Proceedings

Ninth Data Communications Symposium

Whistler Mountain,
British Columbia
September 10-13, 1985

Association for Computing Machinery Special Interest Group on Data Communications (SIGCOMM)

IEEE Computer Society Technical Committee on Computer Communications

IEEE Communications Society Tech. Committees on Data Communication Systems, Computer Communications

IEEE
THE INSTITUTE OF ELECTRICAL AND ELECTRONICS ENGINEERS, INC.

IEEE COMPUTER SOCIETY PRESS

ACM Order Number 533850
IEEE Catalog Number 85CH2137-8
Library of Congress Number 85-60159
IEEE Computer Society Order Number 640
ISBN 0-89791-164-4

Published by IEEE Computer Society Press
1730 Massachusetts Avenue, N.W.
Washington, D.C. 20036-1903

ACM Order Number 533850
IEEE Catalog Number 85CH2137-8
Library of Congress Number 85-60159
IEEE Computer Society Order Number 640
ISBN 0-89791-164-4

Prices (1985) ACM or IEEE Members: $20.00 prepaid
All Others: $40.00 prepaid

Additional copies of the 1985 or prior Proceedings may be ordered prepaid from:

ACM
Order Department
P.O. Box 64145
Baltimore, MD 21264

IEEE Computer Society
Post Office Box 80452
Worldway Postal Center
Los Angeles, CA 90080

IEEE Service Center
445 Hoes Lane
Piscataway, NJ 08854

acm Association for Computing Machinery
IEEE THE INSTITUTE OF ELECTRICAL AND ELECTRONICS ENGINEERS, INC.
IEEE COMPUTER SOCIETY

Message from the General Chairman

In many ways the first of these symposia heralded the beginning of data communications as an identifiable technology. Since then the field has evolved and has begun to mature. Yet in the midst of this process revolution on at least two fronts is still occurring. First, the maturation of existing approaches to data communications has set in motion both standards activities and the integration of data communication into local, nationwide, and worldwide networks. These occurrences promise that data communications will in the not too distant future be as available and as cost-effective as voice communications. This is indeed revolutionary.

As a counterpoint, technology is moving so rapidly that it may outstrip these existing approaches. Computing is demanding ever higher and more ubiquitous performance. Today's personal "workstation" that will be tomorrow's personal computer will wish to casually move large files and high resolution images while at the same time participating in distributed systems requiring submillisecond transmissions. Technology via optical and satellite transmission coupled with subnanosecond VLSI promises data communications that will support these demands. But the form of this "second generation" of data communications is not clear and much research is yet to be done. This is also revolutionary. Maturation in this case does not imply stability.

In the intervening years since the First Data Communications Symposium many other conferences on this topic have emerged. In the midst of this activity, the Data Communications Symposia have remained intentionally small with a single thread of sessions and much opportunity for informal interaction. The goal in selecting Whistler, British Columbia, as the site for the ninth symposium was to provide pleasant yet somewhat remote surroundings with few diversions. Past experience has shown that this atmosphere fosters strong technical interchange. In a sense this symposium is as much a workshop as a formal conference.

Bart W. Stuck, in his role as program chairman, has put together a technical program of the highest caliber. The technical program has always been the key to the success of these symposia; the Ninth Data Communications Symposium carries on this tradition. Moreover, to undertake a symposium of this caliber is also a huge amount of work beyond the program itself. Each member of the symposium committees deserves a great deal of thanks. Jill Miller deserves special thanks for her double duty in publicizing the symposium. Ron Frank, who in the past has so ably worked in publicizing this series and who was sharing the publicity duties with Jill, passed away. Ron is and will be missed. Dawn Williams also deserves special mention for her tireless efforts on local arrangements.

I'd like to acknowledge the support of the ACM/SIGCOMM, the IEEE Communications Society, and the IEEE Computer Society. Also I personally wish to thank all the speakers, panelists, session chairs, members of the program and steering committees, paper reviewers, and the legion of others, many unknown to me, who helped. Finally I wish to thank Joan Maroscia who coordinated many of the administrative details and who made my life a great deal easier.

WILLIAM P. LIDINSKY

Message from the Program Chairman

Welcome to the Ninth Data Communications Symposium, continuing a tradition begun 18 years ago as a meeting of practitioners in the field of data communications. The field continues to mature, but interest is exploding: it can honestly be said that we are living in the midst of a revolution in communications that will have a major impact well into the next century on the daily lives of many people throughout the world. Many feel that this symposium, which attempts to attract the best of the best for informal discussions on technical issues, is the premiere technical data communications conference in the world. The present event hopefully lives up to those expectations by presenting a blend of theory and practice while always looking ahead to the future. As in the past, an informal birds of a feather session will be held to deal with special topics. Watch for notices as to time and place.

This year 85 manuscripts were submitted for review for presentation, while 21 were accepted. The last two symposia each received well over 100 manuscripts with under 30 accepted for presentation at each. In my opinion the drop in submitted manuscripts is due to a shift of much of the pioneering work in data communications from academia and laboratory to industry—from an environment where the end goal is a published paper to one where the end goal is a profitable, proprietary product. The technical program attempts to attract strong participants from both areas.

In an attempt to solicit participants from industry working on the frontier of data communications, seven panels will be presented: space satellite earth stations, DARPA interdomain addressing, packet switching network operational experience, message handling and directory services, remote procedure calls versus problem oriented protocols, integrated services digital network (ISDN), and end user applications.

Papers reflect a mix of theory and practice. The seven sessions will cover transport protocol case studies, theoretical foundations of protocols, communication system performance, packet switching network management, prototype networks, naming and addressing and routing, and application level issues.

Two keynote addresses will be presented. Ed Sussenguth, a lead architect for IBM's Systems Network Architecture (SNA), will examine trends in data communications. Shen Lin, a key designer for AT&T Electronic Tandem Network (ETN) private line networks, will survey network design methodologies.

Two tutorials will be presented. Peggy Karp will cover message handling protocols (e.g., X.400). Stuart Wecker will examine voice/data PBXs.

My most sincere thanks go to the reviewers of the manuscripts, the panel organizers and session chairs, and the technical program committee for bringing this all together. Finally, I sincerely thank Vivan Wu, who helped with many of the administrative aspects of the technical program.

Bart W. Stuck

Ronald A. Frank

1934—1984

Ronald A. Frank, who had a long association with the Data Communications Symposiums, died on December 11, 1984, at his home in New Hampshire. Ron had a distinguished career as a writer and editor in the field of telecommunications. He had been *Computerworld*'s senior technical editor, contributor to *Datamation* and *Telecommunications Products & Technology*, and, most recently, the publisher of his own analytical newsletter, *Telecommunications Times & Trends*. In addition, he served as managing director of The Frank Communications Group.

Ron had been a participant in past technical programs for this series of symposia as an organizer and as a session leader and moderator. Most recently he had directed his talents toward publicizing the symposia. At the time of his death, he was functioning as co-publicity chairperson for the Ninth Data Communications Symposium.

We will miss Ron at Whistler and also in the future.

Sincerely,

William P. Lidinsky

WILLIAM P. LIDINSKY
General Chairman

List of Reviewers

G.A. Agnew
M. Ammar
V. Cerf
F. Chung
E. Cooper
D. Clark
J. Daigle
P. Danzig
M. Decina
M. Dolan
F. Ferrante
D. Ferrari
J. Field
S. Fuhrman
M. Gerla
G. Gopal
R. Gusella
B. Hampson
A. Harvey
P. Humblet
W. Infosino
C. King
D. Knudson
P. Kuehn
K. Kuemmerle
J. Lagarias
W. Leland
S. Li
W. Lidinsky
J. Low
J. Mark
S. Marshall
L.J. Miller
R.J.T. Morris
D. Oran
T. Phinney
G. Ramamurthy
H. Rubin
I. Rubin
A. Salazar
B. Samadi
S. Sechrest
N. Shachum
M. Strohl
N.V. Srinivasan
C. Sunshine
D. Taylor
D. Thompson
D. Unger
A. Vernon
J. Wong
P. Weinberger
H. Yang
O. Yang
S. Zatti
W. Zou

Table of Contents

Session 1

Transport Protocol Case Studies

Chairperson
P. Leach
Apollo Computer

DEVELOPMENT OF A TCP/IP FOR THE IBM/370

Robert K. Brandriff, Clifford A. Lynch, and Mark H. Needleman

University of California
Office of the President and Universitywide Services
Division of Library Automation
186 University Hall, Berkeley, CA 94720

Abstract

This paper describes the design and implementation decisions that have been made in developing software to support the DARPA TCP/IP protocols for the IBM OS/370 environment at the University of California Division of Library Automation. The implementation is designed to support over 100 concurrent TCP connections, all of which are managed by a single program, which acts as a specialized sub-operating system. The system is optimized for line-by-line or screen-by-screen terminal traffic rather than character-by-character traffic. In addition, this TCP is designed to exploit the availability of the large main storage and processor speed available on the IBM/370.

TCP/IP is generally considered to be a mature protocol specification; however, in the course of our implementation we found several parts to be either ambiguous or problematic -- in particular, error handling and notification, ICMP and its relationship to other protocols, and synchronization of data flow with TCP callers.

We also discuss problems encountered in trying to replace hardwired terminals in a public access environment with TCP and TELNET, and some protocol changes that would make these protocols more hospitable to our environment.

Introduction

Implementing a complex and robust telecommunications protocol on an IBM 370 mainframe computer presents some unique design problems and opportunities, particularly when the objectives of the implementation include high performance and the support of a large number of connections. This paper discusses the development of a TCP/IP program at the University of California Division of Library Automation. As background, the first section briefly describes the application for which our particular implementation of TCP/IP was designed and the network environment in which it runs. We then describe the implementation and some of its unusual features. The following sections discuss design issues such as performance optimization, compares our implementation with other "second-generation" TCP/IP protocols such as those for the IBM Personal Computer, and reviews the problems we encountered in our implementation effort.

The University of California Division of Library Automation Network

The Division of Library Automation (DLA) of the University of California is mandated to design, develop and maintain an online library catalog for the nine campuses of the university [1]. This large-scale effort has been in progress since the mid 1970's. The ultimate goal is to allow access by approximately 1600 terminals throughout the state of California to a bibliographic database located at the main DLA computer center in Berkeley. This catalog will eventually contain records for nearly six million different monographs that are held by over 100 libraries on the nine University of California campuses. Stringent requirements were placed on the catalog's design since the application is intended as a public service to be used by untrained library patrons. Great emphasis was placed on user-friendliness and other human factors. In addition, because the system will play a central role in providing library services throughout the university in the future, reliablilty and expandability were also important considerations. A production system, known as the MELVYL™ online catalog, has been in operation for the last two years, and this system is constantly being expanded and improved. It is also expected that the MELVYL system, in the years ahead, will become part of an interconnected network of systems on the UC campuses, as well as part of an interconnected group of systems in libraries around the country, using applications-level protocols currently under development [2].

To support a system of this scale, DLA created a large telecommunications network linking the nine University of California campuses to the computer complex at Berkeley that supports the MELVYL catalog. This complex consists of four IBM/370-compatible machines. The network uses the TCP/IP protocols [3] and packet switching hardware supplied by Bolt Beranek and Newman (BBN), with a variety of transmission media, including leased telephone lines, microwave and

© 1985 ACM 0-89791-164-4/85/0009/0002$00.75

satellite links [4,5]. The TCP/IP protocols were chosen as the backbone telecommunications software for the network for several reasons. The use of datagrams rather than virtual circuits contributes to the reliability and flexibility of the network since each packet is routed individually. In addition, TCP/IP had already been implemented on several computer systems within the UC system (most notably the ubiquitous UNIX systems), thus facilitating the connection of those systems to ours. The internet capabilities of IP were also important to the long-range plans of interconnecting local networks that are being installed on the nine campuses. Finally, TCP/IP is a well-defined, mature, vendor-independent and widely used standard upon which to build distributed library automation and information retrieval systems, which are becoming increasingly important [6]. In particular, the availability of TCP/IP implementations on popular microcomputers, such as the IBM Personal Computer and the Apple MacIntosh, offers the possibility of solving the problem of microcomputer-to-mainframe links, allowing microcomputers to function as peer hosts in a distributed workstation environment.

Having chosen the TCP/IP protocols, we then had to decide whether to implement them in the 370 host, or in special-purpose, front-end minicomputers that would relieve the host of the burden of protocol processing and overhead. We chose to do the former for a variety of reasons. First of all, even if the TCP/IP protocols had been implemented on a front-end processor, some ad hoc mechanism would have been needed to allow the 370 to communicate with the TCP/IP protocols. By implementing the TCP/IP protocols on the 370, we avoided not only the need for this nonstandard interface but also one layer of overhead in the passage of data back and forth in the network. In addition, this approach simplifies multihoming of hosts on several networks and avoids the problems involved in debugging, tuning, managing and maintaining several small, special-purpose computers. Moreover, it promotes consistency since we interface to each network at the same level (in essence, a lowest common denominator). Finally, implementing the TCP/IP protocols on the 370 allowed us to take advantage of the high processing speed, expansive architecture, storage capacity and sophisticated operating system services provided by a large machine, both as a development and operational environment.

Description of the IBM 370 TCP/IP Program

The TCP/IP program at DLA was written in IBM 370 Assembly language. Although the program was specifically optimized for the OS/360 operating system currently in use at DLA, the operating system-dependent routines were kept isolated so that the program could be used on other IBM (OS-type) operating systems (e.g., MVS) with the replacement of only a few routines. Assembly language was chosen because of the speed it provides during program operation and, more importantly, because it permits precise control over critical functions such as storage management and I/O operations.

The TCP/IP program consists of two parts, a single task (from the operating system's viewpoint) and a small, operating-system resident portion, which users communicate with by means of supervisor calls (SVCs). More details on the actual user interface to the supervisor routines and their command structure are given below. The TCP/IP task internally consists of four levels: three protocol levels and a level that contains service functions and routines shared by the other three levels. The service functions consist of initialization and termination routines, the program dispatcher, buffer management routines, error handling and failure logging routines, and the main timing routines. (The various protocol levels make use in their own timing routines of the general services provided by the main timing routines.) The three protocol levels are TCP, IP and the various network drivers. Each one consists of a series of processes that are controlled by the main program dispatcher, which passes control to a process when it has work to do. The process performs its tasks, posting other processes in its own protocol level or other protocol levels as needed, and then returns control to the main dispatcher, which decides where to deliver control next. Note that these processes are invisible to the operating system, which views the entire TCP/IP implementation as a single process. This is discussed in more detail later.

For the most part, our implementation was broken down along protocol boundaries (i.e., IP and TCP reside in different processes), although we made considerable use of shared service routines. The division of functions supporting the different protocol layers into modules was far from rigorous (compare [7]); thus, the protocol layering is mirrored in the software modularization only in the most general way.

The network driver level was designed so that each driver would support one type of network (e.g., 1822, Ethernet, etc.). In the DLA implementation, each driver can support connections to multiple networks of the same type and multiple connections to the same network. The names, addresses and other parameters for a specific network are passed to the driver initialization routines by the main program initialization. A driver-specific control block contains all the information needed by a specific connection. To make it easy to support new types of networks in the TCP/IP task, a common and well-defined interface was provided to each network driver. However, individual drivers can rework this interface, within the confines of their own level, to meet the needs of the specific type of network with which they are communicating. This design facilitates multihoming of an individual host. It also isolates certain network-specific functions that are not well defined in the TCP/IP protocols from higher levels. For instance, internet address conversion routines, which can become quite intricate for networks such as Ethernets [8], can be hidden in the network driver level without affecting the main protocols. A consistent interface to the higher levels is still maintained. Currently, two network driver modules exist in our implementation: an

1822 driver for BBN IMPs [9], which also supports the 1822L logical addressing scheme [10], and an Ethernet driver.

A few other elements of our TCP/IP design are worth mentioning at this point. First of all, the IP protocol recognizes both internal and external callers. An internal caller is one, like TCP, that resides with IP in the same task. Such internal callers are considered to be "trustworthy" and receive a performance advantage since they have variables and pointers that IP can access and modify directly, bypassing considerable validity checking. UDP is also defined as an internal caller. The definition of a category of external callers permits several protocols at the transport level, including protocols that reside in other tasks in the system, to use the services of a single IP. The IP protocol supports multiple concurrent callers, and the SVC structure allows direct calls to be made to IP, bypassing the TCP level. This makes it possible, among other things, to test and debug new versions of TCP/IP without disrupting the production version of TCP, and to implement new protocols that require IP services.

A second important element in our implementation is the support of multiple connections at every level. Before discussing this, some distinctions should be emphasized. In the view of the IBM operating system, the entire TCP/IP program runs as one task. The program handles all multiplexing and multiprocessing internally. One copy of IP manages several host access protocols like TCP that request use of the IP datagram and routing services. As mentioned earlier, the network drivers are designed to handle connections to multiple networks of a specified type. In addition, one copy of TCP (i.e., one set of processes) handles multiplexing for an arbitrary number of connections simultaneously. These may be connections to a single, higher-level protocol like TELNET or to several higher-level application protocols, such as file transfer and mail, which run simultaneously with TELNET. The division of time and resources in the TCP/IP task is managed internally through the TCP/IP program dispatcher and timer management routines. This scheme was used to avoid the overhead that would result if the operating system managed these functions. However, this required us to create, in effect, a miniature operating system, thus reproducing, for efficiency's sake, specialized versions of many of the same services normally provided by the operating system. This design greatly increased the size and complexity of our TCP/IP implementation. It is interesting to note that this same approach is widely used in other large-scale systems, such as job schedulers and database management systems, that have been implemented on IBM hosts.

Finally, we attempted to make the implementation as flexible and adaptive as possible. Many routines use heuristics to dynamically implement different strategies, depending on conditions at the host and out in the networks. For example, the program dispatcher includes priority dispatching provisions. The buffer allocation routines will consider the changing demands in deciding how to allocate buffers. The IP routing routines attempt to respond to changing conditions in the network, especially in the way they keep track of available gateways and their reliability. The network drivers attempt to make the best use of multiple connections into a network if they exist, and to recover dynamically from failures in any network connection. In designing a flexible and adaptive structure, we tried to separate policy-making routines from those routines that carry out policy in order to minimize the effects of changes in the former routines on the latter ones. Consequently, we have been able to test new algorithms and heuristics with minimal impact on the functioning of the main protocol implementation.

User Interface to TCP/IP

In the DLA implementation of TCP/IP, applications-level protocols (e.g., TELNET, file transfer, etc.) run as separate tasks on the same host with the TCP/IP task. These tasks all run as asynchronous processes at the operating system level and communicate with each other via service routines in the operating system. The user/TCP interface is similar to the one described in the DARPA Transmission Protocol Specification [3]. There are six basic calls: OPEN a connection, SEND data, RECEIVE data, CLOSE a connection, ABORT a connection, and get STATUS information about a connection. All these calls make use of parameter blocks to communicate the necessary information. The send and receive calls are asynchronous, and multiple sends and receives may be outstanding on a given TCP connection.

A simplified example will demonstrate how this interface works in the IBM environment. An application protocol task desiring to send a packet on a previously established connection builds up the appropriate parameter block and calls the operating system-resident part of TCP via an SVC instruction. This routine validates the request, places it on the TCP/IP work queue, informs the application protocol that TCP has accepted the request, and informs the TCP/IP task that there is a command to be processed. TCP/IP sends the packet, receives acknowledgement for that packet, and then calls operating system services to inform the calling task that the data have been sent. Because of the asynchronous nature of the interface, both the application protocol and TCP/IP can be doing other work at the same time this SEND command is processed. A separate but analogous user/IP interface supports the external IP users described earlier.

Since TCP/IP knows little about its clients, a major part of the user interface is dedicated to protecting TCP/IP from ignorant or malicious behavior on the part of the users. Without the proper defenses, an application protocol can cause serious problems for the network: it can flood the network with too many packets, provide insufficient space for a receive buffer, or pose erroneously as a common server application. Self protection comes in many forms. For example, any given job attempting to use TCP/IP is restricted to representing specific protocols. It is also restricted to some maximum number of connections and perhaps even to certain foreign sockets. Further protection takes place at the level of individual connections. Any one connection is restricted to a maximum number of allowable outstanding calls. Callers must provide their own send and receive buffers, thus preventing one process from monopolizing all available buffer space in the TCP/IP task. The TCP/IP implementation also makes heavy use of various IBM/OS conventions for protecting itself. Over the years a whole set of standard practices have evolved in the IBM/OS environment for protecting tasks in a multi-tasking environment from each other. Thus, for example, the memory-protection features of IBM/OS are used to ensure that all data areas, parameter blocks, etc., actually belong to the process calling TCP/IP.

If TCP/IP knows little about local clients, it knows even less about remote hosts. In addition to the examples of antisocial behavior mentioned above, other possible problems could occur: a remote TCP could consistently send bad window information, resulting in excessive retransmission on the part of the local TCP, or a remote host could flood the local TCP/IP with "garbage" packets. Protection against remote hosts is more difficult than protection against local users. The ultimate solution, and perhaps the only realistic one, is to ignore packets from "problem" hosts, protocols, or connections. This needs to be done at the lowest possible level to minimize both overhead and the spread of congestion. One can envision a system in which local network drivers simply drop any packets that come from certain hosts. A table of "bad" remote hosts could be maintained, and, periodically, these hosts could be removed from the table to see whether their behavior had improved. Of course, determining whether a host is behaving badly is a difficult problem in itself. It is an example of a complexity inherent in network processes -- the difficult leap from knowing that there is something wrong to knowing precisely what is wrong, an area that is discussed later in this paper.

Optimization and Performance

Given the "half-duplex" nature of its primary applications, the DLA implementation of TCP/IP is optimized for line-by-line, rather than character-by-character, terminal traffic. This optimization is achieved mainly through buffering techniques similar to those described in [11]. Data from both outbound and inbound packets are stored in buffers provided by the applications-level protocols. TELNET, the major application protocol, uses relatively large send and receive buffers. For outbound data, it can buffer up to a full screen of text before calling TCP/IP to transmit the data. Similarly, for inbound data it can indicate to TCP/IP how much data should be contained in a receive buffer before TCP/IP signals that a receive call is complete. Of course, urgent and push pointers override this processing. Nonetheless, the use of fewer, larger packets cuts down on both I/O and CPU cycles, making the network processing faster. The tradeoff for improved performance in this scheme is increased memory requirements; however, this tradeoff is well worthwhile, given the relatively large amount of memory available in the IBM 370.

Generous buffer pools have been exploited throughout our design. For example, any inbound packet that cannot be processed immediately (because it is out of order or because there are no outstanding receive calls for the connection) is stored for a brief period on a "trash" queue until it can be processed. The trash queue is scavenged periodically to remove old packets that have not been claimed, and buffers on the queue are also subject to immediate seizure in situations where the task is running out of buffers.

One of the big bottlenecks in a "structured" implementation of TCP/IP is often the inevitable copying of data as a packet is passed from one protocol layer to another. This bottleneck is even more severe in more recent operating systems such as MVS, where both data and control information must be copied across address spaces. The DLA implementation virtually eliminates data copying and yet maintains separation between the different protocols, since packets are built for output by exploiting a feature of the IBM/370 called chained channel programs [12]. In essence, this permits a group of discontiguous chunks of memory to be written out together, one after another, in a single I/O operation. Thus, user data resident in the send buffer of the applications protocol and the various protocol headers scattered throughout the buffer pools of the TCP/IP task can all be written out together in the appropriate order.

The buffering strategy described above resolves one of the problematic areas in any TCP/IP implementation -- that is, deciding when to inform an application protocol that a packet has been sent. Waiting to inform the application protocol until the remote TCP has acknowledged all the data octets associated with the send buffer offers several benefits. First of all, since the application protocol must provide its own buffers, TCP/IP is protected against buffer overflow due to poor behavior by the application protocol. Memory is also saved; since the application protocol must maintain the integrity of its data buffers, packets need not be saved for possible retransmission. Packets may simply be reformatted from the buffers of the application protocol.

TCP/IP Protocols on Small Machines

It is interesting to compare the DLA TCP/IP implementation with other "second-generation" implementations. One of the most interesting comparisons is with the MIT TCP/IP protocols developed for the IBM Personal Computer [13,14] and since ported to the Apple MacIntosh [15]. Obviously, this implementation is quite different: it is for a small, memory-constrained machine and currently supports only a single TCP connection. It is, however, designed to be performance-oriented and shares several features with our implementation. The TCP/IP protocols in a personal computer still have to deal with the "multi-tasking" necessary to handle the essentially asynchronous activity of inbound and outbound packets. The MIT implementation required the development of multi-tasking support software just as the DLA implementation did, but for an entirely different reason: the personal computer operating system does not have multi-tasking capabilities.

It will be interesting to see how protocol implementations in the personal computer change as the operating systems become more powerful and start to support multi-tasking. The implementations could become more closely intertwined with the operating system, or they might shy away from the multi-tasking features of the operating system for the same reasons that the mainframe DLA implementation did -- cumbersome interfaces and excessive overhead.

Problems Encountered in the Implementation

As might be expected with the implementation of a mature protocol like TCP/IP, relatively few problems were encountered in interpreting the protocol definition. However, several areas are worth discussing, particularly error handling and notification, both in and outside the realm of ICMP. As the protocols are specified, the timeout mechanism is the only way that serious problems on the network (e.g., when a host is unreachable or a local network driver is down) are communicated to TCP and the applications-level protocols. In a system used by the public, a long delay followed by a "connection timed out" message is not a particularly acceptable means of notification. The DLA implementation solves this "error communication" problem in a rather ad hoc manner. The IP protocol passes incoming ICMP packets that contain useful information to the appropriate host-level protocol, be it TCP or UDP. These protocols, in turn, can notify affected applications-level protocols through the user interface described earlier.

An extension of this technique is the creation of "dummy" ICMP messages for error conditions signaled from sources other than ICMP. For example, a local network driver encountering serious I/O errors on the hardware interface to the network could manufacture a fake ICMP packet and place it on the input queue of the IP protocol. This "dummy" packet would percolate through to the higher-level protocols. Of course, the type and code fields of the ICMP header would be different from those used in real ICMP messages.

Another source of information is the local networks themselves. A local network header may contain useful error information. A local network driver receiving information in this manner could build a fake ICMP packet and pass it to the IP protocol. For example, the 1822 host/IMP protocol can communicate to the local network driver such information as that a host is down, why the host is down, that the local IMP is going down, and how long the IMP will be down.

This use of ICMP messages has both benefits and disadvantages. On the one hand, it provides some mechanism for communicating errors. The mechanism provides uniformity of interface and is designed to allow ready incorporation of diagnostics for new error conditions as the need for these diagnostics arises. On the other hand, the use of ICMP messages forces higher-level protocols and applications to become sensitive to the format of ICMP messages -- a decided disadvantage. In addition, it fails to solve some basic problems associated with error handling in layered communications protocols. Knowing that something is wrong is different from knowing precisely what is wrong, and it often takes a substantial amount of case-specific software to advance from the former state to the latter. Additional complexity is introduced by the multiplexing involved as one advances from one protocol layer to the next. Determining which applications, protocols, or entities within a protocol (connections, networks, or whatever) are affected by a particular error is another problem. It is interesting to note that approximately 50 percent of the software in the MIT implementation of TCP/IP is devoted to error processing. Although the percentage is lower in the DLA implementation, it is increasing all the time.

Several additional minor problems are worthy of mention. In the TCP specification, it is unclear when to "push" without a push pointer. The DLA implementation circumvents the problem by forcing the application protocol to decide how many octets of data are to be filled in the receive buffer before it is to be notified.

Some header fields and algorithms in the protocol specifications cannot be implemented in a straightforward manner on a 32-bit machine such as the IBM/370. 16-bit unsigned numbers, such as those used in the check sum routines, length fields, and window sizes, are a nuisance on a 32-bit machine, as are header fields that do not start on octet boundaries, such as offset fields and header lengths. The coding details here are significant since a good deal of time is spent on operations such as check sums and manipulation of these fields. As our debugging progressed, it became clear that little provision has been made in most TCP implementations for diagnosing problems between two implementations of the protocol. It was simple enough for us to incorporate tracing and packet dumping facilities in our local implementation; however, these are often of little use when a remote TCP fails to interoperate properly with ours. We would argue that explicit provision needs to be made for software diagnostics in the protocol definition, either through new forms of ICMP messages or through new flags and fields built into the TCP headers themselves.

Multihoming Support and Gateway Issues

One of the more interesting "gaps" between the TCP/IP specifications and implementation of the protocols involves multihoming and gateways [16]. Because of the throughput and reliability requirements of our application, we expect to multihome hosts on a given network by connecting each to two IMPs. In addition, we are starting to connect our hosts to local networks such as Ethernets. This gives rise to a configuration similar to that shown in Figure 1.

Figure 1

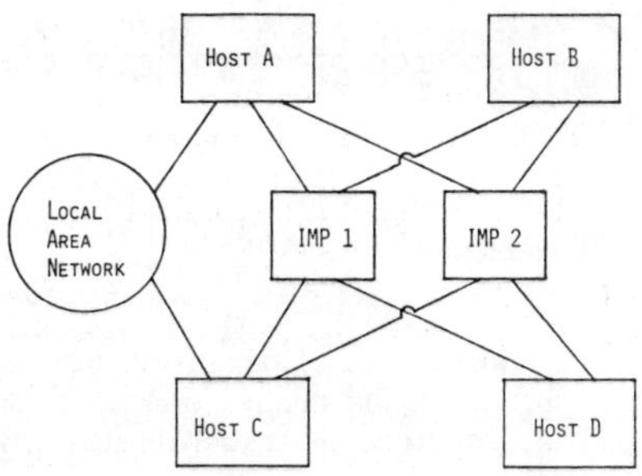

Obviously, Hosts A and C in Figure 1 can function as full internet gateways by implementing GGP [17]; an additional set of routines is simply added to the IP protocol. Along with GGP, which is really a set of network management functions, support is needed to forward packets from one network to another.

A form of gateway processing also comes into play when the IP protocol on a host selects a network over which to send a packet. This is done with a fairly simple heuristic. If the destination network is directly accessible, the packet is sent out over the destination network. If the destination network is not directly accessible, a gateway to the destination network is sought by examining various tables. However, this must be done for each IP packet originating from the host as well as those forwarded by the host, and one must be careful about generating gateway redirect messages. In effect, what occurs is that if the connection between a host and one network fails, the host automatically falls back to operating as a gateway for locally generated traffic.

Multihoming on the IMP network presents particularly difficult problems. In essence, we have had to establish equivalence classes of internet addresses. Traffic directed to a host from either IMP interface is considered to be destined for the same host, despite the fact that it may carry one of two different internet addresses as a destination. Outbound traffic may legitimately bear one of two source internet addresses. We allow the callers at both the IP and TCP level to select which source address will be placed in the IP header. Note that for TCP connections this source address selection persists for the life of the connection. Thus, traffic can pass out to a remote host that does not understand multihoming (such as a BBN TAC) over either IMP, but the remote host will always route inbound traffic to one of the two IMPs. This is not optimal, especially from the point of view of reliability, but it seems to be the best we can do in the absence of explicit support for multihoming in the IP protocol.

Integration of TELNET

TELNET has been implemented external to the main TCP/IP task as part of a program called MILTEN, which serves as a communications monitor. Server TELNET resides in MILTEN as a terminal handler, along with other types of terminal handlers (e.g., for standard ASCII terminals and the IBM 2741). User TELNET is a separate system that users can log on to from MILTEN to make outgoing TELNET connections. We decided to locate the TELNET functions inside existing telecommunications software to provide uniformity in the handling of all types of terminals (both local and remote). We have also explored the possibility of integrating TELNET into the public access application, in order to reduce the path length between the terminals and our applications. At least for the present, we have concluded that the added programming and operational complexities do not seem to be justified.

We have discovered several functionality problems in TELNET. Most of the negotiated options defined in the TELNET protocol specification [18] are not applicable to half-duplex systems like the IBM 370. In a user-friendly application, it is essential to provide the user with a sense of control. One of the key ways this is done in the DLA implementation is by permitting the user to stop the computer from displaying information and regain control by using the "break" key on the terminal. In a packet switched, buffered network, it can take a long time for the break to have any effect, and the user may have to sit through many lines of queued output. DLA has proposed an extension to the TELNET protocol that would allow the two ends in a TELNET connection to synchronize on breaks by using sequence numbers that would indicate which portions of the transmission were covered by the break. The receiving TELNET protocol could then throw away any data until it received a sequence number indicating that the other side had processed the break and was now sending new data no longer covered by the break [19].

Finally, a problem that we have encountered in many TELNET implementations, specifically those designed to run on full-duplex systems like UNIX, is that very small data packets are often transmitted. In a line-by-line system, this is merely a great deal of nuisance overhead. It would be very useful if TELNET negotiation included a choice of line-by-line or character-by-character mode. Currently, some TELNET implementations (e.g., the BBN TAC [20]) offer this option, but the user must know enough about the remote system to request transmission at end of line from his or her local TELNET.

Conclusion

In summary, we have found the implementation of the TCP/IP protocols on the IBM 370 to be a challenging and rewarding experience. The IBM 370 architecture, with its expansive instruction set, high processing speed, large memory size and rich collection of operating system services, provides a hospitable host for the protocols, even though certain portions of the protocols do not fit easily into the 370 environment. However, the environment the architecture provides must be used carefully, especially in the areas of resource management and adequate refereeing among multiple users of a single TCP/IP task.

Acknowledgements

We would like to thank Mary Jean Moore for editorial support.

References

1. University of California, Office of the Executive Director of University-Wide Library Planning, The University of California Libraries: A Plan for Development, 1978-1988 (Berkeley: The University of California, 1977).

2. James Aagaard, Application Level Protocol Development for Library and Information Science Applications (Washington, D.C.: Council on Library Resources, 1982).

3. SRI Network Information Center, Internet Protocol Transition Workbook (Menlo Park, CA: Network Information Center, SRI International, 1982).

4. Edwin Brownrigg and Clifford Lynch, "Development of a Packet Switching Network for Library Automation," in Proceedings of the National Online Meeting, New York, April 1983 (Medford, NJ: Learned Information, Inc., 1983), pp. 67-74.

5. David L. Shaughnessy and Clifford A. Lynch, "Telecommunications for an Online Catalog," Information Technology and Libraries 2, No. 1:73-86 (March 1983).

6. Edwin B. Brownrigg and Clifford A. Lynch, "Beyond the Integrated Library System Concept: Bibliographic Networking at the University of California," Proceedings of the Second National Conference on Integrated Online Library Systems, Atlanta, Georgia, September 1984.

7. David Clark, Modularity and Efficiency in Protocol Implementation, RFC 817 (Cambridge, MA: Laboratory for Computer Science, Massachusetts Institute of Technology, 1982).

8. D. Plummer, An Ethernet Address Resolution Protocol or Converting Network Protocol Addresses to 48-Bit Ethernet Addresses for Transmission on Ethernet Hardware, RFC 826 (Cambridge, MA: Laboratory for Computer Science, Massachusetts Institute of Technology, 1982).

9. Bolt Beranek and Newman, Interface Message Processor: Specifications for the Interconnection of a Host and an IMP, BBN Report 1822 (Cambridge, MA: Bolt Beranek and Newman, Inc., 1981).

10. Andrew Malis, The ARPANET 1822L Host Access Protocol, RF 878 (Cambridge, MA: Bolt Beranek and Newman, Inc., 1983).

11. Eric C. Rosen, Issues in Buffer Management, IEN 182, (Cambridge, MA: Bolt Beranek and Newman, 1980).

12. IBM Corporation, IBM SYSTEM/370 Principles of Operation (New York: IBM Corporation, 1981).

13. MIT Laboratory for Computer Science, PC/IP User's Guide (Cambridge, MA: Laboratory for Computer Science, Massachusetts Institute of Technology, 1984).

14. Jerome Saltzer, David Clark, John Romkey, and Wayne Gramlich, The Desktop Computer as a Network Participant. (Cambridge, MA: Massachusetts Institute of Technology, 1985).

15. Mark Sherman, A Network Package for the Macintosh Using the DoD Internet Protocols, (Hanover, NH: Computer Network Laboratory, Dartmouth College, 1984).

16. Carl Sunshine, "Addressing Problems in Multi-Network Systems," Proceedings INFOCOM 82, (Silver Spring, MD: IEEE Press, 1982), pp. 12-18.

17. Virginia Strazisar, How to Build a Gateway, IEN 109 (Cambridge, MA: Bolt Beranek and Newman, Inc., 1979).

18. SRI Network Information Center, Internet TELNET Proctocol and Options (Menlo Park, CA: Network Information Center, SRI International, 1983).

19. Clifford Lynch, "Extensions to TELNET for Effective Break Support," (1984, unpublished).

20. Robin S. Clifford, TAC Users' Guide. BBN Report 4780 (Cambridge, MA: Bolt Beranek and Newman, Inc., 1982).

PERFORMANCE IMPROVEMENTS FOR ISO TRANSPORT

RICHARD COLELLA - ROBERT ARONOFF - KEVIN MILLS

NATIONAL BUREAU OF STANDARDS
Building 225, Room B217
Gaithersburg, Maryland 20899

ABSTRACT

The NBS Protocol Performance Laboratory is developing enhanced protocol mechanisms for OSI class 4 transport that will improve the throughput efficiency achieved on a satellite channel. A selective acknowledgement mechanism has been shown to improve throughput efficiency by as much as 34%. Several alternative expedited data mechanisms have demonstrated throughput efficiency improvements as great as 38%. Most of the protocol mechanism enhancements considered require only minor changes to the international standard OSI transport protocol.

INTRODUCTION

The ISO transport protocols[3] were designed for effective and efficient communications over a set of network services available from public data networks (PDNs), local area networks (LANs), and combinations of PDNs and LANs; however, satellite networks have not been completely accomodated. This paper proposes and evaluates two minor, optional changes to the ISO class 4 transport protocol that provide substantial performance improvements over transmission paths exhibiting long propagation delay, high bandwidth, and unstable error properties.

First, an option permitting more flexible transmission of expedited data is considered. Use of this variant of expedited data is shown to provide throughput efficiency improvements of up to 38% and delay improvements of up to 61% when compared with the standard ISO expedited mechanism.

Second, an optional selective acknowledgement capability is examined. Use of this optional mechanism is demonstrated to yield throughput efficiency increases of up to 34% when compared with a standard one-for-one acknowledgement scheme; however, the improvement is realized at a cost of increased receive buffer utilization. Both the optional expedited data modification and the selective acknowledgement procedures can be added to the ISO class 4 transport protocol such that implementations of the protocol that do not include the options will be completely interoperable with implementations including the options. Thus, these improvements will increase the range of network media over which the ISO transport protocols can operate efficiently without sacrificing the effectiveness of the protocols.

ISO EXPEDITED DATA MECHANISM

Two forms of the expedited data mechanism are discussed and contrasted here. The first mechanism is that currently specified in the ISO transport protocol specification. The ISO expedited data mechanism specifically precludes sending any new normal data after transmitting an expedited data message until the expedited acknowledgement is received (although retransmission of previously sent data is permitted)[3].

Figure 1 shows a time graph representation of how the ISO expedited data mechanism operates. As expedited data is received from the user it is promoted to the front of the send queue ahead of all normal data. It is then sent by $Transport_A$ to $Transport_B$ which acknowledges the expedited data message, XDT_1, when received satisfactorily. Not until this acknowledgement, XAK_1, is received by $Transport_A$ can $Transport_A$ resume sending normal data.

Thus, as can be seen in Figure 1 where the time dimension progresses down the page, the transport connection is idle between the time of initial sending of the expedited data message, XDT_1, and the time of receipt of the expedited acknowledgement, XAK_1, by $Transport_A$. It is only at that point that normal data transmission can resume. Furthermore, were there additional expedited data messages to be sent, there would be additional delay in the resumption of normal data transmission directly proportional to the number of expedited data messages and the propagation delay on the communications network.

From this brief analysis of the ISO expedited data mechanism, there appears to be an undesirable throttling effect which occurs because all normal data transmissions are suspended until the expedited data message is acknowledged. Thus, the transport connection is idle for the period of time required to send the expedited data and receive its acknowledgement.

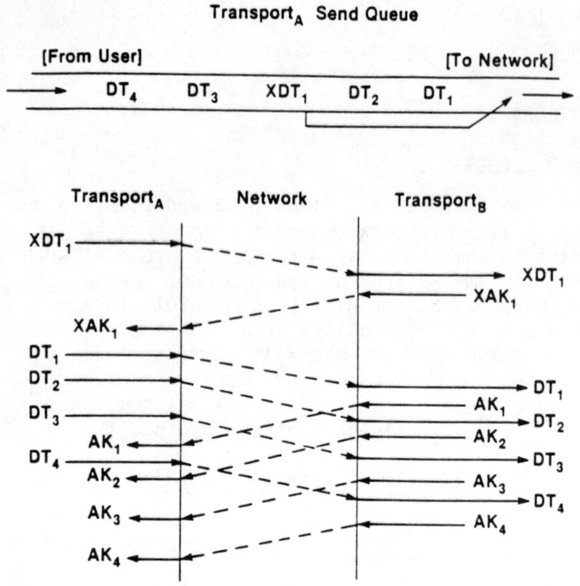

Message Sequencing for ISO Expedited Data Mechanism.
Figure 1

NBS EXPEDITED DATA MECHANISM

The second expedited data mechanism is proposed as part of the NBS transport specification[1], which also satisfies the ISO service specification[4]. The NBS expedited data mechanism may reduce or eliminate the transport connection idle time incurred by the ISO scheme during expedited data transfer. The reduction is accomplished by allowing some normal data to be transmitted while the expedited data acknowledgement is awaited. One restriction applies: the normal data must have been received from the sending user before the expedited data. This satisfies the NBS and ISO transport service specifications [2,4] which guarantee that expedited data received from the user by a transport entity will be delivered to the receiving entity's user before any subsequent normal data received from the sending user. This is achieved because the sending transport entity does not transmit any normal data received from the user after the expedited data until the expedited data has been acknowledged, although data already sent may be retransmitted.

The distinction between the ISO and NBS schemes lies in the signficance attached to the order of data received from the user. With the NBS mechanism, normal data received from the user before the expedited data may be transmitted while awaiting the expedited data acknowledgement (this assumes that expedited data will be placed ahead of normal data in the send queue). With the ISO scheme, the order of receipt from the user is ignored; no data (except retransmissions) may be sent while awaiting an expedited data acknowledgement. The projected improvement in performance for the NBS expedited data mechanism derives from allowing transmission of normal data which had been received from the user up to the time of receipt of the expedited data from the user. By so doing, the normal data does not suffer the full stop-and-wait delay to which the ISO mechanism is subject and which results in idle time on the transport connection. As noted above, this idle time is equal to the difference between the time expedited data is transmitted and the time of receipt of the expedited acknowlegement.

Figure 2 presents a time graph representation of how the NBS expedited data mechanism operates. From this figure we can see the transport connection is being used to send normal data during the time between sending expedited data, XDT_1, and receiving the related expedited acknowledgement XAK_1. More specifically, we can postulate the ideal situation for this mechanism to be one in which there is sufficient normal data in the send queue prior to arrival of the expedited data such that the sending of normal data fills the entire time between sending the expedited data message and receiving the acknowledgement. With such a situation, the NBS expedited data mechanism allows no idle transport connection time.

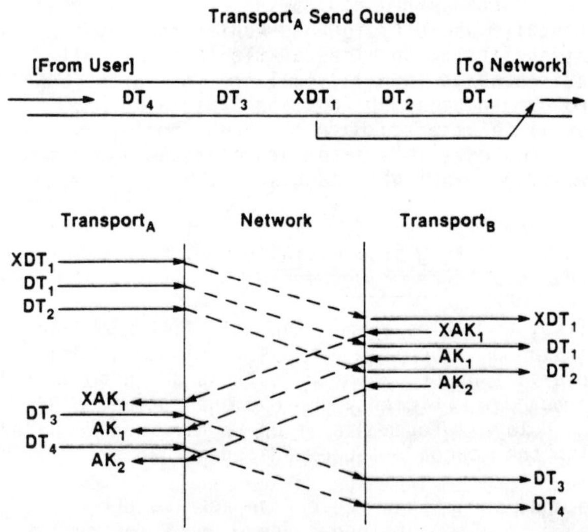

Message Sequencing for NBS Expedited Data Mechanism
Figure 2

This expedited mechanism is easily implemented, requiring no changes in message formats. Normal data are marked with the most recent expedited sequence number, preserving their order of receipt from the user with respect to the expedited data. The transport entity can then determine which normal data is allowed to be transmitted based upon the status of expedited data transmission.

This mechanism, furthermore, can be implemented unilaterally, placing no restriction on interoperation with implementations based on the ISO mechanism. The receiving transport entity assumes that all normal data received in sequence can be delivered. Thus, the receiver relies on the sender to transmit only what is deliverable. It is transparent to the receiver which of the two mechanisms is being used by the sender.

ONE-FOR-ONE ACKNOWLEDGEMENT

The second protocol mechanism studied is the acknowledgement mechanism. In general, an acknowledgement and retransmission mechanism determines the scheme for confirming data that has been received so that the sender can determine which normal data messages, if any, must be retransmitted. Two such schemes are discussed here.

First, consider the acknowledgement and retransmission mechanism scheme currently specified in the NBS transport protocol specification. This is a one-for-one acknowledgement scheme. In a one-for-one acknowledgement and retransmission mechanism, an acknowledgement message is sent for every normal data message received.

According to both the ISO and NBS protocol specifications, the sequence number in the acknowledgement confirms all normal data messages with lower sequence number. If there is a gap in the sequence numbers of received data messages, only those before the gap can be acknowledged. Thus, when a data message is acknowledged, all other messages with lower sequence numbers are also acknowledged. Normal messages with sequence numbers after a gap will be kept by the receiver, if buffer space is available, but cannot be acknowledged until the missing messages are received. Retransmission of the missing data occurs when the retransmission timer at the sender expires.

SELECTIVE ACKNOWLEDGEMENT AND RETRANSMISSION MECHANISM

In order to try to obtain improved performance, we have proposed a variation of the one-for-one acknowledgement mechanism, based upon the selective acknowledgement concept. The selective acknowledgement messages contain one additional optional field not found in the one-for-one acknowledgement messages (Figure 3). When present, this field contains a sequence number for a normal data message which cannot be represented by the regular sequence number field in the acknowledgement message because of an intervening gap in the received data message stream.

For example, assume $Transport_A$ transmits normal data messages with sequence numbers one through five. When message number one arrives at $Transport_B$, an acknowledgement is returned confirming all messages up to and including message one using the normal format shown in Figure 3. Similarly, when message two arrives the acknowledgement confirms message two and all previous messages. If message three is lost in the network and message number four now arrives at $Transport_B$, a selective acknowledgement is returned (Figure 3) where the sequence number confirms messages numbered two and below, as before, and the optional sequence number is set to four, selectively acknowledging that individual message. A similar acknowledgement is returned when message number five is received. Eventually, the retransmission timer for message three will expire and message three will be resent by $Transport_A$. Correct receipt at $Transport_B$ will allow a normal acknowledgement message to be returned for all messages up to five, inclusive.

Using selective acknowledgement, improved performance of the transport protocol may be realized in two ways. First, data messages correctly received, but after a gap, are now acknowledged immediately, allowing the sending transport to release, for other uses, the buffer space occupied by those data messages. Thus, this mechanism may promote more efficient buffer utilization at the sending transport entity.

The second way that selective acknowledgement may improve performance over the one-for-one mechanism is by eliminating unnecessary retransmissions. If a normal data message is lost or damaged in transmission, the retransmission timer for that message will eventually expire causing it to be sent again. However, using the one-for-one scheme, before the retransmitted data message can be received and all received messages confirmed, some number of timers for messages subsequent to the one in error will expire, causing retransmission of correctly received (but unacknowledged) data. These unnecessary retransmissions are eliminated using selective acknowledgement; after a gap at the receiver, data received correctly are confirmed individually using the optional field in the selective acknowledgement message. Transmission bandwidth, therefore, should be used more efficiently.

EXPERIMENTAL APPROACH

The experiments reported in this paper investigate class 4 transport performance for bulk data transfer with particular emphasis on comparing alternative internal mechanisms of the protocol. In order to evaluate and obtain a quantitative comparison of the relative performance characteristics of the contrasted transport mechanisms described above, we used a simulation model of class 4 transport[5], developed at the NBS, to run experiments and generate data for the performance metrics described in Table 1[6].

Acknowledgement Formats

Normal (One-For-One) Acknowledgement

Octet	1	2	3 4	5 6 7 8	9 10
	Length Indication	AK	Destination Reference	Sequence Number	Credit

Selective Acknowledgement

Octet	1	2	3 4	5 6 7 8	9 10	11	12	13 14 15 16
	Length Indication	AK	Destination Reference	Sequence Number	Credit	Parameter Code (Opt. Seq. #)	Parameter Length (4)	Optional Sequence Number

Comparison of Acknowledgement Message Formats Used in the One-For-One and Selective Acknowledgement Mechanisms
Figure 3

TABLE 1

PERFORMANCE METRICS USED IN EXPERIMENTS

<u>Throughput Efficiency</u>	The ratio of user throughput to maximum theoretical throughput. User throughput is the data transfer rate, in octets/sec., as perceived by the user. It is calculated by dividing the amount of user data transferred by the transfer time. Maximum theoretical throughput is the subnetwork access link speed expressed in octets/sec.
<u>Average Per Octet Delay</u>	The delay, as seen by the users, between transmission and receipt of a TSDU (transport service data unit) divided by the number of octets in the TSDU, averaged over all TSDUs sent over the connection. TSDU is the logical unit for data exchange by the users of transport. A TSDU may be broken into multiple normal data messages by the transport entity.
<u>Average Memory Utilization</u>	The average amount of buffer memory used by a transport entity during an experiment.
<u>Maximum Memory Utilization</u>	The maximum amount of buffer memory used by a transport entity during an experiment.

EXPERIMENT 1 (EXPEDITED DATA) -- DESCRIPTION

Experiment 1 explores the performance differences between the two previously described expedited data mechanisms. Table 2 lists the parameters and their values which describe this experiment. The independent variable in this experiment is the expedited data rate. The expedited data rate is the rate of occurrence of expedited data requests per normal data requests at the user interface. The decimal values listed in Table 2 for expedited data rate represent the specific ratios used (i.e., the number of expedited messages divided by the number of normal messages). This variable defines the x-axis of the results graphs. The y-axis represents the particular metric (throughput efficiency or average per octet delay) being used. This experiment is run for each of the two expedited data mechanisms, ISO and NBS.

EXPERIMENT 1 (EXPEDITED DATA) -- RESULTS

The simulation results indicate that the NBS and ISO expedited data mechanisms exhibit very different performance characteristics. Figure 4 maps throughput efficiency as a function of expedited data rate for the NBS and ISO mechanisms. The reason ISO throughput efficiency is lower than NBS efficiency is due to the restriction placed on normal data transmission while awaiting the expedited data acknowledgement.

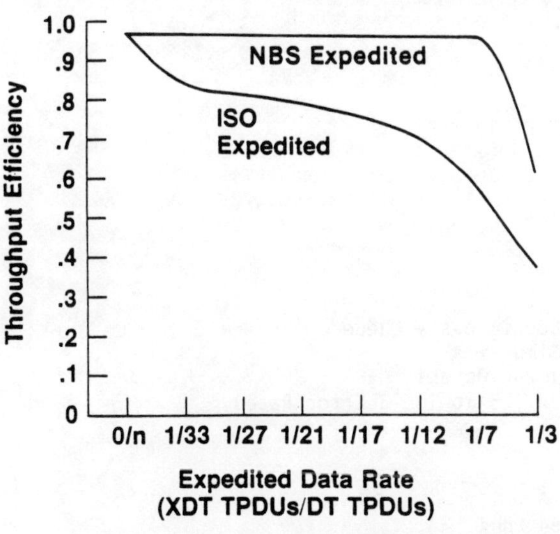

Throughput Efficiency as a Function of Expedited Data Rate for NBS and ISO Expedited Mechanisms
Figure 4

Regardless of the order in which data was received from the user, no normal data may be transmitted until an outstanding expedited message is acknowledged. For each expedited message transmitted, this forces a fixed amount of idle time on the transport connection equal to the round trip time.

The NBS scheme allows normal data received from the user before expedited data to be transmitted after the expedited message is sent, yet before it is confirmed. The idle time on the link is thereby greatly reduced or eliminated. At rates between 0/n and 1/7, any time on the transport connection between transmission and confirmation of expedited data is consumed with normal data transmission. The result is an expeditious transfer of all data.

The throughput efficiency curve for NBS expedited drops significantly as the expedited data rate increases from 1/7 to 1/3. In this region, transport is not receiving sufficient normal data before each expedited message to keep the transport connection fully utilized, resulting in idle time on the connection and lowered throughput efficiency. In effect, the expedited mechanism is imposing an artificial window on normal data flow.

The cost of the ISO mechanism in throughput efficiency is also reflected in delay (Figure 5). Increasing the rate of expedited data causes a concomitant increase in interference in the normal data; the data transfer is stretched out over a longer period of time.

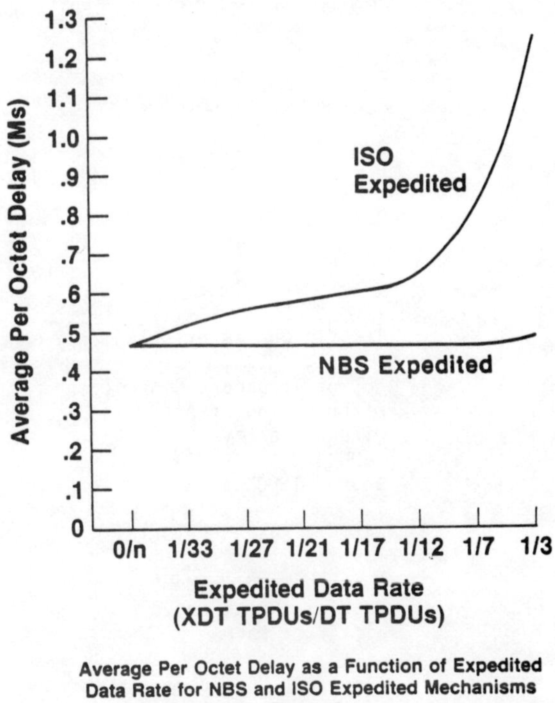

Average Per Octet Delay as a Function of Expedited Data Rate for NBS and ISO Expedited Mechanisms
Figure 5

EXPERIMENT 2 (ACKNOWLEDGEMENT MECHANISMS) -- DESCRIPTION

Table 3 lists the parameters and values describing this experiment, which compares the performance of two acknowledgement schemes. The independent variable in this experiment is the bit error rate. The bit error rate is the ratio of the number of

TABLE 2

EXPERIMENT 1: EXPEDITED DATA MECHANISMS

Experiment Parameters

TRAFFIC CHARACTERISTICS	
No. of Transport Connections	1
Distribution of Arrrivals	Continuously Queued
Directionality	Simplex
TSDU Size	10240 Octets
No. of TSDUS to Send	60
NETWORK CHARACTERISTICS	
Network Window (MSGS)	4
Satellite Link Speed	64 Kbps
Bit Error Rate	10E-9
TRANSPORT CHARACTERISTICS	
Send Buffer Size (Octets)	32K
RCV Buffer Size (Octets)	32K
Window Size (Octets)	12K
Retransmission Timer	2000 ms
Window Timer	6000 ms
Maximum Ack. Delay Timer	Infinite
DT TDPU Size	1024
Expedited Data Rate (expedited/normal)	.33, .14, .08, .0588, .0455, .037, .03, None
Acknowledgement Mechanism	One-For-One
Expedited Mechanism	ISO; NBS

TABLE 3

EXPERIMENT 2: NBS ACK. VS. SELECTIVE ACK.

Experiment Parameters

Traffic Characteristics	
No. of Transport Connections	1
Distribution of Arrivals	Continuously Queued
Directionality	Simplex
TSDU Size	10240 Octets
No. of TSDUS to Send	20 (5 at 10E-3 Error Rate)
Network Characteristics	
Network Window (MSGS)	4
Satellite Link Speed	64 Kpbs
Bit Error Rate	10E-3, 10E-4, 10E-5, 10E-6,
Transport Characteristics	
Send Buffer Size (Octets)	32K
RCV Buffer Size (Octets)	32K
Window Size (Octets)	20K
Retransmission Timer	650 ms
Window Timer	2000 ms
Maximum Ack. Delay Timer	Infinite
DT TPDU Size	128 Octets
Expedited Data Rate	0
Acknowledgement Mechanism	One-for-One; Selective
Expedited Mechanism	NBS

bits received incorrectly to the total number of bits received. For example, a bit error rate of 10E-6 indicates that, on average, a bit will be received incorrectly once in one million bits. This variable defines the x-axis of the results graphs. The y-axis represents the particular metric (throughput efficiency, average and maximum receive memory utilization) being used. This experiment is run for each of the two acknowledgement schemes, one-for-one and selective.

EXPERIMENT 2 (ACKNOWLEDGEMENT MECHANISMS) -- RESULTS

This experiment examines the performance of transport using two different acknowledgement schemes as described earlier, selective and one-for-one. As shown in Figure 6, the benefit of using selective acknowledgement is increased throughput efficiency for the medium to high error rates (E-7 to E-3). The increase is due to a reduction in the number of retransmissions as was anticipated.

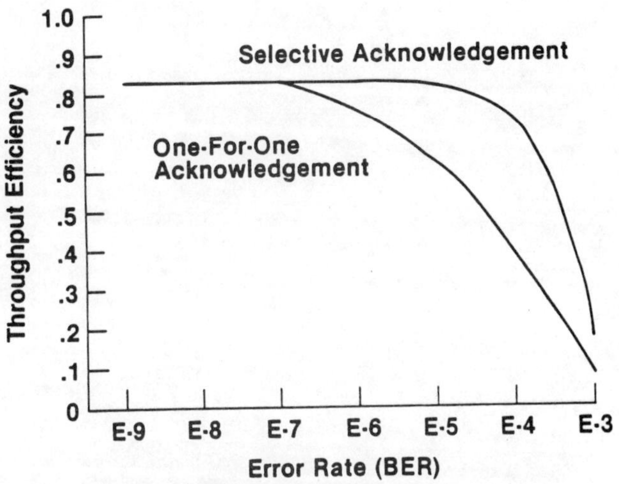

Throughput Efficiency as a Function of Error Rate for One-For-One and Selective Acknowledgement
Figure 6

As one should expect, there is a cost associated with the performance improvement discussed above. In order to maintain a continuous flow of original data messages using selective acknowledgement, the transmit window must be set to twice that required by the one-for-one mechanism. Under the one-for-one scheme, once one retransmission timer expires, a large number will subsequently expire due to the long delay experienced with satellite communications; hence, a major portion of the satellite link bandwidth is consumed by the retransmitted data. Selective acknowledgement, however, eliminates most retransmissions, allowing further original data messages to be sent while awaiting the acknowledgement. The amount of additional data sent is approximately equal to the window size used in the one-for-one mechanism. Therefore, selective acknowledgement requires twice the window size to maintain continuous data flow.

The effect on average and maximum receive memory utilization is shown in Figures 7 and 8. The increase in memory utilization is due to the fact that fewer retransmissions (and therefore more original transmissions) arrive at the receiver per unit time. The received data may not be forwarded to the user except as a sequenced stream of data. Therefore, the receiving transport must buffer the correctly received data, pending the arrival of missing messages. Many more original messages can be sent in a specified period of time when selective acknowledgement is used; thus, many more messages must be buffered by the receiver.

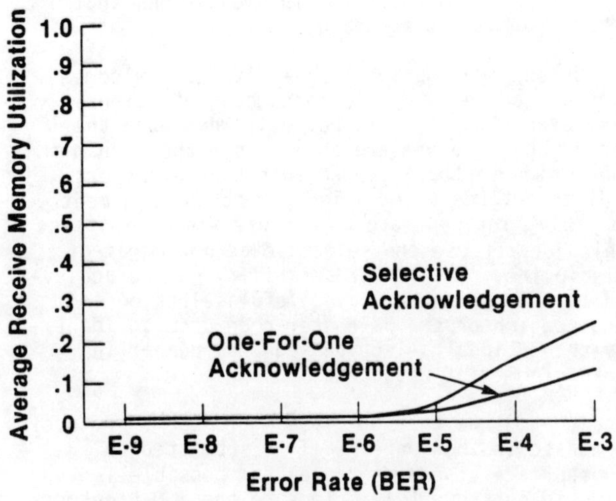

Average Receive Memory Utilization as a Function of Error Rate for One-For-One and Selective Acknowledgement
Figure 7

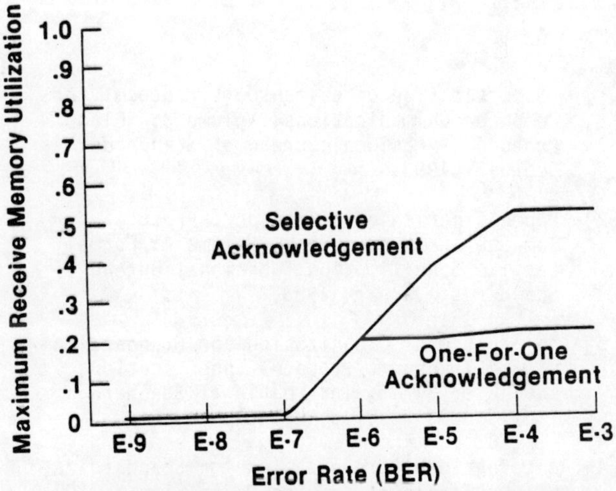

Maximum Receive Memory Utilization as a Function of Error Rate for One-For-One and Selective Acknowledgement
Figure 8

CONCLUSIONS

This paper has shown that two minor, optional changes to the ISO class 4 transport protocol can extend the range of network media over which the ISO transport protocols can achieve efficient operation. A simple modification in the discipline for sending transport expedited data can provide throughput efficiency improvements of up to 38% and delay improvements of up to 61%. The modification in the sending procedures requires no corresponding change in the receiving procedures; therefore, an implementation of the class 4 transport protocol can include the enhanced expedited transfer mechanism with no loss in interoperability with those implementations that adhere to the ISO standard.

Inclusion of an optional selective acknowledgement procedure can yield throughput efficiency increases of up to 34%, but only when both the sender and receiver are able to use the mechanism. The option can be included such that transport implementations using selective acknowledgement can fully interoperate with those implementations which do not use the selective acknowledgement discipline. This interoperability can be achieved in the short-term by careful selection and publication of the parameter code used to identify the optional selective sequence number in acknowledgement messages.

The changes to the ISO class 4 transport protocol proposed in this paper will be submitted to appropriate standards bodies for possible inclusion in subsequent revisions to the ISO transport protocol standard. Adoption of these optional procedures by the standards bodies is necessary to ensure that any short-term agreements to implement the procedures will be protected in the long run by codification within international standards.

REFERENCES

[1] "Specification of a Transport Protocol for Computer Communications, Volume 3: Class 4 Protocol," National Bureau of Standards, February 1983.

[2] "Specification of a Transport Protocol for Computer Communications, Volume 4: Formal Service Specification," National Bureau of Standards, January 1983.

[3] International Organization for Standardization, "Transport Protocol Specification," ISO/TC 97/SC 16, International Standard ISO/IS 8073 rev., June 1984.

[4] International Organization for Standardization, "Transport Service Definition," ISO/TC 97/SC 16, International Standard ISO/IS 8072 rev., June 1984.

[5] Mills, Kevin L., Marnie Wheatley, and Richard Colella, "Simulation of an International Standard Transport Protocol," National Bureau of Standards (pending publication).

[6] Colella, Richard, Marnie Wheatley, and Kevin Mills, "COMSAT/NBS Experiment Plan for Transport Protocol," National Bureau of Standards (pending publication)

AN INTERNODAL PROTOCOL FOR PACKET SWITCHED DATA NETWORKS

D. Drynan and D. Baker

Bell-Northern Research, Ottawa, Canada

This paper describes a proprietary data transmission protocol developed for internodal trunks on the Northern Telecom SL-10™ Packet Switched Data Network. The protocol has been implemented and tested in the new Universal Trunk Processor. Two separate market forces dictated the protocol requirements. The first was a demand for high speed trunks in the T1 (1.54 Mb/s) and above range to cost effectively serve future networks. The second market force was the emergence of satellite data technology with its attendant high bandwidth and lower cost. A protocol with a large sequence number range and the ability to selectively retransmit more than one packet in error without impacting the flow of correctly received packets was required. None of the standard data protocols met these requirements.

Introduction

A typical SL-10 Packet Switched Data Network (PSDN) consists of many nodes interconnected by trunks. Subscriber communication facilities are implemented across the network using three logical layers[1]. At the core, a connectionless subnet provides all internodal communication via the trunks. The protocol described in this paper applies to this connectionless layer. Above this is a virtual circuit (VC) layer, providing the basic end-to-end service. The VC layer guarantees order preserving, non-lossy, non-duplicate packet delivery. Customer access to the network is provided by the outermost network access layer, using standard data communication protocols such as X.25.

There are several advantages in adopting a connectionless subnet layer underneath a VC layer. Because VCs are only set up between the end nodes and not between the intermediate nodes, the packet routing path can be altered dynamically without impacting calls in progress. Normally all calls from one node to another are directed along a fixed path presenting the minimal delay. A trunk outage on this path will trigger a dynamic network change to route around the affected facility. This routing change does not affect the end points of the connection and thus is transparent to the subscriber. A connectionless subnet is also capable of routing packets to their destination with minimal processing delay. This allows economic provisioning of features like fast select, where user data is carried in the call request and the overhead of establishing a VC is not incurred.

Protocol Requirements

SL-10 networks in operation today are experiencing rapid growth, both in the number of nodes and subscriber usage. Some networks are already very large. For example, at year-end 1984 the DATAPAC™ Canadian public PSDN consisted of over 50 nodes and 150 trunks, and was transporting one billion packets a month. As this growth trend continues, currently deployed trunks in the 56 Kb/s to 128 Kb/s range will require upgrading. Higher speed PSDN trunks in the T1 (1.54 Mb/s) and above range, perhaps as high as 44 Mb/s, are required for future growth.

Until recently, most PSDNs relied on terrestrial facilities for internodal communications. This is changing as satellite data technology matures and moves into the data communications marketplace. Satellite facilities are usually available as leased end-to-end links, or as private satellite networks of earth stations. Both offer high bandwidth, high reliability (except during heavy rain or fog), and the potential for substantial cost savings. The existence of facilities can be a factor, since private satellite networks are often the only high bandwidth option in regions not cost-effectively served by terrestrial facilities. Future PSDN trunks must be capable of supporting satellite facilities at high link utilizations.

The average subnet packet size observed in several large SL-10 networks is between 30 and 40 bytes. Assuming a 32 byte packet, 100% trunk utilization, and nodes 1000 km apart, the maximum number of packets in flight at various trunk speeds is shown in figure 1. It is evident that the subnet protocol must accommodate a large packet sequence number range, and be capable of handling several hundred to potentially thousands of packets in transit and awaiting acknowledgment. The most commonly used link protocol, HDLC or LAP-B with extended control field, supports only 127 frames in transit. The most serious drawback, however, is the limitation of only one frame in error outstanding at a time. If a second packet also encounters an error in transmission, all subsequent packets are discarded, whether correct or not, until the outstanding error is corrected. This can impose a significant throughput penalty. At a bit error rate of 10^{-6}, there is a 57% probability of more than one frame in error on a T1 satellite link. If the bit error rate rises to 10^{-5}, the probability rises to over 99%. Therefore, a mandatory requirement for the PSDN trunk protocol is the ability to signal and retransmit multiple packets in error while allowing all correctly received packets to be forwarded [2]. The European Computer Manufacturers Association proposed a HDLC checkpoint mode for selective repeat error recovery [3]. This was examined but rejected because it did not support a variable sequence number field and required frame order preservation.

MAXIMUM PACKETS IN FLIGHT
Packet size is 32 Bytes

Trunk Speed kb/s	Terrestrial Trunk	Satellite Trunk
56	22	124
112	44	249
1544	603	3426
2048	800	4544

FIGURE 1

In an SL-10 only network a proprietary protocol can be optimized for throughput and maximum bandwidth utilization. Such an approach is not possible for multivendor networks where gateways between networks will be based on X.75 or similar VC based protocol.

In summary, a new connectionless subnet protocol was required to meet the following criteria:

- support both terrestrial and satellite facilities.
- handle link speeds from 56 Kb/s to 44 Mb/s.
- allow selective retransmission of multiple packets in error.
- maximize link efficiency for error-free transmission.
- use standard integrated circuits for link handling wherever possible.

Packet Flow

The logical flow of packets using this protocol is shown in Figure 2. The bidirectional trunk is represented by two pipes, with a trunk processor (TP) at each end implementing the protocol. Other processors on the node place packets destined for delivery on this trunk in one (or more) packet queues. The TP determines the order of packet transmission on the trunk by interrogating packet queues according to the priority algorithm specified. The SL-10 algorithm sends network routing information packets first, followed by packets awaiting retransmission, and then user data packets according to the subscribed call priority. Once the packet transmission order is determined each packet, regardless of type, is sent with the next available sequential Sequence Number (SN). The receiving TP first verifies basic packet integrity using the CRC, and then compares the packet's SN to the expected value. When both are correct, the packet is forwarded on, and an explicit acknowledgment (ACK) of that packet's SN is returned to the transmitting TP. The ACK can be returned in a dedicated control frame or be piggybacked unto an outgoing user data packet if available. Upon receipt of the ACK, the transmitting TP will release its local copy of the packet.

LOGICAL PACKET FLOW

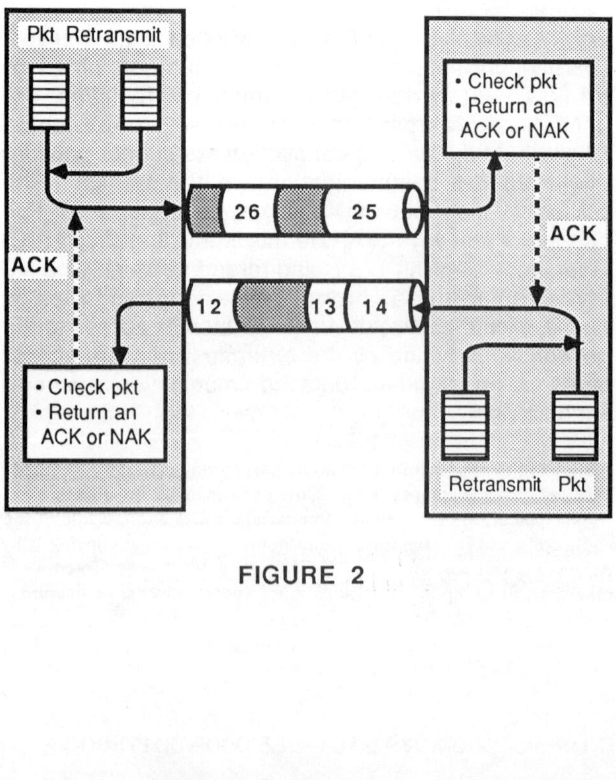

FIGURE 2

If the packet is incorrect, the CRC will fail and the packet is discarded. The SN of the discarded packet or packets can be determined from the gap in subsequently received correct sequence numbers. Explicit Negative Acknowledgments (NAKs) are sent back to the transmitting TP identifying exactly which packets require retransmitting. The packets are retransmitted with the next available SN, not their original SN. It is not necessary or desirable to preserve packet order on trunks, as the VC at each end of the connection will reorder packets. This is seldom necessary in practice as trunks carry packets from many unrelated subscribers, and the probability of disordering a single subscriber's packets is very low.

The two transmission directions are logically independent, although usually implemented in the same TP. The trunk in one direction may be restarting the frame numbering, while the other direction can be transmitting normally. The only interaction between the transmitting and receiving functions within a single TP is the passing of control information, like ACKs, for delivery back to the far-end TP.

Link Framing

The framing chosen for this protocol is from standard level-2 HDLC, because HDLC framing maximizes link bandwidth available for data and is commercially available today in integrated circuits up to the T1 link rate. Within the HDLC flags and CRC, unique control fields are defined to provide the specific capabilities of this protocol. The opening and closing flags and 16-bit CRC are all standard HDLC, as is frame abortion, invalid frames, interframe fill and bit transparency. The HDLC defined idle or out-of-service state, transmitting continuous 1s, is not allowed. This permits trunks to run inverted HDLC for zero code supression. Trunks are either in service or failed, in which case there is no control over what is transmitted on the trunk.

There are two frame types, information frames and trunk control frames. Information frames, as illustrated in figure 3, will be described first. A control field always follows the opening flag, its length depends upon the SN range selected. User data follows the control field and may contain any number of bits up to a defined maximum packet size. The frame is closed by a 16-bit CRC and a standard HDLC flag. The control field consists of the ID, T and E bits, a variable length transmit SN, and an optional acknowledgment field. The ID bit allows fast differentiation between control and information frames without resorting to implicit methods such as length. The T bit indicates the presence of the optional acknowledgment field. This field consists of P1,P2,P3 and a variable length receive SN. When T is set, the acknowledgment field is present and piggybacked in this information frame. The E bit is reserved for future extensions to the control field. The transmit SN of this frame is always present, and can be 5,13,or 21 bits in length. Selecting the appropriate field length is discussed later.

INFORMATION FRAME

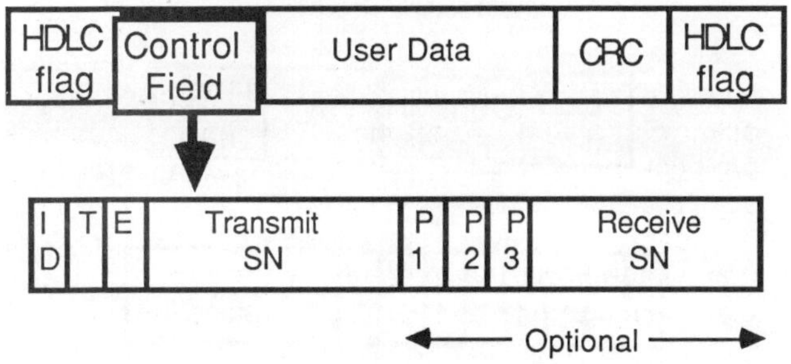

FIGURE 3

Acknowledgments or ACKs of packets travelling in the other direction must be carried back, either in dedicated control frames or piggybacked in information frames. In either case the format is the same. A receive SN uniquely identifies one correctly received frame. P1, P2 and P3 carry the status of the previous three frames, a 0 is an ACK and a 1 is a NAK. Thus, each acknowledgment field carries the status of four consecutive frames. This acknowledgment redundancy allows the protocol to survive the loss or mutilation of up to three consecutive frames. The fourth frame, if received correctly and ACKed, will NAK the previous three frames and cause their retransmission. This feature of the protocol makes the trunk highly robust and capable of continuous data throughput even under high link error conditions.

Trunk control frames as shown in figure 4 contain no user data; they signal information only between TPs. The ID bit identifies a control frame and is always 0. The T bit selects the code field type discussed below. The E bit is reserved for future protocol extensions. The two bit frame length field, L1 and L2, define the size of the SN field to be used between the two TPs. Allowable SN field sizes are 5, 13 and 21 bits, selecting the appropriate size is discussed under link initialization. Three version bits, V1 to V3, uniquely identify the protocol version implemented and allow protocol changes to be gracefully introduced into an existing network. The two communicating TPs must always be executing the same protocol version, but within the network more than one version can coexist.

The code field may contain either a packet ACK or a code value used to signal staging and special commands between the two TPs. When the T bit is set the code field contains an ACK consisting of P1, P2, P3 and a receive SN exactly as described above for an information frame. Separate packet ACK control frames are sent whenever an ACK is queued and there is currently no information frame going out to piggyback on to. This occurs under low traffic conditions or unbalanced traffic flow. The code field length will be 8, 16 or 24 bits long depending upon the length of the packet SN field selected.

CONTROL FRAME

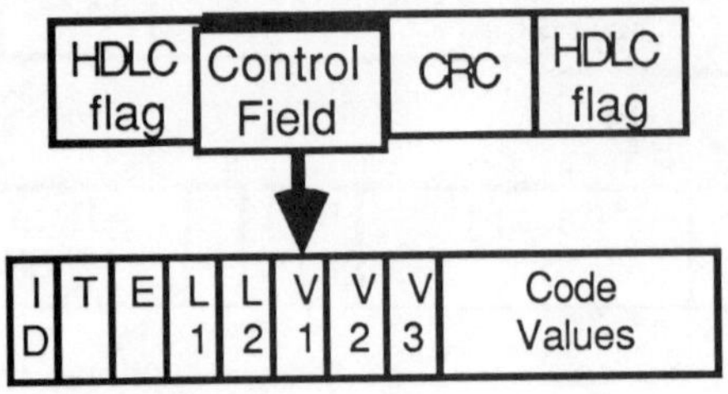

Code Field Values in Order of Precedence

Initialization Command	Decimal Value
UP	3
READY	4
DELAY PROBE	1
DELAY RESPONSE	2
LENGTH CHECK	12
LENGTH OK	13
WINDOW SET	8
WINDOW ACK	9
LOOP EVEN	10
LOOP ODD	11
AYT (Are You There)	5
YIA (Yes I Am)	6

FIGURE 4

When the T bit is not set, the code field contains one of the initialization commands shown in Figure 4. The commands are presented according to their precedence hierarchy, with the highest being the UP command. The four unallocated values are currently illegal and will be rejected. These initialization commands are used to signal the various stages of link initialization between the two TPs, and also to convey special actions under error conditions.

Link Initialization

Link initialization is required whenever a trunk recovers from a facility failure or is first brought into service. Each TP must pass through a self-test, and then establish successful communication with the far-end TP. Link initialization consists of exchanging control information, automatically measuring the facility speed and the round-trip delay, and selecting the appropriate SN field size. The facility can then be used by higher level processes to exchange test packets and measure the facility error rate. If the quality is deemed acceptable, the trunk is placed in service for user packets. The protocol definition deals with initializing the link and handling errors under normal packet flow.

After hardware self-tests are complete and the facility speed is measured, each TP continuously transmits UP control frames. Receipt of an UP control frame at any time always forces a TP to reset the link if it has not already done so. The UP control frame has the highest precedence and overrides any previous state. Once reset, receipt of an UP control frame from the far end directs the TP to send READY control frames and wait for the far end to catch up. Receipt of a READY control frame means both TPs have completed this basic handshake protocol and are now ready to measure the facility delay.

For the round-trip delay measurement a DELAY PROBE control frame is transmitted by each TP. When received, a DELAY RESPONSE control frame is returned as quickly as possible. The processing delay should not exceed 15% of the round-trip facility delay in order that both TPs calculate approximately the same delay. Each TP then independently selects the most appropriate SN field size based on that facility speed and delay. The SN field size must be large enough to accommodate more than the maximum number of packets that could be in flight and awaiting ACK. The three allowed values are 5, 13 or 21 bits corresponding to a maximum SN range of 31, 8192 and 2 million. A short terrestrial trunk would use a 5-bit field, a high speed satellite trunk a 21-bit field. This strategy maximizes bandwidth available for user data by automatically selecting only the necessary overhead for each particular trunk.

Each TP conveys its SN field size choice in a LENGTH CHECK control frame. If the other TP agrees, it returns a LENGTH OK control frame. If both ends do not agree on the same field width, the round-trip delay is remeasured. If disagreement persists after a few attempts, the longer of the two field lengths is adopted. At this point the link is considered initialized and normal packet transmission may occur. The transmit SN is set to 1 and the TP begins to read its packet queues.

Errors and Recovery

Errors can be classified into transient and link affecting. Transient errors force the link to restart frame numbering and retransmit all unacknowledged frames, but the link does not have to be reinitialized and retested. Typically this takes less than one second before normal packet flow commences again. Transient errors can be caused by microwave fades or short periods of interference on the facility. The result as seen by the TP can be any one of the following:

- A gap in the packet ACKs resulting in an unknown status of one or more packets. This only occurs after the loss of at least four consecutive ACKs.
- A timeout waiting for the last information frame to be ACKed.
- Closure of the transmit window for longer than allowed.
- Transmit SN wrapped around without ACK having been received.
- Any other loss of contact with the far-end TP.

Whenever one of the above transient error conditions is encountered, the TP enters the Are You There (AYT) mode. This stops all information frame flow in the affected direction and causes the TP to continuously send AYT control frames. Normal traffic flow may continue in the other direction. The receiving TP will respond with Yes I Am (YIA) control frames interspersed with normal information frames. When a YIA control frame is received by the transmitter, the transmit SN is reset to 0 and all unacknowledged packets are retransmitted. The TP will then resume normal packet flow. If no YIA control frame is received within the specified interval of approximately one second, the link in both directions is declared down.

All other errors are classified as link affecting and result in both TPs going through full self-test and link initialization. Each TP signals this to the far-end by transmitting UP control frames. From the network perspective the trunk is taken out of service and the routing system will choose alternate trunks to carry user packets.

A trunk facility may be accidentally looped, causing each TP to transmit to itself. This condition must be quickly detected and the trunk taken out of service. To achieve this the trunk supervisor process assigns one TP as odd and the other TP as even. Each TP periodically sends the appropriate LOOP ODD or LOOP EVEN control frame. Receipt of the wrong loop control frame causes the TP to immediately reset.

The purpose of flow control at the trunk level is to reduce or stop data flow when the receiver is too busy or can not get memory buffers fast enough. A higher supervisor level may also request flow control. Flow control at the link level is not intended to be used for network congestion control. The maximum number of packets that may be transmitted before the transmit SN window closes is implementation dependent, but must be less than the maximum SN field size allowed. Usually a default value is specified at link initialization. The WINDOW SET and its associated WINDOW ACK control frames allow the TPs to dynamically change the active transmit window without resetting the link. These two control frames carry the new window value to be adopted. The new value must be acknowledged before it becomes active, requiring a delay at least equal to the round trip facility delay. Hundreds of frames may still be received on a satellite link after a new window size is requested, so adequate packet buffering must be available.

Under conditions of light traffic, the last information frame must be acknowledged periodically to maintain TP to TP communication. This and the periodically expected loop-check control frames provide the only means of maintaining contact and ensuring the link is still active.

Conclusion

The protocol described in this paper was implemented in the spring of 1984 as part of the new SL-10 Universal Trunk Processor (UTP). The UTP offers full satellite or terrestrial network trunking at speeds up to 192 Kb/s. Extensive trunk processor and system verification occurred in a laboratory environment at data rates as high as 256 Kb/s. The characteristics of a satellite link were simulated using custom built equipment capable of introducing a round-trip delay of up to 1.2 seconds. Link errors could also be simulated by the same equipment, including frequency and duration of bit inversion and random link fades. Much of the testing was done at full link utilization for periods exceeding 24 hours. A field trial is underway now in the DATAPAC network.

The protocol and the UTP have met all the objectives. The multiple selective retransmission feature coupled with the built-in ACK redundancy allowed both satellite and terrestrial trunks to continuously maintain 100% link occupancy under very high error rates. The ability to automatically select the minimum SN range size has maximized bandwidth available for user data.

References

[1] D. Sproule and F. Mellor, "Routing, Flow, and Congestion Control in the Datapac Network", IEEE Trans. on Communications, Vol. Com-29, No 4, April 1981.

[2] S. Lin, D. J. Costello Jr., and M. J. Miller, "Automatic-Repeat-Request Error-Control Schemes", IEEE Communications Magazine, Vol. 22, No. 12, pp. 5-17, Dec 1984.

[3] European Computer Manufacturers Association (ECMA), "HDLC / CM Elements of Procedure (Checkpoint Mode)", Working Paper ECMA / TC24 / 81 / 38, April 1981.

PROTOCOLS FOR LARGE DATA TRANSFERS OVER LOCAL NETWORKS

Willy Zwaenepoel

Department of Computer Science
Rice University
Houston, TX 77251

Abstract

In this paper we analyze protocols for transmitting large amounts of data over a local area network. The data transfers analyzed in this paper are different from most other forms of large-scale data transfer protocols for three reasons: (1) The definition of the protocol requires the recipient to have sufficient buffers available to receive the data before the transfer takes place; (2) We assume that the source and the destination machine are more or less matched in speed; (3) The protocol is implemented at the network interrupt level and therefore not slowed down by process scheduling delays.

We consider three classes of protocols: *stop-and-wait, sliding window* and *blast* protocols. We show that the expected time of blast and sliding window protocols is significantly lower than the expected time for the stop-and-wait protocol, with blast outperforming sliding window by some small amount. Although the network error rate is sufficiently low for blast with full retransmission on error to be acceptable, the frequency of errors in the network interfaces makes it desirable to use a more sophisticated retransmission protocol. A go-back-n strategy is shown to be only marginally inferior to selective retransmission and is, given its simplicity, the retransmission strategy of choice.

Our results are based on measurements collected on SUN workstations connected to a 10 megabit Ethernet network using 3-Com interfaces. The derivation of the elapsed time in terms of the network packet error rate is based on the assumption of statistically independent errors.

1. Introduction

Recent studies have shown the importance of using large page sizes in order to achieve high performance file access, both locally as well as over a network [10,12,15]. This is due to economies in accessing the disk in large quantities as well as to economies in accessing the network in large quantities. In this paper we study the latter phenomenon. In particular, we study the performance of protocols for transmitting large amounts of data across a local network characterized by a low error rate, low propagation delay and high bandwidth.

By *large* amounts of data, we denote amounts that are one or two orders of magnitude bigger than the maximum network packet size. We show how our analysis can be extended to larger sizes such as those involved, for instance, in remote file system dumps. We study three classes of protocols: *stop-and-wait, sliding window* and *blast* protocols (See Figure 1). With *stop-and-wait* protocols, the source refrains from sending a packet until it has received an acknowledgement for the previous packet. With a *blast* protocol all data packets are transmitted in sequence, with only a single acknowledgement for the entire packet sequence. Different protocols within the category of blast protocols are distinguished by their retransmission strategies (e.g. all packets can be retransmitted, or some form of selective retransmission can be used). With *sliding window* protocols every packet is individually acknowledged but the sender continues to transmit data without waiting for an acknowledgement. In typical sliding window protocols, the sender is silenced when the window "closes". Here we assume that the window is large enough so that it never gets closed.

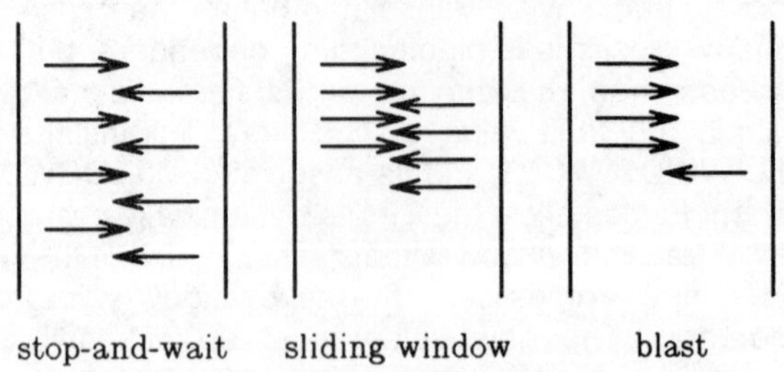

Figure 1: Stop-and-Wait, Sliding Window and Blast Protocols

For error-free transmissions, stop-and-wait protocols do not perform as well as sliding window or blast protocols, because of delays in waiting for the acknowledgement for every packet. Given the low latency and the high bandwidth of local networks, one would expect the difference in performance to be rather small. However, experimental evidence shows that the penalty for using a stop-and-wait protocol on a local network is substantially higher than expected, because of significant extra delays in generating and receiving packets. Sliding window protocols are slightly inferior to blast protocols, because of the overhead involved in handling the extra acknowledgements.

We then consider the performance of the protocols in the presence of transmission errors on the network. Given typical error rates on a local network, the expected elapsed time of a given transmission is almost identical to the error-free transmission time. As a result, under normal local network operating conditions, blast and sliding window protocols outperform stop-and-wait protocols. In fact, *network error* rates are sufficiently rare to make it possible to use full retransmission on error in conjunction with a blast protocol without significant degradation in the expected elapsed times. However, frequent *interface errors* force a more sophisticated retransmission strategy in order to maintain a near-optimal expected time and a small standard deviation.

Our results are based on measurements of error-free transmission times between SUN workstations connected to a 10 megabit Ethernet network using a 3-COM Multibus interface, and on a probabilistic analysis of the performance in the presence of errors. Since our measurements are done in the absence of any substantial network load, contention delays are all but absent from the results. Our conclusions are therefore valid only under low load conditions. Fortunately, such conditions are typical of most local network based systems. We also use delay under low load as a measure of performance, rather than throughput under high load, because low load conditions are so prevalent. In the error analysis, we assume that packet transmissions are statistically independent events with a constant failure probability. In practice, this assumption is a reasonable approximation of reality, although burst errors occasionally occur.

The outline of the rest of this paper is as follows. Section 2 describes measurements of error-free transmissions using each of the three protocols considered. In Section 3, we compare the performance of these three protocols in function of the error rate of the network. In Section 4, we study different retransmission strategies that can be used in conjunction with blast protocols. Related work is covered in Section 5 and conclusions are drawn in Section 6.

2. Error Free Data Transmissions

The large data transfers discussed in this paper occur as part of the interprocess communication functions provided by the V kernel [4,6]. The V kernel is a distributed operating systems kernel, currently implemented on SUN workstations [2] connected to a 10 megabit Ethernet [8] by a 3-Com interface [1]. As part of its interprocess communication facilities, the V kernel provides two operations — **MoveTo** and **MoveFrom** — which allow one process to move an arbitrary amount of data from its address space into the address space of another process, or vice versa. Both operations are *network transparent:* the destination process may or may not be on the same machine as the source process. By definition of the V interprocess communication primitives (see [4,6]), the recipient has sufficient buffers allocated to receive the data prior to the transfer. For instance, when a process wants to read an entire file into its address space, it first allocates a buffer big enough to contain that file. It then sends a message to the file server indicating the starting address of the buffer and its length. If necessary, the file server reads the file from disk, and then uses **MoveTo** to move the file from its address space into that of the client. In the local case, the fact that the client's buffer is already allocated allows the kernel to move the data from the source to the destination address space without an intermediate copy. In the remote case, it allows the kernel(s) to move data from the source address into the network interface of the sending machine, and from the network interface of the receiving machine into the destination address space, again without an intermediate copy.[1]

The total time necessary to execute a **MoveTo** or a **MoveFrom** is the sum of the cost of network communication and the cost of kernel overhead. In Section 2.1, we describe a set of experiments to quantify the cost of network communication. Measurements with the V kernel implementation of **MoveTo** and **MoveFrom**, including both network communication and kernel overhead, are discussed in Section 2.2.

2.1. Network Communication Overhead

On a local network, one would expect a blast or a sliding window protocol to perform only marginally better than a stop-and-wait protocol, because of the low propagation delay and the high bandwidth of the network. For instance, assume that $N = 64$ kilobytes of data have to be transferred over a 10 megabit Ethernet. Assume that we transmit the data in 1 kilobyte packets and that the acknowledgement packets are 64 bytes in size. Based on the 10 megabit data rate of the network, the transmission of a data packet takes $T = 820$ microseconds and the transmission of an acknowledgement packet takes $T_a = 51$ microseconds. The latency of the network τ can be estimated to be below 10 microseconds. If a stop-and-wait protocol is used, and no errors occur, the transfer is complete (including the receipt of the last acknowledgement at the source) after

$$N \times (\ T + T_a + 2 \times \tau\ ) \ \textit{or}\ 57024\ \textit{microseconds}$$

For a sliding window protocol, with a window large

[1] At least with suitable network interfaces.

enough so that it never closes during transmission, the elapsed time becomes

$$N \times (T + T_a) + 2 \times \tau \text{ or } 55764 \textit{ microseconds}$$

If the data is transmitted by a blast protocol, then the resulting elapsed time is

$$N \times T + T_a + 2 \times \tau \text{ or } 52551 \textit{ microseconds}$$

None of these results differ from each other by more than 10 percent. The experiment described below contradicts the above line of reasoning: the stop-and-wait protocol takes about twice as much time as either the sliding window or the blast protocol. We first describe the experiment and its results, and then explain this somewhat counterintuitive outcome.

2.1.1. Experimental Method and Results

In order to quantify the cost of network communication, two standalone programs are run on two different machines connected to the network. One program acts as the source of data and the other as the destination. Data is transmitted from the source, and acknowledgements are returned from the destination as appropriate, according to which protocol is used. Data packets are 1024 bytes in length, while acknowledgements are 64 bytes. For statistical accuracy, the experiment is repeated a number of times and the results are averaged. In all measurements, the network is essentially idle, so no significant contention delay is experienced. The transfers are implemented at the data link layer and device level so that no protocol or process switching overhead appears in the results. In particular, no header (other than the Ethernet data link header) is added to the data, and no provisions are made for demultiplexing packets, or for retransmission. If a transmission error occurs, the experiment halts and is restarted. Also, each of the two programs simply busy-waits on the completion of its current operation, in order to avoid interrupt handling overhead. The experiment therefore provides an accurate approximation of the communication cost involved in the data transfer. Measurement results for the elapsed time of multi-packet transfers in stop-and-wait (SAW), sliding window (SW) and blast (B) mode are reported in Table 1.

Size	SAW	SW	B
1 Kb	4.1 ms.	4.1 ms.	4.1 ms.
8 Kb	32.7 ms.	21.7 ms.	19.8 ms.
64 Kb	261.6 ms.	161.5 ms.	141.1 ms.
512 Kb	2,093.0 ms.	1,284.0 ms.	1,149.0 ms.

Table 1: Standalone Measurements of Error Free Transmissions

2.1.2. Interpretation

Let us first consider data transfers that fit within a single network packet, for instance a 1 kilobyte transfer.[2] The reliable transfer of 1 kilobyte of data, using a 64 byte acknowledgement, takes 4.1 milliseconds (See Table 1). This is significantly more than the transmission time of the data and the acknowledgement packet, which would be 0.87 milliseconds, when computed at the 10 megabit data rate of the network. The difference between the network time, computed at the network data rate, and the measured elapsed time is accounted for almost exclusively by the time necessary for the processors to copy the packets into and out of the interface (See Figure 2).

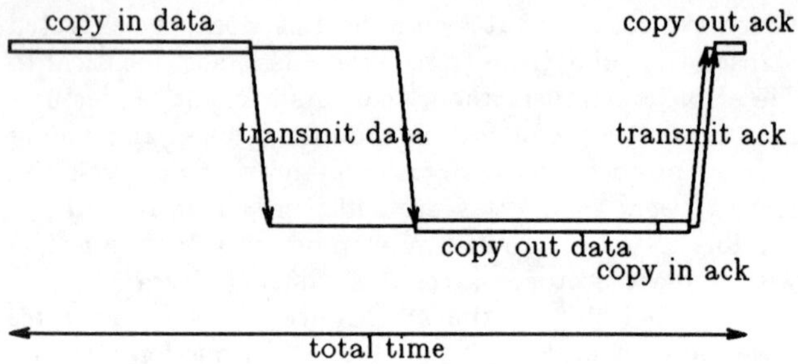

Figure 2: Network Packet Transmission

Table 2 shows a breakdown of the total elapsed time over its various components. As can be seen, of the 4.1 milliseconds total elapsed time, only 21 percent is network transmission time, while 75 percent is copying overhead, the rest (presumably) being network and device latency.

Operation	Time
Copy data into sender's interface	1.35 ms.
Transmit data	0.82 ms.
Copy data out of receiver's interface	1.35 ms.
Copy ack into receiver's interface	0.17 ms.
Transmit ack	0.05 ms.
Copy ack out of sender's interface	0.17 ms.
Total	3.91 ms.
Observed elapsed time	4.08 ms.

Table 2: Breakdown of Transmission Cost over its Various Components

Let us now consider the case where the data transfer requires N packets to be sent from the source to the destination. The reason for the superior performance of the blast and sliding window protocols is explained in Figure 3. Figure 3.a corresponds to the transmission in stop-and-wait mode, Figure 3.b corresponds to the blast transmission, and Figure 3.c depicts the sliding window protocol. The time axis runs horizontally from left to right and the example is for the case of N=3.

---

[2] The maximum packet size on the 10 megabit Ethernet is 1536 bytes.

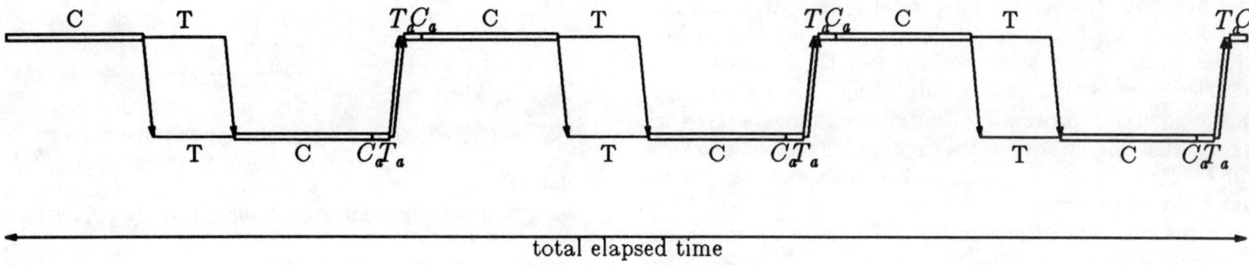

Figure 3.a: Stop-and-Wait

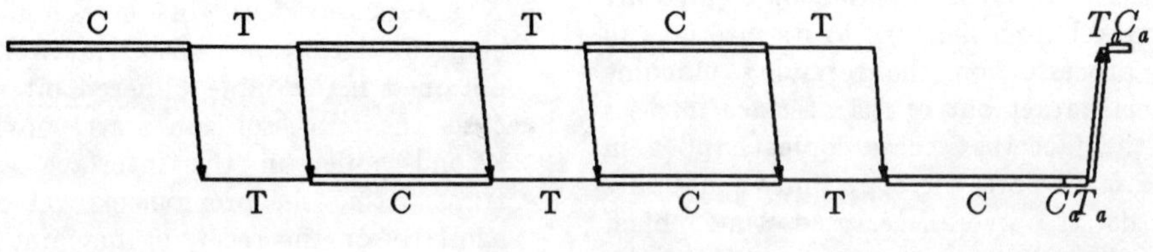

Figure 3.b: Blast Protocol

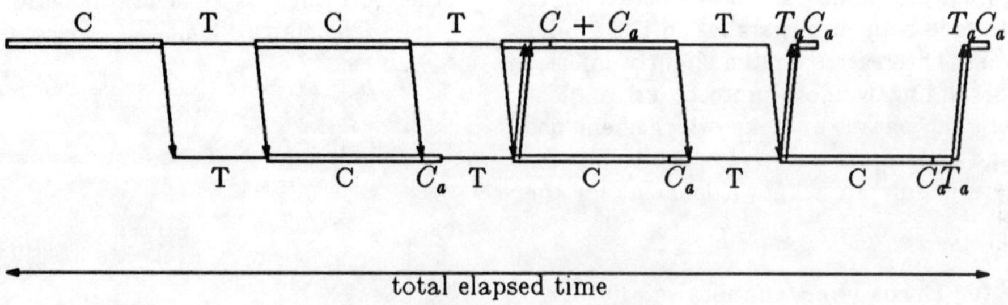

Figure 3.c: Sliding Window Protocol

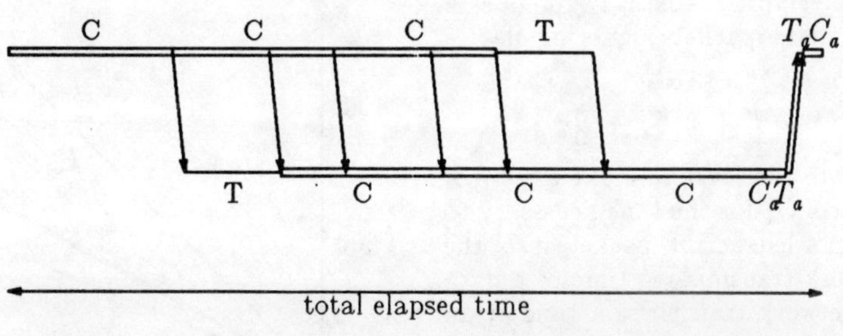

Figure 3.d: Double Buffered Interface with Blast Protocol

Consider first the sequence of events in the case of a stop-and-wait protocol. The sending processor copies a packet from main memory to its interface and then the interface puts the packet on the network. After a time period equal to the network propagation delay the packet arrives at the receiver's interface and then it is copied from the receiver's interface into the receiver's memory by the receiving processor. This process repeats itself in the reverse direction for the acknowledgement packet, and then again for the next packet, and so forth. Note that the two processors are never active in parallel. This is not the case when the transfer is done in blast mode, as shown in Figure 3.b. Due to the very low propagation delay of a local network, the packet is received in the receiver's interface almost completely concurrently with the sender's interface transferring it over the network.[3] Therefore the processor on the sending machine can start copying the next packet from memory to its interface in parallel with the processor on the receiving machine copying the previous packet out of its interface into its memory. Due to the fact that these copies happen in parallel, and, as we saw before, the copy times contribute significantly towards the overall elapsed time, blast transfer results in elapsed times that are substantially lower than those obtained for stop-and-wait transfers, much lower than one would be lead to expect if only transmission time for acknowledgements were taken into account.

Finally, consider the operation of the sliding window protocol (Figure 3.c). Again, the copy operations in and out of the interface happen in parallel on the sender and on the receiver. The reason for the slightly inferior performance of the sliding window protocol vs. a blast protocol is that for each packet an acknowledgement has to be copied in and out of the interfaces, while for the blast protocol, there is only an acknowledgement for the last packet.

2.1.3. Formulas for Error Free Transmissions

From Figure 3, we can derive the following formulas for the elapsed times of multi-packet data transfers in the absence of errors. First, in the case of a stop-and-wait protocol, the total elapsed time T_{SAW} for a transmission requiring N data packets, ignoring device latency, is (See Figure 3.a)

$$T_{SAW} = N \times (\ 2\ C + T + 2\ C_a + T_a\ )$$

where C stands for the time necessary to make a copy of a data packet into interface, C_a for the time necessary to make a copy of an acknowledgement packet into the interface, T for the network transmission time of a data packet, and T_a for the network transmission time of an acknowledgement packet. If a blast protocol is used, then due to concurrent operation of the two processors, the elapsed time T_B becomes (see Figure 3.b)

$$T_B = N \times (\ C + T\ ) + C + 2\ C_a + T_a$$

For a sliding window protocol, the resulting elapsed time T_{SW} equals (see Figure 3.c)

$$T_{SW} = N \times (\ C + C_a + T\ ) + C + T_a$$

Note that the utilization of the network U_n, even when using a blast protocol, is still significantly below 100 percent.

$$U_n = \frac{N \times T + T_a}{N \times T + T_a + N \times C + C + 2\ C_a}$$

For instance, for the 64 kilobyte transfer shown in Table 2, the network utilization is only 38 percent. Better elapsed times and better network utilization can be obtained if a double buffered interface is used. In that case, the processor can start copying a packet into the second buffer in the interface while the interface is transmitting the previous packet over the network, and similarly on the receiving machine . Note that having a third transmission buffer does not provide any further improvement over double buffering, since we assume that both C and T are constant. The value of T is constant as long as there is no significant contention delay. The value of C is also constant since we assume the network transfer is the only activity occupying the processor, and therefore there is no delay in performing the copy operation. The elapsed time $T_{d/b}$ becomes (See Figure 3.d)

$$T_{d/b} = N \times C + T + C + 2\ C_a + T_a \quad (\ T \leq C\ )$$

$$T_{d/b} = N \times T + 2C + 2\ C_a + T_a \quad (\ T \geq C\ )$$

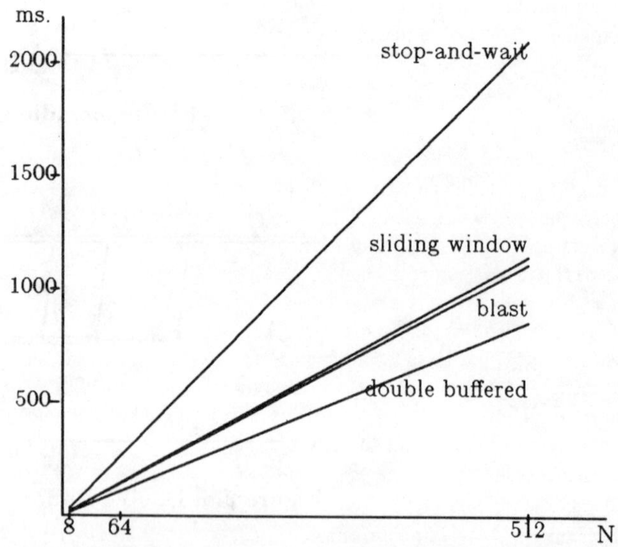

Figure 4: Comparison of Different Protocols

[3] In fact, the propagation delay is far exaggerated in Figures 2 and 3 to make it visible at all: typical propagation delays on a local network are on the order of 10 microseconds while the copy and transmission times depicted in Figure 2 and 3 are on the order of 1 millisecond.

Figure 4 compares the performance of the different protocols in terms of N for values of C, C_a, T and T_a as on the SUN workstation with a 10 megabit Ethernet (See Table 2).

One might wonder whether it is possible to get rid of the copy into the interface altogether, by simply moving the data from main memory onto the network. An interesting design that allows network access without an intermediate copy appears in the Xerox Alto personal computer, where network access is incorporated as an independent task in the processor's micro-engine [20]. The copy can also be avoided by virtual memory techniques (if the origin and the destination of the data are aligned on a page boundary). The Apollo Domain architecture supports this feature [11]. For more conventional architectures, one would like a DMA interface to copy the data from main memory directly onto the network. Most DMA interfaces do not allow such a direct copy. For instance, the Excelan DMA interface first copies the data into on-board buffers before it transmits it on the network [9]. The CMC interface allows the programmer to define the host-board interface [7]. However, the manual strongly recommends that data be copied first into on-board buffers before it is transmitted onto the network, due to limited bus bandwidth and due to the possibility of the host processor timing out on memory access over the bus, while the DMA processor is accessing memory [7,16]. Both interfaces seem to require a copy, albeit that the copy is performed by the interface processor rather than by the host processor. The formulas derived above for the elapsed time therefore remain valid, provided that C and C_a are no longer the time required for the host processor to make the copies, but rather the time required for the DMA processor to make the copies. Our experience with DMA interfaces has been mixed in this respect. With the Excelan board, the copy performed by the 8088 interface processor is much slower than the copy performed by the 68000 host processor into the 3-Com interface. We have no experience to date with the performance of the CMC interface (which contains a 68000 processor).

In summary, it seems that the elapsed time is not significantly improved by using currently available DMA interfaces. The amount of host processor utilization for network access is decreased, since the interface processor can perform the copy, although bus traffic slows down the processor somewhat during DMA operation. In general, the relative importance of the copy operation indicates that memory and bus bandwidth are the critical factors. Therefore, it seems likely that a processor with a fast block move operation, accompanied by very high speed device memory is more promising than any kind of special purpose hardware on the interface.

2.2. Large Data Transfers at the Kernel Level

The protocol used for the set of measurements in the previous section assumes that the network is error-free. When an error occurs, the experiment is terminated and has to be (manually) restarted. A real protocol has to deal with the possibility of errors on the network by some form of retransmission strategy. This introduces overhead, even for error-free transmissions. Additionally, in the experiment above network access is the only task that the processor has to perform. There is no multiplexing, no access rights checking, and the processor busy-waits on the completion of a transmission in order to start the next one. In a real system, the processor has to be shared by a number of different tasks, that have to be protected from each other, and network I/O has to be performed in an interrupt-driven rather than a busy-wait fashion.

All of these requirements have been implemented as part of the V kernel's network interprocess communication. Table 3 gives the results of our measurements of the V kernel's **MoveTo** operation.[4] The results only confirm the measurements of Section 2.1. The extra overhead stems from the transmission of the headers, as well as from access right checking, demultiplexing and interrupt handling. The formulas derived for the elapsed times under various combinations of transmission protocols and network interfaces remain valid, if the extra overhead is added to C, and C_a. For instance, for the blast protocol, the modified values of C and C_a are 1.83 and 0.67 milliseconds, vs. 1.35 and 0.17 milliseconds in the standalone experiments. The relative increase of C and C_a, compared to T and T_a, makes the blast protocol even more advantageous than in the case of a standalone program. In fact, in the case of the V kernel, the extra overhead is relatively minimal compared to other protocol implementations, which suggests that the use of a blast protocol would be even more advantageous for other implementations.

Size	SAW	B
1 Kb	5.9 ms.	5.9 ms.
8 Kb	43.1 ms.	24.8 ms.
64 Kb	340.2 ms.	173.0 ms.
512 Kb	2,719.0 ms	1,370.0 ms.

Table 3: V Kernel Moveto Measurements

3. Effect of Transmission Errors

So far we have considered the performance of the various protocols for the **MoveTo** operation in the case of error-free transmissions. We now turn our attention to their performance in the presence of transmission errors on the network. We first compare stop-and-wait with retransmission of a single packet after a time interval T_r to blast with retransmission of the full sequence of packets after T_r. We show that, for typical local network error rates, the expected time of the blast transmission is

[4] Measurements for the sliding window protocol are not available at the time of writing. We would expect the elapsed times for sliding window to be slightly higher than those for blast, as suggested by the standalone measurements in Table 1.

significantly better than the expected time of the stop-and-wait transmission. We conclude therefore that under normal operating conditions, a blast protocol is superior to a stop-and-wait protocol for use on a local network. We then consider various retransmission strategies that can be used in conjunction with a blast protocol. We do not consider the sliding window protocol in any great detail in this section. Its error characteristics are similar to those of the blast protocol with selective retransmission (See Section 3.2.3).

In this analysis, we assume that packet transmissions are statistically independent events which can fail with probability p_n. This is a reasonable first order approximation of the behavior of the network. Analysis of the performance under other error distributions is beyond the scope of this paper.

3.1. Expected Time

3.1.1. Stop-and-Wait

Denoting by $T(D)$ ($T(1)$) the time necessary for a D-packet (1-packet) transfer, we obviously have

$$T(D) = D \times T(1)$$

The probability p_e of a 1-packet exchange failing is

$$p_e = 1 - (1 - p_n)^2$$

and the probabilities $s(i+1)$ of the exchange succeeding on the (i+1)-th transmission attempt form a geometric distribution with parameter p_e

$$s(i+1) = p_e^i \times (1 - p_e)$$

The expected time for a 1-packet transfer to complete is then

$$T_0(1) + (T_0(1) + T_r) \times (\frac{p_e}{1-p_e})$$

where $T_0(1)$ is the time necessary for a 1-packet exchange without any errors.[5] For D packets, we get for the expected time

$$\mu = D \times [T_0(1) + (T_0(1) + T_r) \times (\frac{p_e}{1-p_e})]$$

3.1.2. Blast

Let us next consider the performance of a blast protocol with full retransmission on error (without a negative acknowledgement). A D-packet transfer succeeds in this case if all D packets reach the destination machine and the acknowledgement packet reaches the source machine. Assuming independent transmissions, the probability of the D-packet transfer failing is then

$$p_e = 1 - (1 - p_n)^{D+1}$$

The probabilities s(i+1) that a transmission succeeds on the (i+1)-th transmission attempt form a geometric distribution with parameter p_e.

$$s(i+1) = p_e^i \times (1 - p_e)$$

The expected time T(D) for a D-packet transfer becomes

$$\mu = T_0(D) + (T_0(D) + T_r) \times (\frac{p_e}{1-p_e})$$

3.1.3. Comparison

Figure 5 compares the two strategies for different values of T_r (The other parameters in the figure are D = 64, $T_0(1) = 5.9$ *msec.* and $T_0(D) = 173$ *msec.*, from Table 3). In addition to these curves we need some idea about the error rate on a 10 megabit Ethernet. Surprisingly enough, very little empirical data is available about the error rates on local networks. Shoch and Hupp report an observed error rate of 1 in 200,000 packets on the experimental Xerox PARC 3 megabit Ethernet [17]. Our measurements on our local 10 megabit Ethernet indicate an error rate of approximately 1 in 100,000 under normal circumstances. However, when one station transmits at full speed to another workstation, the error rates rise an order of magnitude, to approximately 1 in 10,000. We assume that most of the additional errors are due to failures in the 3-COM Ethernet interface (See also [5,13] for additional evidence of large packet losses in interfaces). We therefore operate somewhere in the region between 10^{-4} and 10^{-5} in Figure 5. In comparing the results for the stop-and-wait protocol and the blast protocol, the key observation to make is that $T_0(D)$ — the error-free transmission time for the blast protocol — is significantly smaller than $D \times T_0(1)$ — the comparable value for the stop-and-wait protocol (See Section 2). Consequently, for low error rates where this term dominates, the blast strategy performs significantly better. While the network error rate allows us to operate in the flat part of the curve, the frequency of errors in the interfaces actually forces us partly into the beginning of the knee of the curve. Nevertheless, the expected time of the blast protocol is still notably better than that of the stop-and-wait protocol.

These results also allow us to make a stronger conclusion: since the expected time for a blast protocol with the crudest retransmission strategy — full retransmission on error and no negative acknowledgement — results in a nearly optimal expected time (for the appropriate range of p_n values), no significant improvements in expected time can be achieved by more sophisticated retransmission strategies. In the next section, we show that such strategies can significantly improve the standard deviation.

[5] The second occurence of $T_0(1)$ in this formula should strictly speaking be replaced by $T_0'(1)$, the elapsed of a *failed* transmission. In practice, the difference is minor and can be subsumed by slightly adjusting the value of T_r.

At this point, an observation is in order about the size of the data transfers used in a blast protocol. Clearly as the size of the data transfer increases, errors are more likely and retransmission becomes more costly. For such very large sizes, we suggest the use of ***multiple blasts***, whereby the transfer is broken up in a number of different blasts, each of which proceeds according to the definition of the blast protocol.

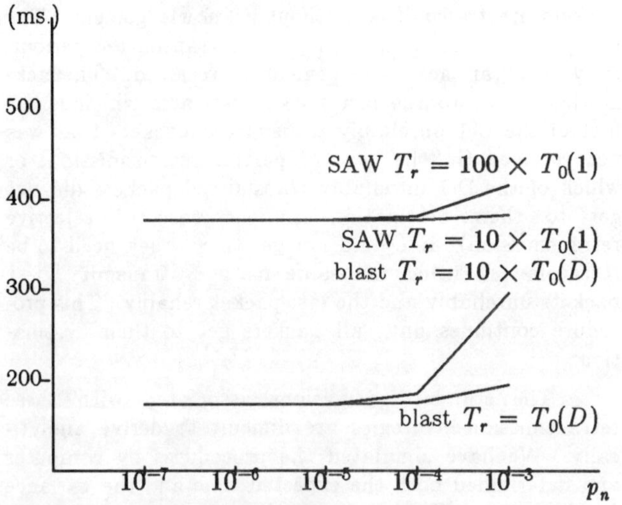

Figure 5: Expected Time for 64 kilobyte Transfers

3.2. Standard Deviation

We now analyze the standard deviation of different retransmission strategies that can be used in conjunction with a blast protocol. We assume that we operate under error conditions such that the expected time of the transfer is nearly identical to the error-free transmission time (i.e. we operate in the flat region of the curves in Figure 6).

Consider a given transmission strategy and denote by $T_0(D)$ the error-free transmission time. Furthermore, let $T_0^k(D)$ be the elapsed time for the k-th transmission attempt, let $T_r^k(D)$ be the interval between the k-th and the (k+1)-th transmission attempt, and finally let $s(i+1)$ be the probability of success on the (i+1)-th transmission attempt. Then, if the transfer succeeds on the (i+1)-th transmission attempt, the total elapsed time for this transfer is

$$\sum_{k=0}^{i} T_0^{k+1}(D) + \sum_{k=0}^{i-1} T_r^{k+1}(D)$$

Assuming we are operating under low error conditions and that thus the expected time is constant and approximately equal to $T_0(D)$, we get for the variance

$$\sigma^2 = \sum_{i=0}^{\infty} [(\sum_{k=0}^{i} T_0^{k+1}(D) + \sum_{k=0}^{i-1} T_r^{k+1}(D))^2 \times s(i+1)] - T_0^2(D)$$

This formula indicates three potentially fruitful avenues for reducing the variance:

(1) Reduce the retransmission intervals $T_r^k(D)$: this can be accomplished either by choosing a small timeout value or by providing a negative acknowledgement when the transfer fails.

(2) Reduce the transmission time $T_0^k(D)$ for retransmissions: this can be done by reducing the number of packets to be sent on retransmission. A negative acknowledgement packet can carry information as to which packets were successfully received.

(3) Reduce the probability of failure of the retransmissions: since we are assuming independent failures, this probability is only dependent on the number of packets transmitted. Thus, here also reducing the number of packets sent has a beneficial effect.

Clearly, a combination of these different approaches is optimal. However, we analyze the different methods in isolation to assess their relative benefits. In particular, we consider the following retransmission strategies:

(1) Full retransmission on error without negative acknowledgement.

(2) Full retransmission on error with a negative acknowledgement after the last packet.

(3) Retransmission from the first packet not received (indicated in a negative acknowledgement)

(4) Selective retransmission of the packets not received (indicated in a negative acknowledgement).

Certain of these retransmission strategies lead themselves to exact analytical evaluation, while others are more easily evaluated by approximation or simulation. In the following, we only present the results of our study. Readers interested in the details of the derivation are referred to [21].

3.2.1. Full Retransmission on Error without Negative Acknowledgement

In the case of full retransmission on error without negative acknowledgement, the standard deviation of the elapsed time is easily shown to be

$$\sigma = (T_0(D) + T_r) \times (p_e^{1/2} \times \frac{(1+p_e)^{1/2}}{1-p_e})$$

$$p_e = 1 - (1 - p_n)^{D+1}$$

This function is set out against p_n for different values of T_r in Figure 6. It is clear from the above formula and from the figure that the value of T_r has a significant effect on σ, even for low error rates.

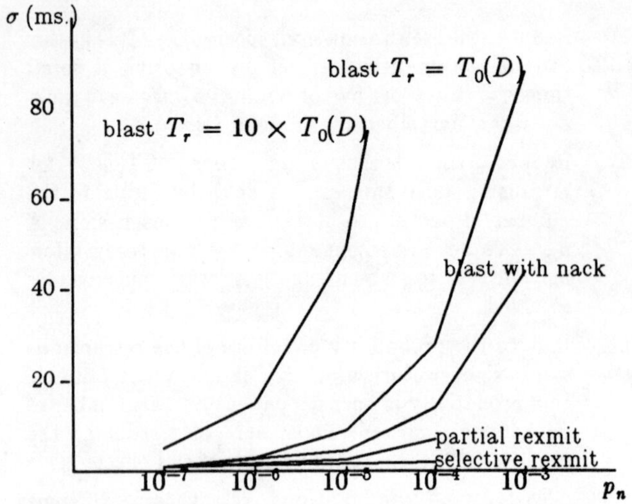

Figure 6: 64 Kilobyte MoveTo: Standard Deviation

3.2.2. Full Retransmission on Error with Negative Acknowledgement

In order to achieve a low standard deviation when using full retransmission, we need to choose T_r small w.r.t. $T_0(D)$. This can be done as shown in Figure 6 by choosing a small value of T_r. Alternatively, a small effective value of T_r can be achieved by using a negative acknowledgement, while still maintaining a much larger physical value of T_r. We use the following strategy:

(1) If the destination receives the last packet, it sends either a positive or a negative acknowledgement depending on whether or not it received all packets in the sequence.

(2) If the sender gets a negative acknowledgement, or if the sender does not receive any acknowledgement within a time interval T_r, it retransmits the whole sequence of packets.

The characteristics of this strategy can be derived by an approximative argument (See [21]). If $p_n \ll \frac{1}{D}$ and $D \gg 1$, then it can be shown that the standard deviation is approximately equal to

$$\sigma \approx T_0(D) \times (p_e^{1/2} \times \frac{(1+p_e)^{1/2}}{1-p_e})$$

This formula indicates that the standard deviation when using full retransmission with a negative acknowledgement is all but independent from the retransmission interval (for low error rates). The values of σ for different values of p_n are set out in Figure 6 for comparison with full retransmission without negative acknowledgement.

3.2.3. Partial and Selective Retransmission

By either choosing a small retransmission interval or by using negative acknowledgements, we have minimized the component of the standard deviation that is dependent on the retransmission interval. The standard deviation is also dependent on the amount of data retransmitted during retransmissions. This component can be minimized using either partial retransmission (from the first packet not received) or by selective retransmission (of the subset of packets not received).

These retransmission strategies are implemented as follows. In order to execute a D-packet transfer, (D-1) packets are transmitted without acknowledgement. The last packet is sent reliably, i.e. it is retransmitted periodically until an acknowledgement is received. The acknowledgement to the last packet indicates which is the first of the D-1 unreliably transmitted packets that was not received (in the case of partial retransmission) or which of the D-1 unreliably transmitted packets did not get to their destination (in the case of selective retransmission). If D' did not get there, they need to be retransmitted using the same method: transmit $D'-1$ packets unreliably and the last packet reliably. This procedure continues until all packets get to their destination.

The standard deviations associated with these retransmission strategies are difficult to derive analytically. We have simulated the procedures by computer and determined both the expected time and the variance from the simulation. Figure 6 shows the standard deviation observed in the simulation.

3.2.4. Standard Deviation: Summary

Figure 6 presents a comparison of the standard deviation of the four retransmission strategies that we consider. Full retransmission without a negative acknowledgement produces unacceptable variations in the elapsed times of the transfers (for realistic retransmission intervals). Use of a negative acknowledgement reduces these variations drastically. Given the presence of such a negative acknowledgement, the extension to partial retransmission starting from the first packet not successfully received is trivial and provides further reduction of the variance. We have chosen not to use selective retransmission, given the additional complexity this introduces in the protocol software and given that the improvement in performance is not very significant. If one were to consider networks with higher error rates or much larger transmissions, selective retransmission might become worthwhile.

4. Related Work

Protocols to support network interprocess communication have been the subject of a number of recent research papers. The protocol supporting the V kernel interprocess communication has been described in [4,6]. Another interesting example appears in Cedar RPC, although this design refrains from using blast protocols for large transmissions [3]. The author first became aware of the name *blast* protocol in conjunction with protocols developed at MIT for downloading screen images from a VAX to an Alto. The idea of blast protocols has been mentioned by various other authors, including Spector who calls it a multi-packet transfer and suggests using an overall software checksum on the entire data segment [18]. Needham mentions the use of a *transmit-and-pray* protocol for file transfers on the Cambridge ring: it is essentially a disk-to-disk (rather than machine-to-machine) blast protocol with full retransmission on error [14].

A large body of work is concerned with the performance of various transmissions strategies such as stop-and-wait, go-back-n and selective retransmission [19]. Most of these analyses assume that the network is a scarce resource, to be shared in an efficient way by a large number of users, and therefore use throughput under high offered load as a measure of performance. The networks studied usually have high error rates (and frequently high latency, such as satellite networks). Although some of them consider the use of cumulative acknowledgements, few consider delaying the acknowledgement altogether until the end of the transmission. Their analysis needs to be reconsidered in a local network environment, where network bandwidth is plentiful, errors are rare, and low delay under low load is more important than high throughput under high load. In fact, most of this work ignores the software cost of generating and receiving the packets, which dominates the transmission cost in a local network environment.

5. Conclusions

The V interkernel protocol has been designed to aggressively take advantage of local network characteristics such as low error rate, high bandwidth and low latency. In order to do so, the software overhead involved in dealing with network interprocess communication must be minimized. As a result, the protocol has been implemented at the network interrupt level and assuming communication partners that are more or less matched in speed. For large data transfers, client buffers are made available prior to the transfer, so that no intermediate copies need to be made.

In such a "tight" implementation of the protocol, we have shown that the overhead of copying data in and out of the network interfaces is a dominating factor in the overall elapsed time. Since blast protocols and sliding window protocols allow these copies to occur in parallel on the source and the destination machine, they perform substantially better than stop-and-wait protocols.

We have also considered the effect of transmission errors on the performance of various protocols. Experimental evidence suggests that network errors are relatively rare, but that interface errors occur more frequently, especially if the devices are driven at full speed. Given the network error rate, it would be acceptable to use full retransmission on error in conjunction with a blast protocol. The frequency of errors in the interface causes such a strategy to have unacceptable variations in the elapsed times. We have argued for a partial retransmission strategy starting from the last packet not received by the destination, since it is simple to implement and not significantly worse than more complicated strategies.

Acknowledgements

The original V kernel protocol was developed by the author at Stanford University under the direction of David Cheriton. I also wish to thank Guy Almes, David Cheriton and Allan Porterfield for their comments on an earlier draft of the paper.

References

[1] 3-COM Corporation, Multibus Ethernet (ME) Controller Model 3C4000, Reference Manual (May 1982).

[2] A. Bechtolsheim, V.R. Pratt and F. Baskett, The SUN Workstation Architecture, Technical Report 229, Computer Systems Laboratory, Stanford University (February 1982).

[3] A.D. Birrell and B.J. Nelson, Implementing Remote Procedure Calls, ACM Transactions on Computer Systems, Vol. 2, No. 1, pp. 39-59 (February 1984).

[4] D.R. Cheriton, The V Kernel: A Software Base for Distributed Systems, IEEE Software, Vol. 1, No. 2, pp. 19-42 (April 1984).

[5] D.R. Cheriton and P.J. Roy, Performance of the V Storage Server: A Preliminary Report, Proc. of the 1985 ACM Computer Science Conference, pp. 302-308 (March 1985).

[6] D.R. Cheriton and W. Zwaenepoel, The Distributed V Kernel and its Performance for Diskless Workstations, Proc. of the 9th Symposium on Operating System Principles, pp. 129-140 (October 1983).

[7] Communication Machinery Corporation, ENP Family Product Brief (February 1984).

[8] Digital Equipment Corporation, Intel Corporation and Xerox Corporation, The Ethernet: A Local Area Network - Data Link Layer and Physical Layer, Specifications, Version 2.0.

[9] Excelan, EXOS 101 Ethernet Front-End Processor.

[10] E.D. Lazowska, J. Zahorjan, D.R. Cheriton and W. Zwaenepoel, File Access Performance of Diskless Workstations, accepted for publication in ACM Transactions on Computer Systems.

[11] P.J. Leach, P.H. Levine, B.P. Douros, J.A. Hamilton, D.L. Nelson and B.L. Stumpf, The Architecture of an Integrated Local Network, IEEE Journal on Selected Areas in Communications, Vol. SAC-1, No. 5, pp. 842-857, November 1983.

[12] M.K. McKusick, W.N. Joy, S.J. Leffler and R.S. Fabry, A Fast File System for UNIX, ACM

Transactions on Computer Systems, Vol. 2, No. 3, pp. 181-197 (August 1984).

[13] J. Nabielsky, Interfacing to the 10 Mbps Ethernet: Observations and Conclusions, Proc. '84 ACM SigComm Conference, pp. 124-131 (June 1984).

[14] R.M. Needham, Systems Aspects of the Cambridge Ring, Proceedings of the Seventh ACM Symposium on Operating System Principles, pp. 82-85 (December 1979).

[15] J. Ousterhout, H. Da Costa, D. Harrison, J.A. Kunze, M. Kupfer, J.G. Thompson, A Trace-Driven Analysis of the Unix 4.2BSD File System, Technical Report, Computer Science Division, University of California at Berkeley.

[16] P.J. Roy, private communication.

[17] J. Shoch and J.A. Hupp, Measured Performance of an Ethernet Local Network, Communications of the ACM, Vol. 23, No. 12, pp. 711-721 (December 1980).

[18] A.Z. Spector, Multiprocessing Architectures for Local Computer Networks, Technical Report STAN-CS-81-874, Department of Computer Science, Stanford University (August 1981).

[19] A.S. Tanenbaum, Computer Networks, Prentice-Hall, 1981.

[20] C.P. Thacker, E.M. McCreight, B.W. Lampson, R.F. Sproull and D.R. Boggs, Alto: A personal computer, Computer Structures: Principles and Examples (Eds. D.P. Siewiorek, C.G. Bell and A. Newell), McGraw-Hill, pp. 549-572, 1982.

[21] W. Zwaenepoel, Protocols for Large Data Transfers on Local Networks, Technical Report TR-85-23, Department of Computer Science, Rice University.

Panel 1

Ku-Band Earth Stations for Data Communications

Chairperson
J. Field
University of Waterloo
Panelists
D. Lyon, *M/A Com Linkabit*
E. Parker, *Equatorial*
E. Yousefzadeh, *Satellite Technology Management*

Session 2

Theoretical Foundations

Chairperson

A. Lyman Chapin
Data General

A NEW TECHNIQUE FOR GENERATING PROTOCOL TESTS

Krishan Sabnani *Anton Dahbura*

AT&T Bell Laboratories
Murray Hill, NJ 07974

ABSTRACT

A novel procedure presented here generates test sequences for checking the conformity of protocol implementations to their specifications. The test sequences generated by this procedure only detect the presence of many faults, but they do not locate the faults. It can always detect the problem in an implementation with a single fault.

A protocol entity is specified as a finite state machine (FSM). It typically has two interfaces: an interface with the user and with the lower-layer protocol. The inputs from both interfaces are merged into a single set I and the outputs from both interfaces are merged into a single set O. The implementation is assumed to be a black box.

The key idea in this procedure is to tour all states and state transitions and to check a unique signature for each state, called the Unique Input/Output (UIO) sequence. A UIO sequence for a state is an I/O behavior that is not exhibited by any other state.

1. INTRODUCTION

In the present communications revolution, public data networks with access protocols such as X.25 [Gr82] have been built in several countries. These networks can interconnect computers and terminals from different vendors. Before any equipment is connected to a public network, the network administrators check the protocol implementations to make sure that they conform to a standard and that they will not cause any problem to the network. This checking procedure is called certification [Ra83], [Bi83].

We give here a novel procedure for generating test sequences for protocols. A test sequence is a series of inputs and outputs. These inputs are applied to an implementation being tested and the outputs are compared with those in the test sequence. If these outputs match, then the protocol implementation conforms to the specification. The procedure given here can be used to automatically generate certifiers. The test sequences generated by this procedure only detect the presence of faults, but they do not locate faults. We assume here that the protocol is specified as a finite state machine (FSM). The implementation can be a piece of hardware, of software, or a hybrid of both.

The key idea in this procedure is to compute a unique input/output sequence for each state in the protocol specification. Then, we use these UIO sequences to generate a test sequence that visits each state and state transition. Two examples are given later to illustrate this procedure.

A test sequence generated by our procedure is similar to a new error-detecting code. Since the test exercises all the state transitions and checks unique signatures (UIO sequences) for all states, it captures a wide range of faults, but not all faults. It can be easily proved that a test sequence generated by this procedure correctly detects the problem in an implementation that has a single fault. A single fault means that the implementation has one incorrect edge label or one edge with an incorrect starting state or an incorrect tail state. We are presently working on the complete characterization of the set of faults that can be detected by a test sequence generated by the procedure given here.

The UIO sequences are not the same as Hennie's distinguishing sequences [He64]. For a test, Hennie's distinguishing sequence is the same for all states and different states are distinguished by distinct outputs. For the UIO sequences, the input portion is normally different for each state. This novel idea allows us to come up with shorter sequences than Hennie's distinguishing sequences. The length of a Hennie's distinguishing sequence, in the worst case, can be exponential in the number of states for the FSM being tested. On the other hand, each UIO sequence grows polynomially with the increase in the number of states in the FSM being tested. Hennie's test sequences check for all faults; but the test sequences generated by the procedure given here check for only some faults.

Even though protocol specification and verification has been an active research area for the past decade [BS83], the test sequence generation for communications protocols is a new area. Recently, some efforts have been made in this area. Piatkowski [Pi80] studied the feasibility of generating tests for the Advanced Data Communication Control Procedures (ADCCP). His conclusions were discouraging and the complete tests in his analysis were too long to be practical. Sarikaya and Bochmann [SB82] used some techniques developed for sequential circuits for certification. Linn and McCoy [LM83] used generative grammars for testing protocols. The major advantage of our procedure is that it is uniquely geared to protocols, while the earlier attempts used the techniques developed in other areas, such as sequential circuits and programming languages, to protocols.

In Section 2, the finite state machine model for the protocol specification is given. Then, we describe the procedure for generating a test-sequence from the specification in Section 3. Some general comments about the protocol testing problem are given in Section 4. Finally, we give some concluding remarks in Section 5.

2. MODEL

A protocol entity is modeled as a deterministic finite state machine (FSM) represented as a directed graph, $G=(V, E)$ where V is a finite non-empty set of vertices and E is a set of edges. The vertices are the states of a finite state machine. Each edge is labeled by an input and output operation. The input operation is an event that causes the state transition and the output operation is typically an externally visible event. The output operation can be null or invisible in some cases. We define I as a set of input operations and O as a set of output operations. Each label is given as (i/o) where i is an element of I and o is an element of O.

We assume that the FSM has an initial state, *IS*. In the directed graph, there exists a path from the initial state *IS* to each state. The machine can be moved from any state to its initial state by a reset input, *ri*. A protocol entity has typically two interfaces: interface with the user (or the upper-layer protocol) and with the lower-layer protocol respectively. The inputs from both interfaces are merged into a single set I and the outputs to both interfaces are merged into a single set O.

We assume that the implementation is a black box with an input port and an output port. The inputs to this black box are given at its input port and the outputs can be observed at its output port. We also assume that the implementation will produce the output to an input in less than a fixed amount of time T. We also assume that the implementation will not produce any output and will remain in its present state if it receives any unspecified input for its present state. This assumption makes our FSM model a fully-specified Mealy machine [Kl72].

During a test, we want to find out whether the input/output behavior of the implementation is the same as that of the specification. For conformance, the output sequences from the specification and the implementation must be the same for any input sequence.

In the rest of this section, we define some quantities used later in this paper.

For each edge e, we define three functions: *Head*(e), *Tail*(e), *Label*(e), where *Head*(e) gives the starting state for the edge e, *Tail*(e) gives the end state of e, and *Label*(e) gives the edge label of e. The UIO sequence for a state s will be represented as *UIO*(s).

A sequence is defined as a series of input/output pairs. In formal terms, a sequence S is given as

$$S= (i_1/o_1)(i_2/o_2)....\ (i_n/o_n)$$

where $i_p \epsilon I$ and $o_p \epsilon O$ with p as a positive integer in the range 1 to n.

We will use two operators, @ and $\cup$, to represent the concatenation of sequences later in our paper. If A and B are two sequences, then their concatenation C is given as

$$C= A@B.$$

Let $A_1, A_2, ..., A_n$ be n sequences. Their concatenation CO is given as

$$CO= \bigcup_{i=1}^{n} A_i.$$

A vector V with k elements is denoted as $V[1{:}k]$. The notation for the ith element of V is $V[i]$.

Now we give two examples of protocol entities specified as finite state machines. Example 1 is a good example for illustrating the procedures described in the next section. Example 2 gives the specification of the transmitter section of the Alternating Bit Protocol (ABP).

Example 1:

The specification of a protocol entity is given in Figure 1. The set of states, S, is given as {*0, 1, 2, 3, 4, 5, 6*}. The initial state *IS* is *0*. The set of inputs is I = {*ri, a, z, x, c*}. The set of outputs is O = { *null, b, f, d*}. The set of edges is given as { *(01), (12), (24), (40), (61), (56), (35), (13), (10), (20), (30), (40), (50), (60)*}. The input *ri* will always bring the FSM to its initial state, *0*. We assume the output to be *null* when the input *ri* is applied to the FSM.

Example 2:

Now we will give an example of the transmitter section of the Alternating Bit Protocol (ABP). The transmitter interacts with a user process (*tuser*), a data medium (*dmd*), a timer process (*timer*), and an acknowledgement medium (*amd*). The transmitter receives messages (*data*) from the user process *tuser*. It sends them to the transmission medium with either a *0* or *1* attached (*data0* or *data1*). From the acknowledgement medium *amd* it receives an acknowledgement with either a *0* or *1* attached (*ack0* or *ack1*). The *timer* signals to the transmitter that it has timed out by sending an *out* signal.

It is modeled as a 8-state FSM as shown in Figure 2. The set of states is {*0, 1, 2, 3, 4, 5, 6, 7*} with *0* as its initial state. The set of inputs is {*tuser?data, amd?ack0, amd?ack1, timer?out, ri, null*}. The set of outputs is {*dmd!data0, dmd!data1, timer!start, null*}. The set of edges is {*(01), (12), (23), (31), (34), (45), (56), (67), (75), (70)*}.

The test-sequences for the two examples given here are computed in the next section.

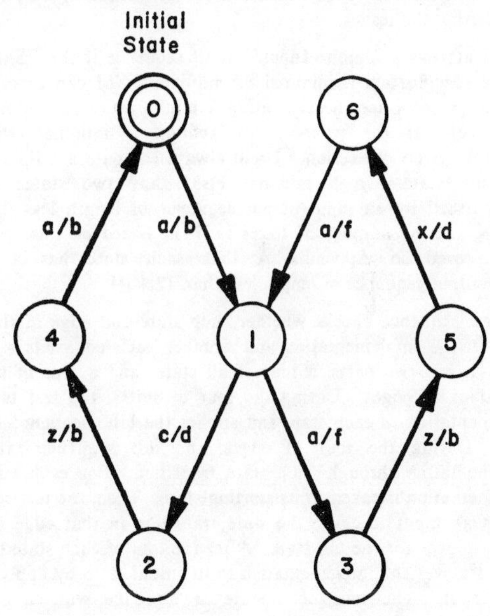

Every state (except for 0) has an edge to the initial state 0 with edge label (ri/null)

FIG. 1 An Example

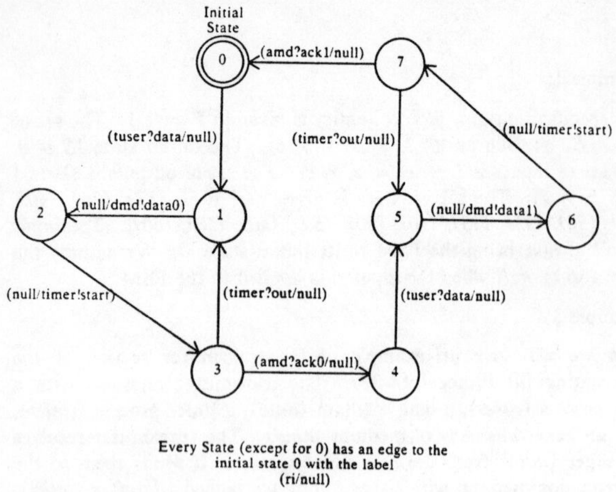

Figure 2. The Transmitter Section of the Alternating Bit Protocol

3. TEST GENERATION PROCEDURE

The main idea here is to compute a unique signature for each state, called the Unique Input/Output (UIO) sequence. A UIO sequence for a state is an I/O behavior that is not exhibited by any other state. It is used to check whether the implementation is in a certain state during the test.

Each state has a Unique Input/Output sequence if the FSM model for the specification is minimal. A minimal FSM can be computed for any protocol specification using a minimization procedure given in [Kl72]. If the protocol specification is minimal, then the algorithm given in Section 3.1 will always compute a UIO sequence for each state. In a minimal FSM, any two states can be distinguished by an input/output sequence of length less than $|V|$, where $|V|$ is the number of states [Kl72]. Based on this result, we have proved in Appendix A that each state has a unique input/output sequence of length less than $(2|V|^2)$.

The test sequence checks whether each state and edge in the FSM exists in the implementation and whether each edge has a correct label. It has two parts: a tour of all states and a tour of all state transitions or edges. During the tour of states, the test takes the implementation to each state and applies the UIO sequence for that state. During the tour of edges, the test sequence takes the implementation through each state transition. For each edge, the implementation is taken to its starting state. Then, the test sequence applies an input to cause the state transition in that edge and the UIO sequence for the tail state. After the tour of each state or edge, the test moves the implementation to its initial state by applying it a reset input. The tour of a state or an edge will be called a subsequence in the rest of this section.

Then, we check if some subsequences are completely contained in the rest. The subsequences completely contained in others are eliminated from the test.

In Section 3.1, we give an algorithm to compute a UIO sequence for each state. Section 3.2 describes an algorithm for computing a test sequence. The length-complexity of the test sequence is given in Section 3.3. Two examples given in Section 3.4 illustrate the application of the algorithms given in Sections 3.1 and 3.2.

3.1 AN ALGORITHM TO COMPUTE A UIO SEQUENCE FOR EACH STATE

We check for uniqueness the input/output sequences of increasing length from each state until a UIO sequence is computed. For each state, all input/output behavior sequences of length 1 are computed and are checked for uniqueness. If there is no unique sequence of length 1, then we repeat the same procedure for all sequences of length 2. This procedure is continued with the sequences of larger length until we find a unique input/output sequence. The details of this procedure are as follows.

Step 1. Let the set of distinct edge labels in the FSM be $ED = \{ e_1, e_2, ..., e_a \}$. For each element $e \in ED$, we compute $Set(e)$, the set of edges with the label e.

Step 2. Let the set of states be $S = \{ s_1, s_2, ..., s_{|V|} \}$ where $|V|$ is the number of distinct states. For each element $s \in S$, we take the following steps. These steps are also given in Figure 3 using an Algol-like language.

Step 3. For each state s, compute the set of outgoing edges, $O_s = \{ o_{s,1}, o_{s,2}, ..., o_{s,k} \}$. For each edge $o_{s,j} \epsilon O_s$, we check whether its edge label $Label(o_{s,j})$ is unique. It is done by checking whether the set $Set(Label(o_{s,j}))$ has a single element. If any edge $o_{s,j} \epsilon O_s$ has a unique label, then its label $Label(o_{s,j})$ is the UIO sequence for state s. Otherwise, we compute the sequences of larger length in the following steps until we find a UIO sequence for s.

Step 4. The input/output sequences of length l are computed from those of length $(l-1)$. The sequences of length $(l-1)$ are stored in the vector *Opattern[1:OND]* where *OND* is the number of distinct sequences of length $(l-1)$. The tail state for each sequence *Opattern[i]* is stored in the ith element of the vector *Oend[1:OND]*, *Oend[i]*. The tail state is the state of the FSM after the application of the sequence *Opattern[i]* starting in the state s. The states other than s that have the same input/output sequence *Opattern*[i] along with the tail states for these states* are stored in the ith element of the vector *Oled*[1:*OND*], *Oled[i]*.

For initializations (upon the first entry to Step 4), we set *OND* to k and l to 2. Then, we store the labels of all edges $o_s \epsilon O_s$ in the vector *Opattern[1:OND]*. The tail states for each edge $o_s \epsilon O_s$ are stored in the vector *Oend*[1:*OND*]. Then, for each edge $o_{s,j} \epsilon O_s$, *Oled*[j] is set to $Set(Label(o_{s,j})) - \{ o_{s,j} \}$, which is the set of edges with the label $Label(o_{s,j})$ and with a head state other than s.

Similar information about the sequences of length l is stored in the vectors *Pattern*, *End*, *Led*. These vectors are assumed to be sufficiently large to accommodate all sequences of length l. The input/output sequences of length l are stored in the vector *Pattern*, the tail states in the vector *End*, and the other states with the similar input/output state in the vector *Led*. The running count of the sequences of length l is stored in *ND*. Before initiating the computation of the first sequence of length l, we initialize *ND* to 1.

In Step 5, for each sequence of length $(l-1)$, *Opattern*[i], we compute some sequences of length l and check their uniqueness. Each one of them has *Opattern*[i] as its leading subsequence and is an input/output sequence for State s. If we are unsuccessful in locating any unique sequence of length l in Step 5, then we go to Step 6.

Step 5. For the tail state *Oend*[i] for the sequence *Opattern*[i], we compute the set of outgoing edges $O_{Oend[i]} =$

---

* The tail state for a starting state x and the input/output sequence *Opattern[i]* is the one that the FSM will enter if the input portion of *Opattern*[i] is applied.

Figure 3. Procedure to compute the shortest UIO Sequence for the state s

```
/* l = Length of the input/output sequences
      being computed.  */
/* OND = Number of the Sequences of length (l−1). */
/* Opattern[1:OND] = A vector to store all sequences
   of length (l−1). */
/* Oend[1:OND] = A vector for storing the tail states
   for the sequences stored in Opattern. */
/* Oled[1:OND] = A vector storing the sets of two-state
   tuples. Such a two-state tuple carries a starting state
   with the I/O sequence Opattern[i] and the
   corresponding tail state. */
/* ND = A running count of the sequences of length l. */
/* Pattern = A vector for storing the sequences
      of length l. */
/* End = A vector for storing the tail states
   for the quences stored in Pattern. */
/* Led = A Vector for storing the sets of two-state
   tuples similar to those stored in Oled. */
/* k = Number of outgoing edges from s.*/

/* Steps 3 and 4: Compute all
   sequences of length 1 starting from s */

UIO[s]:=0;
l:=1;
OND:=k;
for each o_i ε O_s do
 begin
   Opattern[i]:=Label(o_i);
   Oend[i]:=Tail(o_i) ;
   Oled[i]:=Set(Label(o_i))− { o_i};
   if ( Oled[i] = null) then
   /* Success! Label(o_i) is the UIO sequence. */
   begin
     UIO[s]:=Label(o_i);
     exit;
   end
 end

/* If no sequence of length 1 is unique, then compute
longer sequences and check them for uniqueness.*/

while ( UIO[s]=0 ) do
 begin
   l:=l+1;
   ND:=0;
   for i:=1 to OND do
    begin
   /* Step-5*/
     compute the set O_out of outgoing
     edges from node Oend[i];
     for each edge o_out ε O_out do
      begin
        ND:=ND+1;
        Pattern[ND]:=Opattern[i]@Label(o_out);
        End[ND]:=Tail(o_out);
        Led[ND]:=nullset;
        for each two-state tuple (h,t)εOled[i] do
         begin
          for each edge o leaving t do
          if Label(o)=Label(o_out)
            then Led[ND]:=Led[ND]∪ (h,Tail(o));
          if Led[ND]=null then
           begin
             /* Success! The sequence is unique */
             UIO[s]:=Pattern[ND];
```

Figure 3 (Continued)

```
             exit;
           end
         end
      end
    end

/*Step 6: Transfer all sequences of length l to the
   vectors Opattern, Oend, Oled. */

OND:=ND;
for i=1 to ND do
 begin
   Opattern[i]:=Pattern[i];
   Oend[i]:=End[i];
   Oled[i]:=Led[i];
 end
end
```

Table I. List of Edges for Each Label (Example 1)

Edge Label	List of Edges
a/b	(40), (01)
z/b	(24), (35)
c/d	(12)
a/f	(61), (13)
x/d	(56)

Table II. UIO Sequences for Example 1.

State	UIO Sequence
0	a/b, c/d
1	c/d
2	z/b, a/b
3	z/b, x/d
4	a/b, a/b
5	x/d
6	a/f, c/d

Table III. UIO Sequences for Example 2.

State	UIO Sequence.
0	tuser?data/null, null/dmd!data0
1	null/dmd!data0
2	null/timer!start, timer?out/null, null/dmd!data0
3	timer?out/null, null/dmd!data0
4	tuser?data/null, null/dmd!data1
5	null/dmd!data1
6	null/timer!start, timer?out/null, null/dmd!data1
7	timer?out/null, null/dmd!data1

$\{o_{out,1}, o_{out,2}, \cdots, o_{out,k}\}$. For each edge $o_{out} \in O_{Oend[i]}$, the following steps are taken.

We construct an input/output sequence of length l, $trans = Opattern[i] @ Label(o_{out})$ and store it in the NDth element of the vector *Pattern*, *Pattern*[ND]. The tail state of the edge o_{out}, $Tail(o_{out})$, is stored in the NDth element of the vector END, $End[ND]$. Then we use the following procedure to check whether the sequence *trans* is unique.

For each element of the set $Oled[i]$, we check whether the tail state has any outgoing edge with the label $Label(o_{out})$. If no element of $Oled[i]$ has such an edge, then *trans* is the UIO sequence for the state s. Otherwise, the tail states of such edges along with the starting states from the corresponding element of $Oled[i]$ are stored in $Led[ND]$. Then, ND is incremented by 1.

Step 6. OND is set to ND. Then, we transfer all sequences of length l stored in the vector *Pattern*, to the vector *Opattern*. We also store their tail-states, elements of *End*, in *Oend*, and the other states with the same input/output sequences along with their tail states, elements of *Led*, in *Oled*. We also increment l by 1. Finally, we go back to Step 4 to compute the sequences of length l and check them for uniqueness.

For a minimal FSM, the time-complexity of the algorithm given above in the worst case is $O((m_{max})^{2|V|^2})$ where m_{max} is the largest out-degree of any state (or the largest number of outgoing edges from any state) and $|V|$ is the number of states in the machine. If the FSM for the specification is not minimal, then the algorithm given above will never terminate. It is so because at least two states of the non-minimal FSM will have the same output behavior for all input sequences. The algorithm given above is illustrated by two examples in Section 3.4.

3.2 AN ALGORITHM TO GENERATE A TEST SEQUENCE

This algorithm has three major steps. In the first step, we generate a test to check whether each state is reachable from the initial state and whether it has the unique signature, the UIO sequence, computed using the algorithm given in Section 3.1. In the second step, a test is generated to check whether each edge has the correct head state, the correct tail state, and the correct label. In the third step, we concatenate the tests from the first two steps and optimize the combination to reduce the test length.

Step 1. From the initial state, compute shortest paths to all states using Dijkstra's Algorithm [AHO76]. For a state $s \in S$, the shortest path will be represented as $P(s)$.

Step 2. This step generates the portion of the test sequence that tours all states. From the initial state, the test takes the implementation to each state s and applies the UIO sequence for s, $UIO(s)$. The test sequence for the tour of state s is $ri/null @ P(s) @ UIO(s)$. Finally, the test sequence for checking all states, TS_1, is given as

$$TS_1 = \bigcup_{s \in S} ri/null \ @ \ P(s) \ @ \ UIO(s)$$

Step 3. This step generates the portion of the test sequence that tours all edges. The tour of each edge requires the following test sequence

$$T(e) = P(Head(e)) \ @ \ Label(e) \ @ \ UIO(Tail(e)).$$

The tour of all edges requires the following test sequence:

$$TS_2 = \bigcup_{e \in E} ri/null \ @ \ T(e)$$

Step 4. The unoptimized test sequence is the concatenation of the two test sequences generated in Steps 2 and 3 : $TS_1 @ TS_2$. It has $(|V|+|E|)$ subsequences, where $|V|$ = number of states and $|E|$ = number of edges. In this step, we check if some subsequences are completely contained in others. The subsequences completely contained in others are eliminated from the test sequence. Two examples given later illustrate that many subsequences are completely contained in the rest.

The time-complexity of the algorithm given above is $O(|V|.|E|)$. Two examples given in Section 3.4 illustrate the algorithm given in this section.

3.3 LENGTH OF TEST SEQUENCES

The length of an unoptimized test sequence is $O(|V|^2.|E|)$ if the specification is a minimal FSM. For a minimal FSM, the UIO sequence for each state has a length less than $2|V|^2$. After optimization (Step 4 of Section 3.2), the test sequence will be much smaller than the unoptimized sequence as illustrated by the following examples.

3.4 EXAMPLES

We illustrate the algorithms given in Sections 3.1 and 3.2 by applying them to the two examples given earlier in Section 2.

For Example 1, the algorithm given in Section 3.1 is used to generate a UIO sequence for each state. For each edge label, we compute the list of edges with that label[Table I]. These lists are further used to compute a UIO sequence for each state [Table II]. Then, we use the procedure given in Section 3.2 to compute an unoptimized test sequence with 21 subsequences [Figure 4]. This sequence consists of a tour of all states and of all edges. From this sequence, the subsequences contained in the others are eliminated. Notice that 13 subsequences are completely contained in others. After these eliminations, we get the optimized test sequence consisting of 8 subsequences [Figure 5].

Next we use the procedure given in Section 3.1 to compute the UIO sequence for each state for Example 2 [Table III]. Each UIO sequence given in Table III has at least one output other than *null*. A *null* output gives us no indication of the machine's state and the machine may be a unpredictable state. Some state transitions in Example 2 produce a *null* output.

From these UIO sequences, we derive the test sequence given in Figure 6. This unoptimized sequence has 25 subsequences. For optimization, 15 subsequences that are completely contained in others are eliminated from the unoptimized sequence. The optimized sequence has ten subsequences as shown in Figure 7.

4. SOME OBSERVATIONS ABOUT THE TESTING PROBLEM

In this section, we will give some lower bounds on the complexity of the testing problem. The basic problem is at least as tough as solving a graph isomorphism problem. During a test, we must check whether the minimal FSM for the implementation contains a graph isomorphic to the minimal machine for the specification. Further, it must check whether the edge-labels in the minimal FSMs for the implementation and specification match.

The implementation, according to our earlier assumptions, is a black box. Even, if we know what is inside this black box, still we have to solve at least a graph isomorphism problem. Graph isomorphism is an open problem. It is not known whether this graph problem falls in the class P or class NP-complete [GJ79]. We expect the testing problem to be exponential or worse in the number of states.

The equivalence of two finite state machines with no outputs and with final states can be checked by a polynomial algorithm [AHU76]. But those results do not apply here.

Assume that there is no upper bound on the number of states in the implementation. Then, it is impossible to check the conformity of an implementation to a specification with a cycle. We illustrate it by a simple example. Consider a finite state machine with two states: *0* and *1* (Figure 8). Assume that the implementation has *2n* states (Figure 9). Unless the test sequence has more than *2n* steps, it will not detect the problem in the implementation. But, if we can assume that the implementation has the same or less number of states than the specification, then the problem described above will not occur.

5. *CONCLUSIONS AND FUTURE WORK*

Design of the certifiers is an important networking problem which has received little attention from theoretical computer scientists. The main component of a certifier is a good test sequence for the protocol.

We give here a novel procedure for generating protocol tests. This procedure is based on an intuitive idea that the test must visit each state and each state transition and must check a unique signature, called the UIO sequence, for all states. Our experience until now with this procedure shows that it has a good fault coverage. We are working on a complete characterization of the fault set. This procedure will also be applied to a real-life protocol in the future.

The ideas given here can be further extended to the ones for designing protocols that are easy to test. For example, we can design a protocol with a UIO sequence less than 2 or 3 for each state. Such protocols will be easy to test. We can also alter designs of the existing protocols so that the UIO sequence for each state is smaller than 2 or 3.

REFERENCES

AHU76 "*The Design and Analysis of Computer Algorithms,*" A. V. Aho, J. E. Hopcroft, and J. D. Ullman, pp. 143-145, 207-209, Addison-Wesley Publishing Company, 1976.

Bi83 "Methodology and tools for qualitative protocol validation," J. F. Billiard, *Protocol Specification, Testing, and Verification,III,* R. Rudin and C. H. West (editors), Elsevier Science Publishers B. V., IFIP 1983.

BS83 "A Survey of Formal Methods," Gregor V. Bochmann and Carl A. Sunshine, pp. 561-578, Chapter 20, "*Computer Networks and Protocols,*" Paul E. Green (Editor), May 1983, Plenum Press.

GJ79 "*Computers and Intractability: A guide to the theory of NP-Completeness,*" M. Garey and D. Johnson, Freeman and Co., San Francisco, 1979, Page 285.

Gr82 "The Structure of Computer Networks," Paul E. Green, Jr., Chapter 1, pp. 3-32, *Computer Network Architectures and Protocols,* Edited by Paul E. Green, Plenum Press, 1982.

He64 "Fault-Detecting Experiments for Sequential Circuits," F. C. Hennie, Proceedings of the Fifth Annual Symposium, Princeton University, Princeton, NJ, November 11-13, 1964, pp 95-110.

Kl72 "*Introduction to the Methodology of Switching Circuits,*" George J. Klir, D. Van Nostrand Company, 1972, pp. 264-325.

LM83 "Producing Tests for Implementations of OSI Protocols," R. J. Linn and W. H. McCoy, *Protocol Specification, Testing, and Verification, III,* R. Rudin and C. H. West (editors), Elsevier Science Publishers B. V., IFIP 1983.

Pi80 "Remarks on the Feasibility of Validating and Testing ADCCP Implementations," T. F. Piatkowski, IEEE Proceedings of Trends and Applications, May 1980.

Ra83 "Towards an objective understanding of conformance," D. Rayner, *Protocol Specification, Testing, and Verification, III,* R. Rudin and C. H. West (editors), Elsevier Science Publishers B. V., IFIP 1983.

SB82 "Some Experience with Test Sequence Generation for Protocols," B. Sarikaya and G. V. Bochmann, *Protocol Specification, Testing, and Verification, II,* C. Sunshine(ed.), North-Holland Publishing Company, 1982.

APPENDIX A

Theorem. Each state in a minimal fully-specified Mealy machine with an initial state (defined in Section 2) has a UIO sequence of length shorter than $2|V|^2$ where $|V|$ is the number of states.

Summary of the Proof. Let us focus on one state s_i of the machine. For each state s_j (where j is not the same as i), two input/output sequences $IO(i,j)$ (for s_i) and $IO(j,i)$ (for s_j) exist with lengths less than |V| [Kl72] which can distinguish between s_i and s_j. The input portions of $IO(i,j)$ and $IO(j,i)$ are the same, but the output portions are different. If the input portion of the sequence $IO(i,j)$ is applied to the machine in states s_i and s_j separately, the output sequences from the machine will be different.

Let us consider the sub-sequence $ri/null@\ P(i)@\ IO(i,j)$ where $P(i)$ is the shortest path from the initial state to the state s_i. This test-sequence checks whether the machine when applied the sequence $ri/null@P(i)$ enters a state that has different input/output behavior than that of the state s_j and gives an input/output behavior expected from s_i. The following sequence $LD(s_i)$ when applied to the state s_i will uniquely distinguish it from all other states.

$$LD(s_i) = IO(i,1)@ \bigcup_{k=2}^{|V|} ri/null@\ P(i)@\ IO(i,k) \quad \text{(A1)}$$

where k is not equal to i and i is not equal to 1. For state s_1, one can easily compute a similar expression for its UIO sequence. The maximum length of $P(i)$ and of $IO(i,k)$ for all values of k is less than $|V|$. Therefore, the upper bound on the length of $LD(s_i)$ is $2|V|^2$.

Figure 4. Unoptimized Test Sequence for Example 1.

First Part(Tour of States)

[ri/null, a/b, c/d],
[ri/null, a/b, c/d],
[ri/null, a/b, c/d, z/b, a/b],
[ri/null, a/b, a/f, z/b, x/d], [ri/null, a/b, c/d, z/b, a/b, a/b],
[ri/null, a/b, a/f, z/b, x/d],
[ri/null, a/b, a/f, z/b, x/d, a/f, c/d],

Second Part(Tour of Edges)

[ri/null, a/b, c/d], [ri/null, a/b, c/d, z/b, a/b],
[ri/null, a/b, a/f, z/b, x/d], [ri/null, a/b, c/d, z/b, a/b, a/b],
[ri/null, a/b, c/d, z/b, a/b, a/b, c/d],
[ri/null, a/b, a/f, z/b, x/d],
[ri/null, a/b, a/f, z/b, x/d, a/f, c/d],
[ri/null, a/b, a/f, z/b, x/d, a/f, c/d],
[ri/null, a/b, ri/null, a/b, c/d],
[ri/null, a/b, c/d, ri/null, a/b, c/d],
[ri/null, a/b, a/f, ri/null, a/b, c/d],
[ri/null, a/b, c/d, z/b, ri/null, a/b, c/d],
[ri/null, a/b, a/f, z/b, ri/null, a/b, c/d],
[ri/null, a/b, a/f, z/b, x/d, ri/null, a/b, c/d]

Figure 5. Optimized Test Sequence for Example 1.

[ri/null, a/b, c/d, z/b, a/b, a/b, c/d],
[ri/null, a/b, a/f, z/b, x/d, a/f, c/d],
[ri/null, a/b, ri/null, a/b, c/d],
[ri/null, a/b, c/d, ri/null, a/b, c/d],
[ri/null, a/b, a/f, ri/null, a/b, c/d],
[ri/null, a/b, c/d, z/b, ri/null, a/b, c/d],
[ri/null, a/b, a/f, z/b, ri/null, a/b, c/d],
[ri/null, a/b, a/f, z/b, x/d, ri/null, a/b, c/d]

Figure 6. Unoptimized Test Sequence for Example 2.

First Part(Tour of States)

[ri/null, tuser?data/null, null/dmd!data0],
[ri/null, tuser?data/null, null/dmd!data0],
[ri/null, tuser?data/null, null/dmd!data0, null/timer!start,
timer?out/null, null/dmd!data0],
[ri/null, tuser?data/null, null/dmd!data0, null/timer!start,
timer?out/null, null/dmd!data0],
[ri/null, tuser?data/null, null/dmd!data0, null/timer!start,
amd?ack0/null, tuser?data/null, null/dmd!data1],
[ri/null, tuser?data/null, null/dmd!data0, null/timer!start,
amd?ack0/null, tuser?data/null, null/dmd!data1],
[ri/null, tuser?data/null, null/dmd!data0, null/timer!start,
amd?ack0/null, tuser?data/null, null/dmd!data1, null/timer!start,
timer?out/null, null/dmd!data1],
[ri/null, tuser?data/null, null/dmd!data0, null/timer!start,
amd?ack0/null, tuser?data/null, null/dmd!data1, null/timer!start,
timer?out/null, null/dmd!data1],

Second Part(Tour of Edges)

[ri/null, tuser!data/null, null/dmd!data0],
[ri/null, tuser!data/null, null/dmd!data0, null/timer!start,
timer?out/null, null/dmd!data0],
[ri/null, tuser!data/null, null/dmd!data0, null/timer!start,
timer?out/null, null/dmd!data0],
[ri/null, tuser!data/null, null/dmd!data0, null/timer!start,
timer?out/null, null/dmd!data0],
[ri/null, tuser!data/null, null/dmd!data0, null/timer!start,
amd?ack0/null, tuser!data/null, null/dmd!data1],
[ri/null, tuser!data/null, null/dmd!data0, null/timer!start,
amd?ack0/null, tuser!data/null, null/dmd!data1],
[ri/null, tuser!data/null, null/dmd!data0, null/timer!start,
amd?ack0/null, tuser!data/null, null/dmd!data1, null/timer!start,
timer?out/null, null/dmd!data1],
[ri/null, tuser!data/null, null/dmd!data0, null/timer!start,
amd?ack0/null, tuser!data/null, null/dmd!data1, null/timer!start,
timer?out/null, null/dmd!data1],
[ri/null, tuser!data/null, null/dmd!data0, null/timer!start,
amd?ack0/null, tuser!data/null, null/dmd!data1, null/timer!start,
timer?out/null, null/dmd!data1],
[ri/null, tuser!data/null, null/dmd!data0, null/timer!start,
amd?ack0/null, tuser!data/null, null/dmd!data1, null/timer!start,
ack1/null, tuser!data/null, null/dmd!data0]
[ri/null, tuser?data/null,
ri/null, tuser?data/null, null/dmd!data0]
[ri/null, tuser?data/null, null/dmd!data0,
ri/null,tuser?data/null, null/dmd!data0],
[ri/null, tuser?data/null, null/dmd!data0, null/timer!start,
ri/null,tuser?data/null, null/dmd!data0],
[ri/null, tuser?data/null, null/dmd!data0, null/timer!start,
amd?ack0/null, ri/null, tuser?data/null, null/dmd!data0],
[ri/null, tuser?data/null, null/dmd!data0,
null/timer!start, amd?ack0/null, tuser?data/null,
ri/null,tuser?data/null, null/dmd!data0],
[ri/null, tuser?data/null, null/dmd!data0, null/timer!start,
amd?ack0/null, tuser?data/null, null/dmd!data1,
ri/null, tuser?data/null, null/dmd!data0],
[ri/null, tuser?data/null, null/dmd!data0, null/timer!start,
amd?ack0/null, tuser?data/null, null/dmd!data1, null/timer!start,
ri/null, tuser?data/null, null/dmd!data0]

Figure 7. Optimized Sequence for Example 2.

[ri/null, tuser!data/null, null/dmd!data0, null/timer!start,
timer?out/null, null/dmd!data0],
[ri/null, tuser!data/null, null/dmd!data0, null/timer!start,
amd?ack0/null, tuser!data/null, null/dmd!data1, null/timer!start,
timer?out/null, null/dmd!data1],
[ri/null, tuser!data/null, null/dmd!data0, null/timer!start,
amd?ack0/null, tuser!data/null, null/dmd!data1, null/timer!start,
ack1/null, tuser!data/null, null/dmd!data0]
[ri/null, tuser?data/null,
ri/null, tuser?data/null, null/dmd!data0]
[ri/null, tuser?data/null, null/dmd!data0,
ri/null,tuser?data/null, null/dmd!data0],
[ri/null, tuser?data/null, null/dmd!data0, null/timer!start,
ri/null,tuser?data/null, null/dmd!data0],
[ri/null, tuser?data/null, null/dmd!data0, null/timer!start,
amd?ack0/null, ri/null, tuser?data/null, null/dmd!data0],
[ri/null, tuser?data/null, null/dmd!data0,
null/timer!start, amd?ack0/null, tuser?data/null,
ri/null,tuser?data/null, null/dmd!data0],
[ri/null, tuser?data/null, null/dmd!data0, null/timer!start,
amd?ack0/null, tuser?data/null, null/dmd!data1,
ri/null, tuser?data/null, null/dmd!data0],
[ri/null, tuser?data/null, null/dmd!data0, null/timer!start,
amd?ack0/null, tuser?data/null, null/dmd!data1, null/timer!start,
ri/null, tuser?data/null, null/dmd!data0]

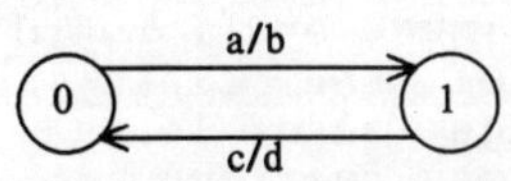

Figure 8. Two-state Finite State Machine

Figure 9. Implementation with 2n states.

An Algorithm for Distributed Computation of a Spanning Tree in an Extended LAN

Radia Perlman

Digital Equipment Corporation
1925 Andover St., Tewksbury MA 01876

Abstract

A protocol and algorithm are given in which bridges in an extended Local Area Network of arbitrary topology compute, in a distributed fashion, an acyclic spanning subset of the network.

The algorithm converges in time proportional to the diameter of the extended LAN, and requires a very small amount of memory per bridge, and communications bandwidth per LAN, independent of the total number of bridges or the total number of links in the network.

Algorhyme

I think that I shall never see
A graph more lovely than a tree.

A tree whose crucial property
Is loop-free connectivity.

A tree which must be sure to span
So packets can reach every LAN.

First the Root must be selected.
By ID it is elected.

Least cost paths from Root are traced.
In the tree these paths are placed.

A mesh is made by folks like me
Then bridges find a spanning tree.

Introduction

Local area networks are limited in geography, traffic, and number of stations. A single local area network will often not meet the needs of an organization for these reasons. Conventional LAN interconnection mechanisms, for instance XNS [1], Sytek[2], IBM1[3], IBM2[4], DNA[5] require cooperation from the stations with compatible protocols layered above the protocol necessary to connect to a single LAN.

An approach that is transparent to stations, and thus allows a station to participate in an extended LAN with no modification, is presented in [6],[7], and [8]. In this approach, a bridge connected to two or more links will listen "promiscuously" to all packets transmitted on each of its links, and forward packets received on one of the links onto the others. A bridge also learns of the location of stations relative to itself, so that it will not forward traffic for a station onto a link unnecessarily.

This approach assumes that the topology is a tree (loop-free). However, requiring a topology to be loop-free means there are no backup paths in the case of bridge or LAN failures. Also, because the technology allows network growth so easily, it might be difficult to prevent someone from adding a bridge and creating a loop. A loop in the topology might cause severe performance degradation in the entire extended network due to congestion caused by infinitely circulating packets. It is undesirable to have a network that can be brought down so easily, merely by plugging a cable into the wrong place.

Thus we have designed an algorithm that allows the extended network to consist of an arbitrary topology. The algorithm is run by the bridges, and computes a subset of the topology that connects all LANs yet is loop-free (a spanning tree). The algorithm is self-configuring. The only a priori information necessary in a bridge is its own unique ID (MAC address), which we are assuming can be attained in some manner, for instance with a hardware ROM containing the value. [9]

Terminology

1. station -- a node in the extended LAN that does not forward packets. It is connected to the extended LAN solely for the purpose of communicating with other stations. It does not participate in the spanning tree algorithm.

2. bridge -- a node connected to two or more LANs, for the purpose of forwarding packets between the LANs.

3. link -- a connection from a bridge to a single LAN.

4. extended LAN -- the collection of LANs connected by bridges.

5. Root--the bridge chosen to be the root of the tree to be formed. Note that there is exactly one Root in the network, chosen dynamically.

6. Designated Bridge--the bridge on a LAN closest to the Root, with a tie-breaker to enforce uniqueness. Thus there is exactly one Designated Bridge per LAN in the network, chosen dynamically.

Goals of the Spanning Tree Algorithm

- Allow interconnection of LANs compatibly with LAN standards, so that stations need not know whether they are attached to a single LAN or to a network of LANs.

- Allow redundant bridges and LANs in an extended LAN, for instance so that connectivity can be preserved after bridge or LAN failures.

- Be self-configuring, so that each bridge need only know its own ID and a well-known generic group address.

- Memory requirements of a bridge should not grow with the number of LANs or bridges, so that a network cannot be misconfigured.

- The communications bandwidth consumed by the algorithm on any particular LAN should be a constant (and small) amount, regardless of the total number of bridges or LANs in the net.

- The algorithm should stabilize as quickly as possible to a deterministic spanning tree (no loops, complete connectivity) in any size network. Assuming no lost messages, it should stabilize in some small multiple of the round trip delay across the network.

- The algorithm should allow bridges to keep caches of station membership (a forwarding data base), to cut down on traffic.

- No permanent loops should ever form. If no messages are lost, no transient loops should form (in any legal sized network). The algorithm should minimize the probability of transient loops, given the possibility of lost messages and the possibility of a network larger than designed for. Similarly, transient behavior should not cycle a link (turn it off, then back on again) in a legal sized network if no messages are lost.

- The algorithm should be self-stabilizing, once any malfunctioning equipment is repaired or disconnected.

- The algorithm should break loops even if some of the bridges in the loop do not implement the algorithm. (Of course it cannot break a loop in which none of the bridges implement the algorithm.)

- The algorithm should require only a connectionless service (LLC type 1) for any LANs on which it is run.

- The algorithm should allow tuning of the configuration by adjustment of parameters. This tuning of the configuration may help performance, but the algorithm should still work acceptably with default values.

Link States

Each bridge has two or more links. For each LAN to which a bridge is attached, the bridge computes a state for that link. In all states, the algorithm continues to run. The only difference between these states is whether data packets are forwarded to and from the LAN.

- FORWARDING -- Forward Data Packets
- BACKUP -- Do not forward data packets
- PRE_FORWARDING -- Do not forward data packets yet; however, unless an event reverts the link to BACKUP, the state will shortly become FORWARDING.
- PRE_BACKUP -- Continue forwarding data packets; however, unless an event reverts the link to FORWARDING, the state will shortly become BACKUP.

HELLO Messages

The Designated Bridge on each LAN periodically transmits a HELLO Message, with period HELLO_TIME, broadcast to the well-known group address "all bridges". The contents of the HELLO Message are:

1. Transmitting bridge ID
2. ID of bridge assumed to be Root
3. Length of best known path to Root
4. Age of the HELLO (time since information from the Root has propagated on this path)
5. Link identifier (of local significance to the transmitting bridge)
6. MAX_AGE (a parameter, passed in HELLOs so that the network will agree on timer values)

If the Root or a Designated Bridge along the path to the Root fails, some time threshold will be exceeded in which no HELLO messages were received on that path. This is detected by keeping track of the age of the HELLO message. When the age exceeds MAX_AGE, the information is discarded, triggering bridges to recompute a new Root or new path.

The AGE field is included in HELLO messages so that a bridge other than the Root can initiate sending a HELLO message, without making the Root's last HELLO seem more recent than it should be. When a HELLO is initiated by the Root, the age will be 0. As the information about the HELLO is held in memory, the age is increased. If a Designated Bridge (other than the Root) initiates transmission of a HELLO, the age field in the transmitted HELLO is filled in to be the age of the stored HELLO. When a HELLO is received, the age is set to the received value initially, and then increased as the information is held.

The Algorithm

As the algorithm is initially presented, primary/secondary status is ignored. In the section "Extensions" the modifications necessary to provide such designation is described.

Electing the Root and Designated Bridges

The Root is the bridge in the network with the smallest ID. When a bridge receives a HELLO Message, it compares the ID of the Root in the HELLO Message with the currently best known Root. If the Root in the HELLO message has a lower ID than the currently known Root, then information about the current Root is discarded and overwritten by information about the newly discovered Root.

The Designated Bridge on a LAN is the bridge with the shortest path to the Root, on that LAN. In case of ties, the bridge with the smallest ID is the Designated Bridge on that LAN. The Root is distance 0 from itself, and claims 0 as the length of its path to the Root in its transmitted HELLO Messages. Other bridges compute their distance from the Root as 1 greater than the minimum distance received from a HELLO Message from the Designated Bridge on any attached LAN.

A bridge B initially attempts to become Root (and therefore Designated Bridge on each of its LANs). When a Root with a lower ID is discovered in a received HELLO message, B then computes its distance from that Root, and attempts to become Designated Bridge on each of its links. If a HELLO Message is received on a link from a bridge that is closer to the Root than B, or equally close with a transmitting bridge's ID lower than B's (as a tie breaker to ensure a unique Designated Bridge per LAN), B defers to the other bridge and stops sending HELLO messages on that link.

The Root is Designated Bridge on each of its LANs, since it is the only bridge which is 0 from the Root. Any other bridge B is some distance K from the Root. B derived its distance K by receiving a HELLO from the Designated Bridge on one of its links containing HOPS of K-1. Thus bridges other than the Root can not be Designated Bridge on all of their LANs -- one of the LANs has to contain a bridge closer to the Root.

Transmission of HELLO Messages
The Root periodically broadcasts a HELLO on each of its links, with period HELLO_TIME. Other bridges only broadcast a HELLO on a link if they are Designated Bridge on that LAN. Since there is exactly one Designated Bridge on a LAN, exactly one bridge on each LAN will periodically broadcast a HELLO message.

The HELLO packet has an age in it, indicating the age of the information contained therein. Usually a Designated Bridge will transmit a HELLO only upon receipt of a HELLO on the link towards the Root, with a younger age. However, a Designated Bridge will also transmit a HELLO on a particular link upon receipt of a HELLO from a different bridge on that LAN that should not be Designated Bridge.

If the age field exceeds some maximum, the information is discarded. In this way failure of the Root or Designated Bridge will be detected, and the other bridges will be triggered to compute a new Root or a new path to the Root.

Link States and Databases
Each bridge, A, keeps a per-link database for each its links, of the state of that link, the information from the latest HELLO Message from the Designated Bridge on that LAN, whether A should send a HELLO on that link, and a timer, activated in states PRE_FORWARDING and PRE_BACKUP, whose expiration will move the link from PRE_FORWARDING to FORWARDING state, or from PRE_BACKUP to BACKUP state.

State Transitions

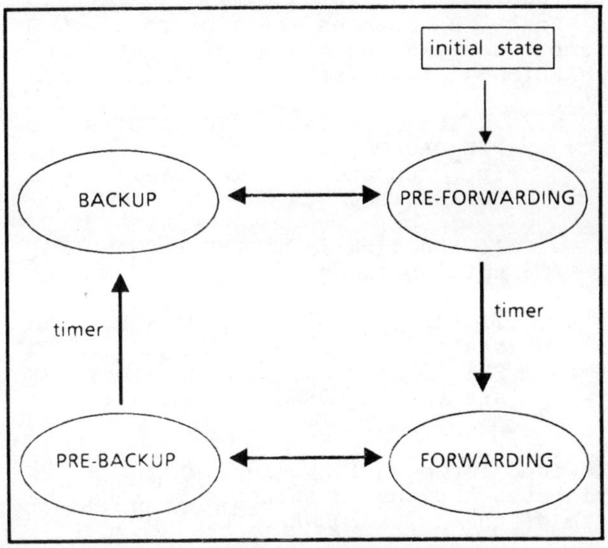

Upon startup, bridge B initializes each link to be in the PRE_FORWARDING state, with the link database for each link initialized to claim B as the Root, B as the Designated Bridge on that link, B's distance from the Root to be 0, and the timer set to PRE_FORWARDING_DELAY.

The events that cause the algorithm to run are receipt of a HELLO Message, and a timer pulse. If a HELLO Message is received that supercedes what is stored in the link database, the information in the link database is overwritten. The timer pulse triggers the Root to transmit a new HELLO message (if HELLO_TIME has transpired since the last HELLO message), triggers each bridge to update the age of the information in the link database for each link (causing the information to be discarded if MAX_AGE is exceeded), and triggers links to transit from PRE_FORWARDING or PRE_BACKUP states (if the delay time has transpired).

When an event causes information in the link database to change (either receipt of a different and superior HELLO message, or expiration of the stored HELLO information), bridge B recomputes its own distance from the Root, and examines each link to determine if B should take over as Designated Bridge. If all information about the current Root has expired, then B will attempt to become the new Root, and therefore Designated Bridge on each of its LANs.

If B is not the Root, it selects a single link which is closest to the Root. If more than one link is equally closest to the Root, B chooses one arbitrarily. B attempts to make the state of that link FORWARDING as follows:

- If the link is in FORWARDING or PRE_BACKUP, the state of the link is changed immediately to FORWARDING.
- If the link is in PRE_FORWARDING, nothing is done.
- If the link is in BACKUP, the state is changed to PRE_FORWARDING, and the timer is set to PRE_FORWARDING_DELAY.

Likewise, for each link for which B is Designated Bridge, B attempts to make the state of that link FORWARDING.

For each other link (links other than the one closest to the Root or for which B is Designated Bridge), B attempts to make the state BACKUP as follows:

- If the link is in BACKUP or PRE_FORWARDING, the state of the link is changed immediately to BACKUP.
- If the link is in PRE_BACKUP, nothing is done.
- If the link is in FORWARDING, the state is changed to PRE_BACKUP, and the timer is set to PRE_BACKUP_DELAY.

Handling Data Traffic

Bridges maintain a forwarding data base of stations. An entry in the forwarding data base consists of a station address, together with the link which is in the direction from which traffic from that station was seen.

Assuming for the moment that all links are in FORWARDING state, the use of the forwarding data base is as follows. The forwarding data base is constructed based on the source field of packets. Traffic is forwarded based on the destination field. If a packet is received on link L, with destination D and source S, then an entry is made in the forwarding data base that S resides in the direction of link L. If no entry for D exists, then the packet is forwarded back onto all links except for L. If an entry for D exists, with link L, then the packet is not forwarded. If an entry for D exists with some other link Q, then the packet is forwarded onto Q.

But links can be in other states than FORWARDING. The rules are as follows:

- Data traffic is ignored when received from links in the BACKUP state.
- Data traffic is examined for source address, for the purpose of updating the forwarding data base, from links in the other states (FORWARDING, PRE_FORWARDING, and PRE_BACKUP).
- Data traffic is forwarded to and from links in the FORWARDING and PRE_BACKUP states.

Interaction with Simple Bridges

Simple bridges are ones that do not participate in this algorithm, but pass HELLO messages through transparently. Thus LANs connected with simple bridges will look to this algorithm like a single LAN.

As long as there is more than one bridge participating in the algorithm in every loop, the algorithm will not even detect the simple bridges. However, if there is only one bridge in a loop that participates in the algorithm, the effect is that it will look to that bridge as though it has multiple attachments to the same LAN. This topology will be detected by the algorithm because a bridge will receive its own HELLO message back on a different link.

The HELLO message should include a link number, of local significance to the Designated Bridge. If the Designated Bridge B receives a HELLO from itself on link L1, that claims to have been transmitted on L2, then B should place the link with the larger link number into PRE_FORWARDING state, with timer PRE_FORWARDING_DELAY. Each time such a HELLO is received, the timer should be reset to PRE_FORWARDING_DELAY. If multiple links are involved in the loop, then only one of them will remain in the FORWARDING state. All others will remain in the PRE_FORWARDING state, and the loop will not exist for data traffic.

Timer Values

To discuss the appropriate timer settings below, let us define the following:

- MaxDelay == the maximum one-way delay across the extended LAN
- MaxLostMsgs == the maximum number of consecutive messages that (within an acceptable probability) can get lost
- MaxPropTime == MaxDelay + (HELLO_TIME * MaxLostMsgs)

MAX AGE -- MAX_AGE is a parameter that should be greater than or equal to MaxPropTime. If the age of a received HELLO message exceeds MAX_AGE, the information is discarded, and the bridge which discarded the information will attempt to take over as Designated Bridge, based on a different path towards the Root, if a HELLO Message that has not timed out is stored for some other link. If all information about the current Root has timed out and been discarded, the bridge will attempt to take over as the new Root, until and unless it hears about a better Root in a subsequent received HELLO message.

The reason MAX_AGE must be at least MaxPropTime is so that a Root that is currently operational will not be mistakenly timed out. The Root can transmit one HELLO that can make it through the extended LAN virtually immediately. (Technically, the Root's HELLO does not propagate through the net, but is received only by the bridges attached to the same LANs as the Root. However, those neighbor bridges transmit HELLOs as a result of receiving a HELLO, so the effect is similar.) The next few HELLOs can get lost (up to MaxLostMsgs). Then the next HELLO might encounter maximum delays, and take the MaxDelay time to reach the periphery of the net.

If MAX_AGE were shorter, then the bridges on the periphery would have timed out the Root before the second HELLO were received, in worst case behavior.

PRE BACKUP DELAY -- To prevent transient partitions, a bridge keeps a link in state PRE_BACKUP for a time PRE_BACKUP_DELAY before switching the link from state FORWARDING to BACKUP.

In state PRE_BACKUP, if newer information arrives suggesting the link should be in FORWARDING state, the link is switched to state FORWARDING without delay.

PRE_BACKUP_DELAY should be greater than or equal to MAX_AGE + MaxPropTime.

The reason it needs to be that large is so that sufficient time will have elapsed so that all bridges have received information about the current topology before any bridges stop forwarding.

The worst case is when the Root fails. As described above, it can take a MaxPropTime from the time the final HELLO was transmitted by the Root to the time it is received on the periphery of the net. Bridges near the old Root (in the worst case) will have timed out the Root MaxPropTime before bridges on the periphery. MAX_AGE after receipt of the last HELLO, a bridge will have timed out the Root. If the bridge destined to be the new Root is maximally far from the old Root, it will issue its first HELLO as the Root MAX_AGE + MaxPropTime after transmission of the final HELLO by the old Root. If the bridge destined to be the new Root is maximally far from the old Root, news of the new Root can take MaxPropTime to reach the rest of the net. Thus following the failure of the old Root, it can take MAX_AGE + 2 * MaxPropTime for news of the new Root to reach all portions of the net.

Thus to be completely safe, bridges should wait 2 * MaxPropTime after timing out the Root to ensure they will be making decisions based on the new topology. Thus PRE_BACKUP_DELAY should be greater than or equal to 2 * MaxPropTime.

For simplicity of parameter setting, PRE_BACKUP_DELAY = 2 * MAX_AGE.

PRE FORWARD DELAY -- To prevent transient loops, a bridge keeps a link in state PRE_FORWARDING for a time PRE_FORWARD_DELAY before switching the link from state BACKUP to state FORWARDING.

While in state PRE_FORWARDING, if newer information arrives suggesting the link should be in BACKUP state, the link is switched to state BACKUP without delay.

Since loops are far more serious than temporary partitions, it is highly desirable that after a topology change all bridges that should not be forwarding in the new topology will have stopped forwarding before any bridges start forwarding.

News of the new topology will have spread to all bridges within a window of MaxPropTime. Thus MaxPropTime after PRE_BACKUP_DELAY, all links that will be turning off in the new topology will have turned off. At this point it is safe to start turning links on. Thus PRE_FORWARDING_DELAY should be greater than or equal to PRE_BACKUP_DELAY + MaxPropTime.

For simplicity of parameter setting, PRE_FORWARDING_DELAY = 3 * MAX_AGE.

Setting Parameters

A network management facility, such as discussed in [8], allows parameters to be set at a node.

To allow tuning of a particular configuration, parameters HELLO_TIME and MAX_AGE are network management settable (the other timers are derived from MAX_AGE).

HELLO_TIME can differ from bridge to bridge, provided that the Root's HELLO_TIME not be so large as to violate the inequality that MAX_AGE be greater than or equal to MaxPropTime. Thus HELLO_TIME can be set independently at each bridge, with an appropriate default so that performance would be adequate without the necessity of setting this parameter.

However, the network must agree on MAX_AGE. To ensure consistency, MAX_AGE (from which the other parameters are derived) is passed in the HELLO messages and the value passed by the Root is used.

Analysis

Exactly one Root in the net

Suppose there are two Roots, A and B, with A having the numerically smaller ID. If there is physical connectivity between A and B, there will be some LAN with a path to both A and B. On that LAN, as soon as a bridge issues a Hello with information about A, the other bridges on that LAN will overwrite their information about B. This will propagate eventually to B, which will stop attempting to be Root.

Exactly one Designated Bridge per LAN

Each bridge has a unique ID. Of the set of bridges on a LAN that have the smallest hop count from the Root, only one of them can have the smallest ID. Thus there is a unique Designated Bridge per LAN.

A bridge will attempt to become Designated Bridge on a LAN unless a different bridge which has a "better" HELLO message (closer to the Root, or equally close, with lower ID) transmits.

Thus if any bridges are attached to a LAN, one of them will become Designated Bridge.

All LANs are included in the formed graph

If there is physical connectivity between a LAN and the bridge that is Root, then news of the Root will have spread to that LAN, and the Designated Bridge will chosen because of its path to the Root. A bridge which is Designated Bridge on a LAN will be in the formed graph, and will include that LAN in the formed graph.

Since every LAN has a Designated Bridge, every LAN (to which there is physical connectivity from the Root) is included in the formed graph.

Amount of Memory

The amount of memory required for finding an acyclic spanning subset of the topology is just the amount needed for the per link database.

To keep the IDs of the Root and the Designated Bridge requires theoretically the log of the number of nodes, but in practice the fixed value of 48 bits for each is sufficient.

To keep the distance from the Root requires the log of the network diameter, but in practice a fixed value of 16 bits or even 8 bits would suffice.

The other pieces of information are of fixed length.

Bridge B must keep this information for each link attached to B, not each link in the network.

Communications Bandwidth

Maintenance of the algorithm when topology has not recently changed requires one HELLO per HELLO_TIME to be transmitted on each LAN by that LAN's Designated Bridge. In the absence of topology changes, only the Designated Bridge on a LAN will transmit HELLOs, and the only event to cause that bridge to transmit a HELLO is receipt of a HELLO message on its upstream LAN (or for the Root, the periodic HELLO-TIME trigger).

Thus regardless of the total size of the network, only one control message will be sent each HELLO_TIME on each LAN, unless the topology changes.

Following a topology change, a few extra HELLO messages might be transmitted, for instance by a bridge that has just come up, or by all bridges on a LAN following information having timed out. Additionally, the Designated Bridge issues a HELLO in response to a "worse" HELLO, to more quickly synchronize bridges that have just come up. However, there is no need to do this more frequently than HELLO_TIME, so a hold down is instituted to prevent a bridge from transmitting more than one HELLO per HELLO_TIME. Thus the maximum amount of control traffic on a LAN even during topological changes is limited by the total number of bridges on that LAN, not by the total number of bridges in the network.

Note that the overhead of the algorithm is independent of the characteristics of the stations' data traffic (e.g. factors such as frequent connection establishment).

The formed graph is a Tree

The properties of a tree are:

1. It has a unique root.
2. Each node other than the root has a unique predecessor closer to the root.

Analysis of the algorithm is simplified by assuming that the formed graph is bipartite, i.e. that there are two types of nodes:

1. Bridges, and
2. LANs.

The ancestor node of a Bridge is the LAN in the direction of the Root. The ancestor node of a LAN is the Designated Bridge. Thus all nodes have a unique ancestor. Likewise, there is a unique Root, which is the bridge with the lowest ID.

Deterministic Behavior

Given a particular physical topology, behavior of the algorithm is completely predictable. The routes chosen do not depend on the order in which messages were received, or on the order in which bridges were brought up, or on the history of previous topologies.

There are many advantages to deterministic behavior:

1. reproducibility -- It is easier to maintain a network if conditions are reproducible. If routes depended on chance occurrences such as the order in which bridges were brought up, or the order in which messages were received, then it is difficult to diagnose problems, since behavior is not reproducible.

2. configuratiblity -- It is easier to configure a network if behavior of a particular topology can be calculated.

3. predictability -- Users will be more satisfied with a particular level of performance, if the same conditions always produce the same level of performance. However, if sometimes performance is noticeably better, due to chance occurrences, users will not be satisfied with the usual level of performance. Similarly if sometimes performance is noticeably worse, due to chance occurrences under what appears to the user as identical conditions, the network will not be viewed as satisfactory.

Extensions

It might be desirable to influence the topology that is computed by the spanning tree algorithm. One facility that might be desirable is the ability to designate some bridges as "primary" bridges, and others as "secondary" bridges. The algorithm should compute a topology in which no secondary bridges appear, unless no topology exists consisting solely of primary bridges. Also, it might be desirable to minimize through-traffic on some LANs, for instance those that might be lower speed or more congested with local traffic. Also, it might be desirable to configure the network so that topological changes will have minimal impact.

All of these facilities are trivial extensions to the algorithm. These facilities can be provided as follows. The bridge ID should actually consist of a priority field, settable via network management, concatenated as the most significant byte onto the unique 48 bit ID assigned by the hardware.

If the topology is mostly a set of tree-like structures off a backbone, then the priority should be set to be the number of levels away from the backbone LAN that the bridge is. In this way, a bridge will be chosen as Root that is closest to the actual backbone, which is preferable, since it minimizes through-traffic on local segments, and minimizes actual topology changes that would occur due to changing Roots. (If the topology doesn't change, then the endnode caches are still valid, and the network is not partitioned temporarily while nodes switch over to other routes. If the backup Root is very near the old Root, then the topology will not change significantly when the old Root dies, and the net will experience no disruptions due to Root switchover.)

If the only switch is "primary" or "secondary", then a "1" will be the high order bit of the ID if "secondary", and a "0" if the bridge is designated "primary".

HOPS should be a double length field, with a high order part and a low order part. If a bridge is designated "secondary", instead of adding 1 to HOPS to get its distance from the Root (incrementing the low order part of HOPS), it will increment the high order part of HOPS.

To minimize traffic on some LANs, links can be assigned a cost. Then, instead of adding 1 to the distance to the Root given by the Designated Bridge on that LAN to obtain this bridge's distance from the Root, it would add the cost of the link. This will tend to place links with higher costs towards the leaves of the tree, which will minimize through traffic.

<u>One Way Connectivity</u>

Connectivity between bridges can be one-way if one transmitter is broken, or the other's receiver is broken, or some channel hardware works in only one direction. One-way connectivity might be a common hardware failure mode, and it is desirable that the algorithm prevent loops in the face of one-way links.

This algorithm can prevent one-way links from causing loops. A Designated Bridge will detect a problem if another bridge on the same LAN persists in sending HELLO messages, because that would indicate that the other bridge is not receiving the Designated Bridge's HELLOs. Note that if the other node disagrees about the identity of the Root, there is no loop. If the other node agrees about the Root, there is a loop.

If there is no loop, then no harm is done by continuing to forward packets to and from the LAN. Thus if the other node disagrees about the identity of the Root, no further action is taken (other than perhaps logging the condition).

If the other node does agree about the Root, the Designated Bridge must stop forwarding packets to and from the LAN.

<u>Summary</u>

The algorithm presented here maintains a spanning acyclic subset of a general mesh topology. It requires a very small, bounded amount of memory per bridge, independent of the total number of LANs or the total number of bridges. It requires a very small, bounded amount of communications bandwidth on each LAN, independent of the total number of LANs or the total number of bridges. It tolerates lost messages and efficiently utilizes the broadcast nature of multiaccess LANs. It requires no effort on the part of stations. The computed topology converges in at most twice the round trip delay across the extended network. The computed topology is deterministic. Bridges implementing this algorithm can coexist with simpler bridges that do not implement this algorithm, and loops will still be broken, provided that no loop exists composed solely of bridges that do not implement the algorithm.

References

[1] Boggs, Shoch, Taft, and Metcalfe, "PUP: An Internetwork Architecture," IEEE Transactions on Communications, April 1980.

[2] C. Sunshine, D. Kaufman, G. Ennis, and K. Biba, "Interconnection of Broadband Local Area Networks", Eighth Data Communications Symposium, Massachusetts, 1983.

[3] Norman Strole, "A Local Communications Network Based on Interconnected Token-Access Rings: A Tutorial", IBM J. Res. Develop, Vol 27, No 5, Sept., 1983.

[4] Kian-Bon Sy, Daniel A. Pitt, Robert A. Donnan, "An Architecture for Interconnecting LAN Segments", IBM Corporation, Technical submission to IEEE 802 LAN standards committee, July 13, 1984.

[5] Radia Perlman, "Incorporation of Multiaccess Links Into a Routing Protocol", Eighth Data Communications Symposium, Massachusetts, 1983.

[6] Bill Hawe, Alan Kirby, Anthony Lauck, "An Architecture for Transparently Interconnecting IEEE 802 Local Area Networks", Paper presented at the IEEE 802 meeting in San Diego, CA on October 1984

[7] Bill Hawe, Alan Kirby, Bob Stewart, "Transparent Interconnection of Local Networks with Bridges", Journal of Telecommunication Networks, June 1984.

[8] George Varghese and Bill Hawe, "Extended Local Area Network Management Principles," Paper presented at the IEEE 802 meeting in San Diego, CA on October 1984

[9] Yogen Dalal and Robert Printis, "48-bit Absolute Internet and Ethernet Host Numbers", Seventh Data Communications Symposium, 1981.

MODELING PHYSICAL LAYER PROTOCOLS USING COMMUNICATING FINITE STATE MACHINES

Mohamed G. Gouda and Khe-Sing The

Department of Computer Sciences
The University of Texas at Austin
Austin, Texas 78712

Abstract

We illustrate the usefulness of communicating finite state machines in modeling a number of physical layer protocols that include (i) an asynchronous start-stop protocol and (ii) a protocol for synchronous transmission with modems. Each protocol is modeled as a network of four finite state machines that communicate by exchanging messages over unbounded, FIFO channels. (Two machines are used to model the protocol itself, while the other two are used to model its interface to the upper data link protocol in the protocol hierarchy.) We outline a methodology to verify communication boundedness and progress for each protocol model. The methodology is based on three techniques that were proposed earlier to verify networks of communicating finite state machines; they are network decomposition, machine equivalence, and closed covers.

1. Introduction

The physical layer is the lowest layer in a protocol hierarchy [1,7,15]; its function is to transmit frames of data characters or bits across a physical line between two adjacent computers in a computer network [1,7,14]. Most of the available literature on physical layer protocols concentrate on defining their electrical specifications and pin descriptions, and ignore their procedural characteristics and interface behaviours. Although the procedural characteristics of these protocols may seem simple and straightforward at first, the task of modeling them formally and proving their correctness can become quite complicated.

The objective of this paper is three-fold:

i. Provide formal models based on communicating finite state machines to two important and widely used physical layer protocols: (a) an asynchronous start-stop protocol, and (b) a protocol for synchronous transmission with modems. We believe that by going through these formal models, one can appreciate the function of these protocols and their different ways of achieving this function.

ii. Illustrate the usefulness of communicating finite state machines as a convenient formal model for defining communication protocols. (For more examples in using communicating finite state machines to model communication protocols we refer the reader to [2,3,4,5,6,16,17].)

iii. Demonstrate the utility of recent verification methodologies [8,10,11,12,13] for networks of communicating finite state machines in establishing the correctness of "real" and nontrivial protocol models.

Following the introduction, the paper is organized as follows. Networks of communicating finite state machines are briefly presented in Section 2. Then, in Section 3, we discuss how to model a physical layer protocol as a network of four communicating finite state machines: two machines model the protocol itself and the other two model the protocol's interface with the data link protocol. In Section 4, we use this model to formally define two examples of physical layer protocols. A verification methodology for these protocol models is outlined in Section 5. Finally, concluding remarks are in Section 6.

2. Networks of Communicating Finite State Machines

A *network* W is a tuple $[M_1,...,M_n]$, where $n \geq 2$, and for $i=1,...,n$, M_i is a finite state machine that communicates with other finite state machines in W by exchanging messages via one-directional, unbounded, FIFO channels. Next, we define communicating finite state machines.

A *communicating finite state machine* M_i in a network $W=[M_1,...,M_n]$ is a labeled directed graph with three types of edges, namely sending, receiving, and internal edges:

- a *sending edge* in M_i is labeled with $-g/M_j$,
- a *receiving edge* in M_i is labeled with $+g/M_j$, and
- an *internal edge* in M_i is labeled with the empty sequence E,

where g is a *message* from a finite set G of messages, and M_j is a machine, other than M_i, in W. One of the nodes in M_i is identified as its *initial node*, and each node in M_i is reachable by a directed path from the initial node. For convenience, we assume that M_i is "nonterminating," i.e., each node in M_i has at least one outgoing edge.

A *state* s *of a network* $[M_1,...,M_n]$ is an $n \times n$ matrix, where an entry $s_{i,j}$ in the i-th row and j-th column is defined as follows:

i. If $i=j$, then $s_{i,i}$ is a node in machine M_i.

ii. If $i \neq j$, then $s_{i,j}$ is a sequence of messages from the set G; in this case, $s_{i,j}$ is called the *content of the channel from* M_i *to* M_j *in* state s.

The *initial state* s^0 of a network $[M_1,...,M_n]$ is a state of the network whose entries $s^0_{i,j}$ are defined as follows:

i. If $i=j$, then $s^0_{i,i}$ is the initial node in M_i.

ii. If $i \neq j$, then $s^0_{i,j}$ is the empty sequence E.

Let s and s' be two states of a network $[M_1,...,M_n]$, and let e be an edge from node v to node v' in machine M_i. State s' is said to *follow* s *over* e iff one of the following three conditions is satisfied:

i. e is a sending edge labeled $-g/M_j$, and each entry in s is identical to its corresponding entry in s' except (possibly) for $s_{i,i}$ and $s'_{i,i}$, and for $s_{i,j}$ and $s'_{i,j}$ whose values are defined as follows:
$s_{i,i}=v$, $s'_{i,i}=v'$, and $s'_{i,j}=s_{i,j}.g$, where "." is the usual string concatenation operator.

ii. e is a receiving edge labeled $+g/M_j$, and each entry in s is identical to its corresponding entry in s' except (possibly) for $s_{i,i}$ and $s'_{i,i}$, and for $s_{j,i}$ and $s'_{j,i}$ whose values are defined as follows:
$s_{i,i}=v$, $s'_{i,i}=v'$, and $s_{j,i}=g.s'_{j,i}$.

iii. e is an internal edge labeled E, and each entry in s is identical to its corresponding entry in s' except (possibly) for $s_{i,i}$ and $s'_{i,i}$ whose values are defined as follows:
$s_{i,i}=v$ and $s'_{i,i}=v'$.

Let s and s' be two states of a network W. State s' is said to *follow* s iff there exists an edge e in one of the machines in W such that s' follows s over e.

Let s and s' be two states of a network W. s' is *reachable from* s iff either s=s' or there exist a sequence $s^0,s^1,...,s^r$ of states of W such that $s^0=s$, $s^r=s'$, and for $i=0,...,r-1$, s^{i+1} follows s^i. A state of a network is *reachable* iff it is reachable from the initial state of the network.

The communication of a network $W=[M_1,...,M_n]$ is said to be *bounded by* K iff for every reachable state s of W, and for every $i=1,...,n$ and every $j=1,...,n$, if $i \neq j$ then $|s_{i,j}| \leq K$, where |x| is the number of messages in sequence x. The communication is *bounded* iff it is bounded by some positive integer K.

A state s of a network W is called an *unspecified reception state for machine* M_i in W iff $s_{i,i}$ is a receiving node in machine M_i, and for every outgoing edge e of node $s_{i,i}$, if e is labeled with $+g/M_j$, then $s_{j,i}$ is a nonempty sequence of the form g'.x, where $g \neq g'$. The communication of W is said to be *free from unspecified receptions* iff no reachable state of W is an unspecified reception state (for some machine in W).

Let s be a state of a network W, and let M_i be a machine in W, and $\{M_a,...,M_b\}$ be a set of machines in W. M_i is said to be *blocked by* $\{M_a,...,M_b\}$ *at* state s iff the following condition holds for each outgoing edge e of node $s_{i,i}$ in machine M_i. Edge e is labeled with some $+g/M_j$ such that

(i) M_j is in the set $\{M_a,...,M_b\}$, and

(ii) either $s_{j,i}$=E (the empty sequence) or $s_{j,i}$=g'.x, where g'≠g.

A set D of machines in W is called a *deadlock set at state* s iff each machine in D is blocked by a subset of D at s. A state s of a network W is called a *deadlock state* iff there exists a set of machines in W that constitutes a deadlock set at state s. The communication of W is said to be *free from communication deadlocks* iff no reachable state of W is a deadlock state.

The communication of a network W is guaranteed to *progress indefinitely* iff the communication of W is free from unspecified receptions and communication deadlocks. The communications of the two protocol models presented in this paper can be shown to progress indefinitely; an outline of the verification methodology is given in Section 5.

3. Modeling Physical Layer Protocols

The physical layer is the lowest layer in a protocol hierarchy; it serves as a communication medium to the data link layer [1,7,15]. As shown in Figure 1, the data frames sent by the data link layer at one site are first delivered to the physical layer at that site, then transmitted as a stream of characters over the communication line between sites, and finally assembled into frames before delivery to the data link layer at the other site. This arrangement can be modeled as a network of four communicating finite state machines [P,M,N,Q], where M and N model the physical layer protocol, and P and Q model the interface of the data link layer to the physical layer (see Figure 2).

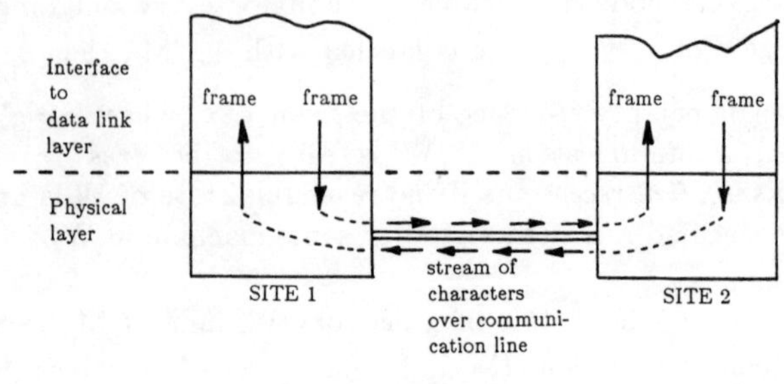

Figure 1: Physical layer.

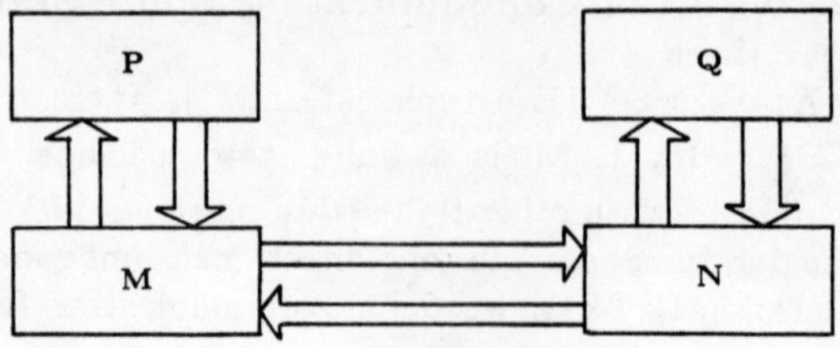

Figure 2: Modeling a physical layer protocol as a network of four machines.

In Section 5, we model various physical layer protocols as networks [P,M,N,Q] that differ in the way the communication between M and N is governed, but provide a uniform interface through P and Q. This means that the interface machines P and Q are the same for all the protocol models discussed in Section 5, but M and N change from one protocol model to another. The interface machines P and Q are defined in this section, and the different versions of machines M and N are discussed in Section 5.

The chosen interface between the data link layer and the physical layer satisfies the following requirements:

i. *Basic full duplex operation:* The interface allows the data link layer at each site to send frames to the other site, and to receive frames from the other site, possibly at the same time.

ii. *Simple operation:* For the data link layer at one site to send a frame, it simply sends the frame to the physical layer at its site. For it to receive a frame, it simply sends a "next" message to the physical layer at its site, then waits for the frame.

iii. *Friendly operation:* Before the data link layer at one site can send one frame to the other site, it must also be ready to receive one frame from the other site.

iv. *No error detection:* The interface does not guarantee that all sent frames will be delivered correctly. It is the function of the data link layer, not the physical layer, to detect frame corruption and loss [1,14,15].

Based on these requirements, the communicating finite state machine P can be defined as shown in Figure 3. (Machine Q is identical to P except that it communicates with N instead of M.) Machine P exchanges the following messages with machine M.

- "frame" denotes a data frame,

- "next" denotes a request from P to M to send the next data frame, and

- "noframe" denotes a reply of "no frame is available" from M to P.

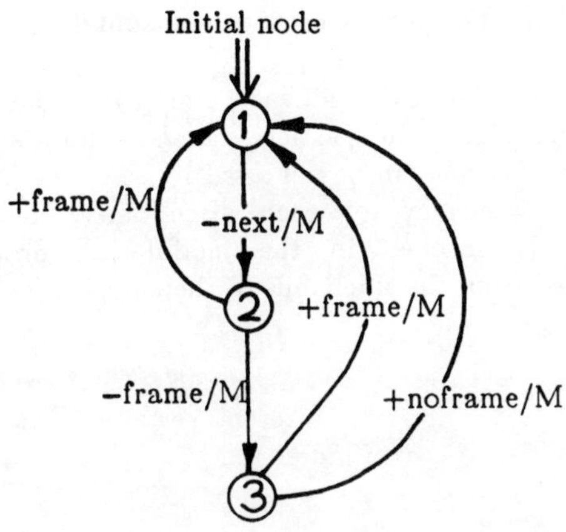

Figure 3: Machine P.

Starting from node 1, P can only send a request "next" to M. It then either waits to receive the requested frame from M and returns to node 1, or sends a data frame to M. As discussed later, M is expected to forward this data frame to N. If in the mean time M receives a frame from N, it sends it to P as a reply to its request; otherwise it sends a "noframe" reply to P. In either case P returns to its initial node.

4. Examples of Physical Layer Protocols

In this section, we discuss formal models of two physical layer protocols: an asynchronous start-stop protocol, and a protocol for synchronous transmission with modems. For convenience, we adopt the following two abbreviations in drawing the communicating machines in this section:

1. An edge label of the form $-g/M_i$ $(+g/M_i)$ can be written as -g (+g), when M_i can be implicitly understood from g. For example, the edge label -frame/P in machine M will be written as -frame, because M sends "frame" only to P.

2. Two or more edges from node v to node v' can be drawn as one edge from v to v' with multiple labels.

4.1. An Asynchronous Start-Stop Protocol

This protocol is used to transmit frames of data characters over a single full duplex communication line [14]. A data frame consists of an integral number of contiguous data characters, where a character has a fixed length of bits. The bits are transmitted serially, hence, there should be an agreement between the sender and receiver on the duration of a bit over the communication line, so the state of the line can be properly sampled.

The protocol provides a mechanism to delineate frames and characters. A data character consists of a START bit (usually a "0" bit), fixed number of data bits, and a STOP bit (usually a "1" bit). When there are no data characters to transmit over the line, idle characters are transmitted instead. An idle character consists of all idle bits (usually "1" bits). Two successive data frames should be separated by at least one idle character. The receiver can detect the beginning of a frame by sensing the START bit of the first data character in the frame, and the end of a frame by the idle character following the last data character.

Formal modeling: Figure 4 shows machine M. (Machine N is identical.) M and N exchange the following messages:

- "idle" denotes an idle character,

- "start" denotes the START bit of a data character,

- "data" denotes the sequence of data bits in a data character, and

- "stop" denotes the STOP bit of a data character.

The initial node, node 1, of M is the "not ready" node, since at this node M has not yet received a "next" from P, and so it cannot accept any data characters from N. If at node 1 M receives a "next" message from P, it proceeds to node 2 and becomes ready to communicate with N. On the other hand, if at node 1 M receives a data character from N, it proceeds to node 20, the "error" node, and discards the incoming data characters as indicated by the dashed receiving edges. M proceeds from node 20 to node 2 when it receives the first "idle" message from N after a "next" message from P.

Node 2 is the "ready" node, where M keeps sending and receiving "idle" messages until it receives a frame from P (then goes to node 3), or a data character from N (then goes to node 4).

Nodes 3 and 16 are "data sending" nodes, where M sends data characters to N, and receives idle characters from N. If at node 3 M finishes sending the characters of the current frame, it recognizes that it has received no data characters from N, so it sends a "noframe" message to P and returns to the initial node. On the other hand, if at node 3 M receives a data character from N, it proceeds to node 8.

Node 4 is the "data receiving" node, where M receives data characters from N, and sends idle characters to N. If at node 4 M receives an "idle" indicating the end of the received frame, it immediately delivers this frame to P. On the other hand, if at node 4 M receives a data frame from P, it proceeds to node 8.

Node 8 is the "data sending and receiving" node, where M and N exchange data characters. From this node M can proceed to node 4 (16) whenever it sends (receives) an "idle" message. Notice that when M receives an "idle" message indicating the end of the current received frame at node 15, it sends the received frame to P.

4.2. A Protocol for Synchronous Transmission with Modems

This protocol is used to transmit frames of data characters over a full duplex communication line with a pair of modems, one at each end [14]. The data bits are modulated in carrier signals, and full duplex communication can be achieved by selecting a different carrier frequency for each direction. The clocking signal is encoded in the modulating process and decoded from the sidebands of the carrier signal.

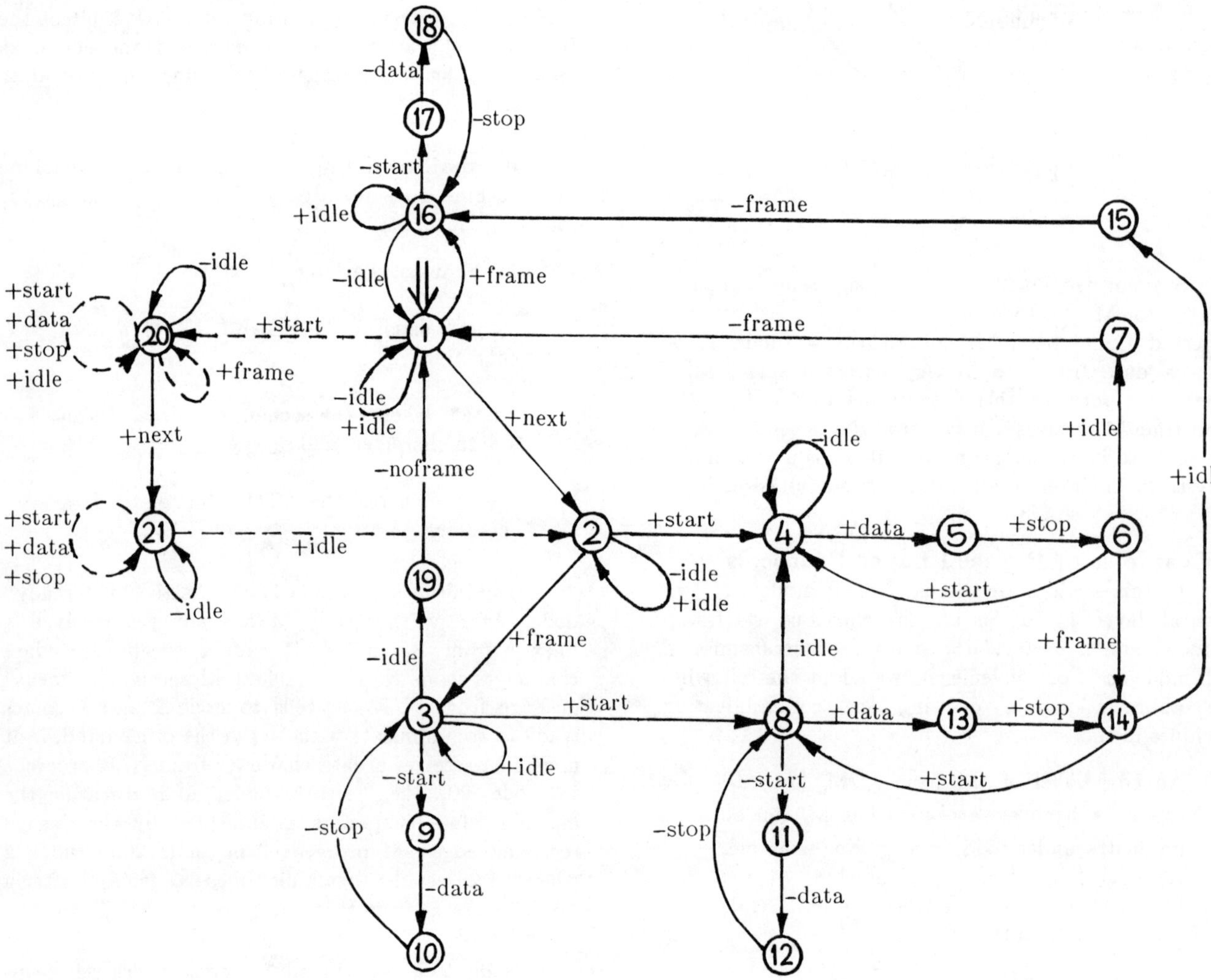

Figure 4: Machine M of the asynchronous start-stop protocol.

The existence of a clocking signal in this protocol necessitates a special synchronizing character (denoted by "sync") to be sent whenever either site has no data characters to send. A data frame consists of integral number of contiguous data characters and can be preceeded or followed by "sync" characters.

Unlike the previous protocol, here a connection should be established between the two sites before either site can send a data frame. A connection is indicated by the presence of carriers in both directions. During the interval from establishing to clearing a connection, at most one frame can be sent in each direction. After clearing a connection, new connections can be established to send more frames.

To establish a new connection, a site keeps on sending connect requests until it receives the incoming carrier, it then propagates its own carrier and the connection is established. Either site can request to clear the connection simply by ceasing to propagate its carrier, and the connection will be cleared after the incoming carrier has been turned off.

Formal modeling: The communicating finite state M (or N, which looks identical to M) of this protocol model is shown in Figure 5. The interpretation of messages exchanged between M and N are as follows:

- "conn" denotes a connect request.
- "car" is a message that represents the existence of carrier signals. ("-car" should be interpreted as starting to propagate the carrier, and "+car" should be interpreted as sensing an incoming carrier.)
- "nocar" is a message that represents the lack of carrier signals.
- "data" denotes a data character.
- "sync" denotes a "sync" character.

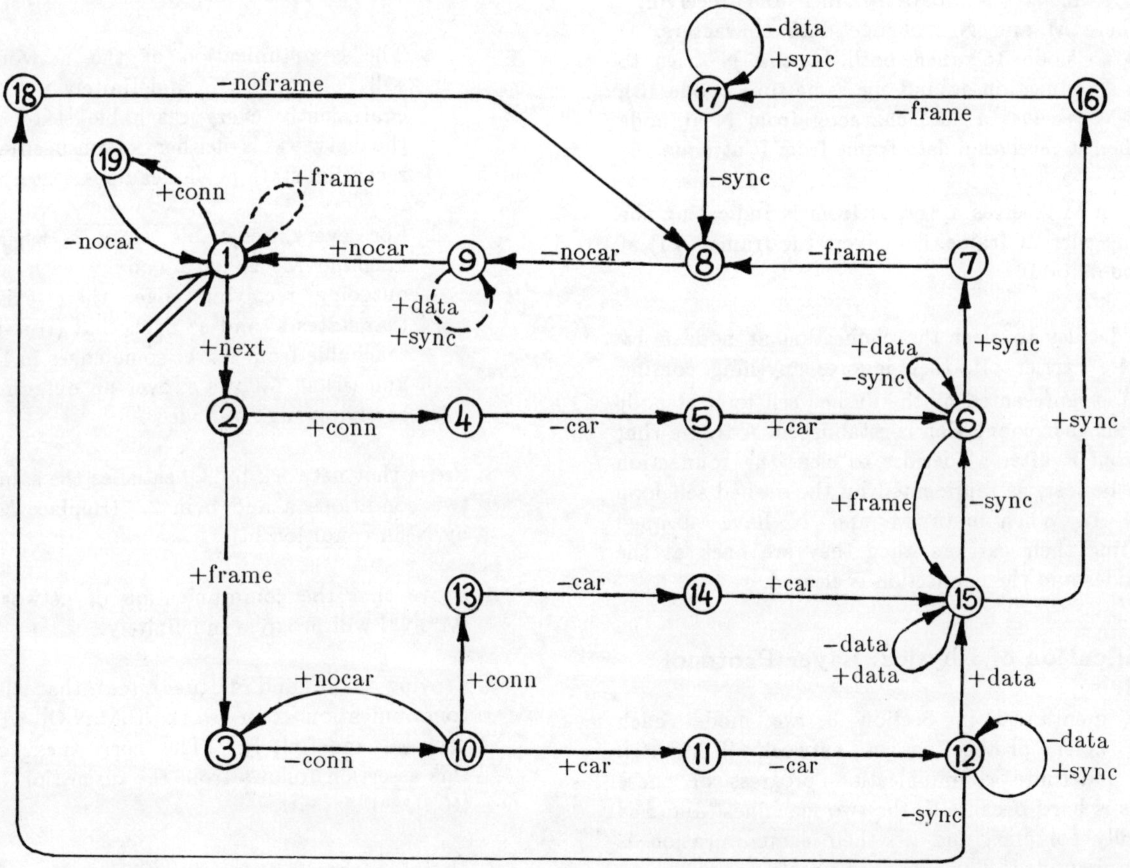

Figure 5: Machine M of the synchronous transmission protocol.

At node 1, the initial node, M has not yet received a "next" from P, and so it will reject any "conn" request from N. If M receives a "next" at node 1, it proceeds to node 2 where it becomes ready to accept a connect request and later starts receiving data characters at node 6, or to accept a frame from P and then tries to establish a connection (node 3).

Nodes 12 and 17 are "data sending" nodes, where M sends data characters to N, and receives "sync" characters from N. If M finishes sending the current frame while it is at node 12, it recognizes that it has received no data characters from N, so it sends a "noframe" message to P and prepares to clear the connection.

Node 6 is the "data receiving" node, where M receives data characters from N, and sends "sync" characters to N. M will come to this node if P does not have a frame to send, or if M has finished sending the frame from P to N but still has to receive data characters from N.

Node 15 is the "data sending and receiving" node, where M and N exchange data characters. M proceeds to node 15 when both M and N tries to establish a connection around the same time (node 13), or when M receives a data character from N at node 12, or when it receives a data frame from P at node 16.

When M receives a "sync" from N indicating the end of the current frame, it delivers the frame to P at either node 7 or 16.

M decides to clear the connection at node 8 by turning its carrier off, then ignores anything coming from N (as represented by the dashed self loop at node 9) until another connection is established. A frame that comes from P after M decides to clear the connection will also be lost, as represented by the dashed self loop at node 1. When both M and N have stopped propagating their carriers, then they are back at the initial nodes and the connection is cleared.

5. Verification of Physical Layer Protocol Models

As mentioned in Section 3, we model each physical layer protocol as a network [P,M,N,Q]. Proving indefinite communication progress of these networks is hard because (i) the two machines M and N are usually complex, and (ii) their communication is usually unbounded. To carry such proofs, we use three techniques: network decomposition, machine equivalence, and closed covers. These techniques are summarized next, but interested readers can find more details and examples about these techniques in [8], [10], and [11].

5.1. Network Decomposition

To prove that the communication of network [P,M,N,Q] in Figure 2 will progress indefinitely, we follow the next procedure.

1. Decompose the network into three networks, with two communicating machines each, [P,M'], [M",N"], and [N',Q], where

   M' (N') is identical to M (N) except that each sending edge to N (M) and each receiving edge from N (M) is replaced by an internal edge, and

   M" (N") is identical to M (N) except that each sending edge to P (Q) and each receiving edge from P (Q) is replaced by an internal edge.

2. Prove that network [P,M'] satisfies the following two conditions:

   a. The communication of the network will progress indefinitely, or equivalently every reachable state of the network is neither an unspecified reception state nor a deadlock state.

   b. For every reachable state s where machine M' is at a node v with an outgoing receiving edge, there exist two states s' and s" such that (i) s' is reachable from s over some edges in P, and (ii) s" follows s' over an outgoing receiving edge of node v.

3. Prove that network [N',Q] satisfies the same two conditions a and b in 2. (Replace M' by N' in condition b.)

4. Prove that the communication of network [M",N"] will progress indefinitely.

5. Proving 2, 3, and 4 guarantees that the communication of network [P,M,N,Q] will progress indefinitely. (The correctness of this assertion follows from the discussion in [10].)

Notice that the function of network decomposition in this procedure is to reduce the problem of proving some property for a four-machine network into proving some related properties for three two-machine networks; it is a *divide and conquer* technique.

The proof needed in Step 2 of the above procedure is straightforward. This is because the communication of network [P,M'] is bounded; hence all reachable states of the network can be generated and examined to verify that conditions a and b are satisfied. Similarly, the proof needed in Step 3 is straightforward, since the communication of [N',Q] is also bounded.

Unfortunately, the proof needed in Step 4 is not straightforward since the communication of [M",N"] is usually unbounded. Therefore, to prove indefinite progress for the communication of network [M",N"], we resort to the technique of closed covers [8]. But before we use this technique, it is convenient to remove the internal edges from M" and N"; this requires a third technique known as machine equivalence.

5.2. Machine Equivalence

Let W be a network of communicating finite state machines, and let M be a machine that have some internal edges in W. Assume that each internal edge in M is in some directed cycle. Then, the equivalence-preserving transformations defined in [11] can be applied to transform M into an "equivalent" machine EM that has no internal edges. The equivalence is such that the communication of network W is guaranteed to progress indefinitely iff the communication of network EW is guaranteed to progress indefinitely, where EW is the same as W except that M is replaced by EM.

This result can be used to remove the internal edges from the two machines in [M",N"]. In particular, these equivalence-preserving transformations can be applied to the two machines M" and N" yielding the equivalent machines EM" and EN", respectively, such that (i) EM" and EN" have no internal edges, and (ii) the communication of [M",N"] is guaranteed to progress indefinitely iff the communication of [EM",EN"] is guaranteed to progress indefinitely. It remains now to discuss how to prove that the communication of [EM",EN"] will progress indefinitely. For that purpose, we need the technique of closed covers.

5.3. Closed Covers

As discussed in [8], to prove that the communication of a network will progress indefinitely, it is sufficient to exhibit a closed cover of the network. A *closed cover* C of network [EM",EN"] is a finite set of state schemas

$$\begin{bmatrix} v_1 & Y_1 \\ X_1 & w_1 \end{bmatrix}, \ldots, \begin{bmatrix} v_r & Y_r \\ X_r & w_r \end{bmatrix},$$

such that the following three conditions are satisfied.

i. Each state schema

$$\begin{bmatrix} v_i & Y_i \\ X_i & w_i \end{bmatrix}$$

in C is such that
(a) v_i is a node in EM",
(b) w_i is a node in EN", and
(c) X_i and Y_i are two (possibly infinite) sets of message sequences.
Each state schema can be viewed as a set of network states. A state

$$\begin{bmatrix} v_i & y_i \\ x_i & w_i \end{bmatrix}$$

is *in* some state schema

$$\begin{bmatrix} v_i & Y_i \\ X_i & w_i \end{bmatrix}$$

in C iff the message sequences x_i and y_i are in the sets X_i and Y_i, respectively.

ii. The initial state of [EM",EN"] is in some state schema in C.

iii. For every state s that is in some state schema in C, there exist two states s' and s" such that
(a) s" is in a state schema in C, and
(b) either (s' follows s over an edge in EM" and s" follows s' over an edge in EN") or (s' follows s over an edge in EN" and s" follows s' over an edge in EM").

For more details on how to apply these three techniques (network decomposition, machine equivalence, and closed covers) to the verification of the protocol models discussed in this paper, we refer the reader to [9] and [16].

6. Concluding Remarks

We have presented a formal model to define physical layer protocols and a methodology to verify the correctness of the resulting protocol models. We have applied this methodology to model and verify an asynchronous start-stop protocol and a synchronous transmission protocol with modems. In [9], we use the same formal model and its accompanying verification methodology to model and verify other physical layer protocols that include the EIA RS-449 Standard and the CCITT X.21 Recommendation.

We hope that this experience can inspire others to utilize these or other formal techniques in modeling communication protocols instead of resorting to informal descriptions that are ambiguous and usually invite errors or imprecision.

REFERENCES

[1] H. V. Bertine, "Physical interfaces and protocols," in *Computer Network Architectures and Protocols*, P. E. Green, Ed. New York: Plenum Press, 1982, pp. 57-83.

[2] G. V. Bochmann, "Finite state description of communication protocols," *Comput. Networks,* vol. 2, pp. 361-371, 1978.

[3] G. V. Bochmann and C. Sunshine, "Formal methods in communication protocol design," *IEEE Trans. Commun.*, vol. COM-28, pp.624-631, Apr. 1980.

[4] D. Brand and P. Zafiropulo, "On communicating finite-state machines," *Journal ACM*, vol. 30, pp. 323-342, Apr. 1983.

[5] C. H. Chow, M. G. Gouda, and S. S. Lam, "An exercise in constructing multi-phase communication protocols," in *Proc. SIGCOMM '84 Symposium*, June 1984.

[6] C. H. Chow, M. G. Gouda, and S. S. Lam, "A discipline for constructing multi-phase communication protocols," *ACM Trans. Comput. Syst.*, to appear Nov. 1985.

[7] R. J. Cypser, *Communications Architecture for Distributed Systems.* Reading, MA: Addison-Wesley, 1978.

[8] M. G. Gouda, "Closed covers: to verify progress for communicating finite state machines," *IEEE Trans. Software Eng.,* vol. SE-10, pp. 846-855, Nov. 1984.

[9] M. G. Gouda and K. S. The, "On modeling and verification of physical layer protocols," Tech. Rep., Dep. Comput. Sci., Univ. Texas, Austin, TX, in preparation.

[10] M. G. Gouda, K. S. The, and C. K. Chang, "Verification of distributed synchronization systems via decomposition," Tech. Rep., Dep. Comput. Sci., Univ. Texas, Austin, TX, in preparation.

[11] M. G. Gouda and C. H. Youn, "On the notion of equivalence for communicating finite state machines," *TR-84-14*, Dep. Comput. Sci., Univ. Texas, Austin, TX, May 1984. Revised Feb. 1985.

[12] B. T. Hailpern and S. S. Owicki, "Modular verification of computer communication protocols," *IEEE Trans. Commun.*, vol. COM-31, pp. 56-68, Jan. 1983.

[13] S. S. Lam and A. U. Shankar, "Protocol verification via projections," *IEEE Trans. Software Eng.,* vol. SE-10, pp.325-342, July 1984.

[14] J. E. McNamara, *Technical Aspects of Data Communication.* Maynard, MA: Digital Equip. Corp., 1977.

[15] A. S. Tanenbaum, *Computer Networks.* Englewood Cliffs, NJ: Prentice-Hall, 1981.

[16] K. S. The, "A framework for formal modeling and verification of physical layer protocols," Master's thesis, Univ. Texas, Austin, TX, 1985.

[17] P. Zafiropulo, C. H. West, H. Rudin, D. Brand, and D. Cowan, "Towards analyzing and synthesizing protocols," *IEEE Trans. Commun.*, vol. COM-28, pp. 651-661, Apr. 1980.

A GRAMMAR-BASED METHODOLOGY FOR PROTOCOL SPECIFICATION AND IMPLEMENTATION

David P. Anderson
Lawrence H. Landweber

Computer Sciences Department
University of Wisconsin - Madison
Madison, Wisconsin

ABSTRACT

A new methodology for specifying and implementing communication protocols is presented. This methodology is based on a formalism called "Real-Time Asynchronous Grammars" (RTAG), which uses a syntax similar to that of attribute grammars to specify allowable message sequences. In addition RTAG provides mechanisms for specifying data-dependent protocol activities, real-time constraints, and concurrent activities within a protocol entity. RTAG encourages a top-down approach to protocol design that can be of significant benefit in expressing and reasoning about highly complex protocols. As an example, an RTAG specification is given for part of the Class 4 ISO Transport Protocol (TP-4).

Because RTAG allows protocols to be specified at a highly detailed level, major parts of an implementation can be automatically generated from a specification. An *RTAG parser* can be written which, when combined with an RTAG specification of a protocol and a set of interface and utility routines, constitutes an implementation of the protocol. To demonstrate the viability of RTAG for implementation generation, an RTAG parser has been integrated into the kernel of the 4.2 BSD UNIX operating system, and has been used in conjunction with the RTAG TP-4 specification to obtain an RTAG-based TP-4 implementation in the DoD Internet domain.

1. INTRODUCTION

Data communication protocols serve to enhance the utility and reliability of communications media. In recent years a great deal of effort has been devoted to understanding the services and mechanisms of such protocols. The ISO, in cooperation with other international organizations is currently in the process of developing protocols corresponding to the layers of its reference model. Earlier work by the U.S. DoD as well as by various computer manufacturers (e.g., IBM's SNA and DEC's DECNET) has resulted in the specification of protocol architectures which are functionally similar to but incompatible with each other and with the proposed ISO protocols.

The need for machine-independent specifications of standard communication protocols has encouraged the definition of a number of formalisms for this purpose. In this paper, we define a new methodology which is rich enough to concisely describe real protocols, is "easy and intuitive" to use, and which we believe can eventually serve as the basis of a software system for automated generation of quality protocol implementations.

Our methodology, "Real-Time Asynchronous Grammars" (RTAG), is based on an extension of attribute grammars in which terminal symbols correspond to messages sent and received. It provides formal constructs for specifying concurrent protocol activities and real-time constraints. RTAG facilitates the expression of most protocol mechanisms, and encourages a top-down approach to designing and specifying protocols.

We have developed a software system consisting of a *grammar analyzer* and an *RTAG parser*. Applied to an RTAG specification of a protocol, these programs provide a major part of an implementation of the protocol (essentially only operating system interface routines and utility routines for functions such as packet assembly need be added).

The protocol implementor writes an RTAG description of a protocol. This description is then processed by the grammar analyzer, which produces a set of "grammar description tables". The grammar description tables for a protocol, plus user supplied interface and utility routines for a particular operating system, are combined with the RTAG parser to obtain an implementation of the protocol. The parser operates by doing a top-down parse on the grammar in response to external events (e.g., arrival of a packet or a request from a user).

If a number of protocols have been specified by RTAG, and implementations of these protocols are desired for a new computer system, it suffices to implement the RTAG parser and the utility routines on the system. Conversely, experimentation with a protocol running on a heterogeneous network of systems all running the RTAG parser can be done by changing the RTAG specification rather than by rewriting many hand-coded protocol implementations.

We have implemented an RTAG parser under 4.2 BSD UNIX and have written an RTAG specification for the ISO Class 4 Transport Protocol (TP-4). The RTAG parser has been installed in the UNIX kernel and interfaced with the other components of the 4.2 BSD UNIX networking system, thus obtaining an RTAG-based implementation of TP-4. The goal of this component of our work is to study the feasibility of generating production implementations in the manner described above.

Section 2 of this paper provides an overview of related work in protocol specification techniques. The remainder of the paper covers various aspects of RTAG. Sections 3 and 4 describe the syntax and informal semantics of RTAG. Section 5 describes the structure of a simple RTAG parser, and Section 6 discusses RTAG software tools and their integration into environments such as operating systems and prototyping/debugging systems. Section 7 summarizes RTAG, contrasts RTAG with related work, and suggests directions for further work.

2. FORMALISMS FOR PROTOCOL SPECIFICATION

The general goal of a protocol specification is to describe the desired behavior in abstract terms, i.e., without dependence on a particular programming paradigm. Besides their use in protocol verification and performance study, specifications aid in implementation by providing formal or informal guidelines for programmers and/or by allowing some degree of automated implementation. The remainder of this section briefly surveys some existing protocol specification formalisms.

Protocols are often modeled by associating a finite-state automaton with each entity. This approach allows verification by various state-space exploration techniques. Pure FSA cannot easily model data-dependent aspects of protocols, such as the use of large sequence numbers spaces. To deal with this an extension called "augmented FSA" (AFSM) or "abstract machines" (see Sun81) has been defined.

Blumer and Tenney (Blu82) describe the use of AFSM in a project whose goals include automatic implementation and verification. Their paper describes software tools for converting an AFSM description to a C program which implements the protocol. A later paper by Sidhu and Blumer (Sid83) describes tools for automated testing of protocols described by AFSM's; the connection establishment phase of TP-4 is used as an example.

An automated implementation project undertaken by IBM in connection with their Systems Network Architecture (SNA) is described in several papers (Sch80, Poz82, Nas83). It is based on the Format and Protocol Language (FAPL) and has been used to develop a specification of SNA from which a partial implementations can be automatically generated. FAPL is based on a subset of PL/1 and provides AFSM's as a programming abstraction.

Harangozo (Har78) uses a grammar-based formalism to specify a portion of the HDLC data-link protocol. An attribute-like mechanism, in which symbols and productions are indexed by integers, is used to represent sequence numbers. The grammars used are all regular grammars and hence are equivalent to FSA's. Teng and Liu (Ten78) use context-free grammars to specify some simple protocols. Their objectives include implementation and limited verification.

Yemeni and Nounou (Yem83) discuss a "protocol development environment" in which several formalisms (FSA, Petri nets, high-level language) are available for protocol specification. They propose tools to convert these forms into a "canonical semantic model" based on Milner's Calculus of Communicating Systems (CCS) (Mil80).

CCS also serves as the basis for LOTOS (Bri85), a protocol specification language intended for use in specifying "real" protocols. LOTOS is similar to RTAG in that it allows complex protocols to be broken into processes which can be composed in different ways. RTAG and LOTOS differ in that LOTOS is oriented towards verification (and has a better-understood semantic model), but its utility in generating implementations has not been shown.

3. OVERVIEW OF RTAG

Real-time asynchronous grammars (RTAG) is a protocol specification language oriented towards specifying and implementing complex protocols. An RTAG specification is based on an underlying context-free grammar. Most terminal symbols correspond to an *event*, that is, to the sending or receipt of a particular message. Event symbols are called *input* or *output* symbols depending on the direction of the message relative to the protocol entity being specified.

RTAG grammars describe event sequences allowed by the protocol. For example, the production

<goal> : [net-data] [user-data] <goal>.

might be part of a trivial protocol which responds to messages from the network layer (the **[net-data]** input event) by relaying them to an upper layer (the **[user-data]** output event).

Each RTAG symbol has an associated set of attributes. The attributes of an input or output symbol correspond to the fields of the associated message or packet. A production may have a boolean *enabling condition*, involving attributes of its symbols. This enabling condition must evaluate to **True** for the production to be applied. Each production can also have an associated set of *attribute assignments* which are performed when the production is applied.

Each RTAG symbol can be thought of as representing a "subprotocol". Productions can combine subprotocols in sequence:

<x> : <y> <z>.

or, using curly brackets, in parallel:

<x> : {<y> <z>}.

In the first case, the events derived from **<y>** must precede (in real time) the events derived from **<z>**, whereas in the second case the events may be interleaved.

Real-time constraints, such as timeouts, can be represented using a special terminal symbol **/timer/**, whose single attribute is the length of a real-time interval in clock ticks.

The semantics of RTAG can be explained in terms of an *RTAG parser* which maintains the current parse tree. When an input event occurs, the parser attempts to left-derive the corresponding input event symbol from some nonterminal symbol which is a leaf in the parse tree. For example, suppose the grammar contains the production

<goal> : [net-data] [user-data] <goal>.

If the current parse tree is

```
                  <goal>
                    |
 ----------------------------------------
 |                  |                   |
[net-data]      [user-data]          <goal>
```

and the **[net-data]** event occurs, the above production would be applied to the lower instance of **<goal>.** If the production has attribute assignments, these would result in input data received from the network being introduced into the parse tree. As a side-effect of applying the production, the output event symbolized by **[user-data]** would be performed.

If an input symbol cannot be derived, it is discarded and ignored. This is appropriate, for example, with a duplicate connection request or confirm packet, where such an occurrence does not signify a protocol error. In other cases it might be preferable to regard unparsed packets as fatal errors and to abort the protocol, or to log them for debugging purposes.

4. RTAG SYNTAX AND INFORMAL SEMANTICS

Because of their role in automated implementation, RTAG grammars are written as machine-readable files obeying syntactic rules. This section develops in parallel the syntax and informal semantics of RTAG. An RTAG specification consists of two parts: a list of symbol definitions and a list of productions.

4.1. Symbols, Attributes, and Names

Each symbol has an associated set of attributes, each of which has a type (e.g. integer, boolean, pointer to data). Symbol names obey the following rules:

(1) Nonterminal symbols are delimited by angle brackets.

(2) Event symbols (both input and output) are delimited by square brackets. Names should suggest whether the symbol is an input or output symbol, and what other entity it involves. For example, in our TP specification **[U-CR]** is a connection request input from the user layer, and **[N-DT]** is a data packet output to the network layer.

(3) Special symbols are delimited with slashes. There are currently three special symbols: **/timer/**, **/remove/**, and **/freedata/.**

Here is an example symbol declaration section:

```
goal     <connection>

nonterm <send>
         int      seqno

input    [U-CR]
         int      refno
         int      address
```

```
output |N- DT|
        int      refno
        dataptr  data
```

4.2. Productions

The syntax for simple productions with a common LHS symbol is:

```
<x>  :     S1.
           (optional enabling condition)
           (optional attribute assignments)
      |    S2.
           (optional enabling condition)
           (optional attribute assignments)
           ...
;
```

where **<x>** is any nonterminal and S1, S2, ... are strings of symbols. To simplify the parsing task, two restrictions are imposed: 1) input symbols must be leftmost in a RHS, and 2) left recursion is not allowed.

Each symbol instance is initially "inactive" and is later "activated". A nonterminal cannot be expanded if it is inactive, and the output event represented by an output symbol is performed only when its symbol is activated. When a production is applied, the leftmost symbol of the RHS is activated, and each remaining RHS symbol is activated as soon as all its left siblings have been fully expanded. This rule connects left-to-right syntactic order with real-time order. For example, in a production of the form

```
<x> : <y> <z>.
```

all of the terminal symbols derived from **<y>** will strictly precede (in real time) all those derived from **<z>.** This corresponds to sequential composition of subprotocols. The ability to compose subprotocols in parallel is also needed: for example, the send and receive portions of a transport protocol can be specified separately but must be interleaved in real time. RTAG uses a curly bracket notation to express this parallel composition. A group of RHS symbols surrounded by curly brackets will all be activated as soon as the leftmost symbol in the group is activated. For example, in

```
<x>  :     {<a> <b>} <c> {<d> <e>}.
```

<a> and **<b>** will be activated when the production is applied, and their subtrees will be expanded concurrently. When they are both fully expanded **<c>** will be activated, and when it is fully expanded **<d>** and **<e>** will be activated.

4.3. Symbol and Attribute References

Symbol references can be *local* or *non-local*. A local reference is of the form $n, and refers to the nth symbol of the production; $0 is the LHS and $1 is the leftmost symbol of the RHS. A nonlocal reference is written as **sym1/sym2.../symn.** The semantics are as follows: find the closest ancestor named *sym*1 of the symbol instance serving as the LHS of this production; find the leftmost of its children named *sym*2; the leftmost of that symbol's children named *sym*3 and so forth. A reference to a nonexistent symbol (detected at runtime) is considered a fatal error.

Attribute references are of the form *symbol-reference.attrname* and refer to the named attribute of the symbol instance specified by *symbol-reference.*

4.4. Expressions, Enabling Conditions, and Attribute Assignments

Expressions are formed from constants, attribute references, integer operators (+, −, mod), logical and relational operators (written as in C), and external functions (see sec. 4.7.1).

A production can have an "enabling condition", a boolean-valued expression that must evaluate to **True** for the production to be performed. For example,

```
<x>  :  |N -DT| <z>.
        if ($0.size > 0) && ($1.number == $0.number)
;
```

specifies that the production can be applied only if the *size* attribute of **<x>** is positive and the *number* attributes of **<x>** and **|N-DT|** are equal.

A production can have zero or more "attribute assignments" of the form *attribute-reference = expression*, which are used to transfer information between symbols. When a production is applied, attributes of newly-created symbol instances (except input symbols) are initially undefined. Assignments are then performed (meaning that the RHS expression is evaluated and the value is stored in the LHS attribute) in the order in which they appear in the grammar. *Persistent* attribute assignments (denoted with =* instead of =) are performed whenever a value on which the RHS depends is changed. For example,

```
<x>  :  |N -DT| <y>.
        $0.data = $1.data
        $2.ready =* $0.ready
```

specifies that the *data* attribute is to be copied from **|N-DT|** to **<x>**when is applied, and the *ready* attribute is to be copied from **<x>** to **<y>** both when the production is applied and whenever thereafter **<x>**.ready is modified; some other enabling condition or attribute assignment may depend on **<y>**.ready.

4.5. Special Symbols

Special symbols are terminal symbols which represent internal actions of the parser. There are currently three special symbols: **/timer/**, **/remove/**, and **/freedata/**.

The **/timer/** symbol, which has a single integer attribute *interval*, represents the passage of that many ticks of real time. Productions in which **/timer/** is the leftmost symbol of the RHS are called *timed productions*. As an example,

```
<x>  :  /timer/ |N- AK|.
        $1.interval = 10
     |  |U -DT| |N- DT|.
;
```

specifies that if the **|U-DT|** event occurs within ten ticks of the time an instance of **<x>** is activated then the second production will be applied to **<x>**, and **|N- DT|** will be performed; otherwise the first production will be applied ten ticks after **<x>** is activated, and **|N- AK|** will be performed as a result. This is accomplished by having the parser start a timer when **<x>** is activated. It cancels the timer and performs the second production if **|U -DT|** occurs before the timeout, otherwise it performs the first production when the timer times out.

The **/timer/** mechanism contributes greatly to the convenience of RTAG for expressing timing constraints because it removes the problem of timer names and the necessity of explicitly starting and canceling timers.

The **/remove/** symbol has an attribute *where* of type symbolptr. When an instance of **/remove/** is activated, the subtree rooted by the symbol instance pointed to by this attribute is removed from the parse tree, and the symbol is considered to be fully expanded. This is used, for example, in handling abrupt closure of connections.

The **/freedata/** symbol has an attribute *data* of type dataptr. When an instance of **/freedata/** is activated, the memory area pointed to by this attribute is freed. This is needed, for example, to specify memory management in protocols that do retransmission.

4.6. External Functions

External functions are used to perform calculations which are not easily expressed within RTAG or which are installation-dependent. These are notated as #name#(expr-list). For example, our TP specification uses external functions for computing the cyclic "between" relation on sequence numbers, and to check the acceptability of a connection request or confirmation.

4.7. Key Attributes

The work done in processing an input event consists largely of finding where to parse it. Most protocols which handle multiple connections have a "reference number" mechanism for associating input events with connections. Rather than enforce this at the leaf level (by having a reference number equality clause in the enabling condition) we have added a mechanism which allows the parser to search only the subtree of the correct connection.

This is done as follows: an attribute name can be designated as *key*. If an input symbol has the key attribute, it can be derived at the end of particular branch only if there is a symbol on the branch having the key attribute, with the same value as that of the input symbol. A second attribute name can be designated as *key−intro*; this is used by input symbols such as connection requests which introduce new key attribute values. If an input symbol has the key-intro attribute, and some nonterminal has the key attribute with the same value, then the input symbol can only be derived below such a nonterminal.

4.8. Parsing at Multiple Places

A single input event may be relevant to several subprotocols. For example, in our specification of the sending component of TP, each packet corresponds to a different subtree, and a single acknowledgement event may apply to several outstanding packets. Therefore, RTAG allows a single input event to be derived at several places in the parse tree.

4.9. Example: an RTAG Specification of TP-4

An RTAG specification has been written for the ISO level 4 transport protocol (TP-4), using the following guidelines: 1) The protocol is recursively divided into logically distinct subprotocols, and each subprotocol is placed in a subtree of the parse tree. 2) Attributes, which are used for reference numbers, sequence numbers, window sizes, etc., are moved to the root of the appropriate subtree. The following diagram shows the top-level decomposition of the protocol.

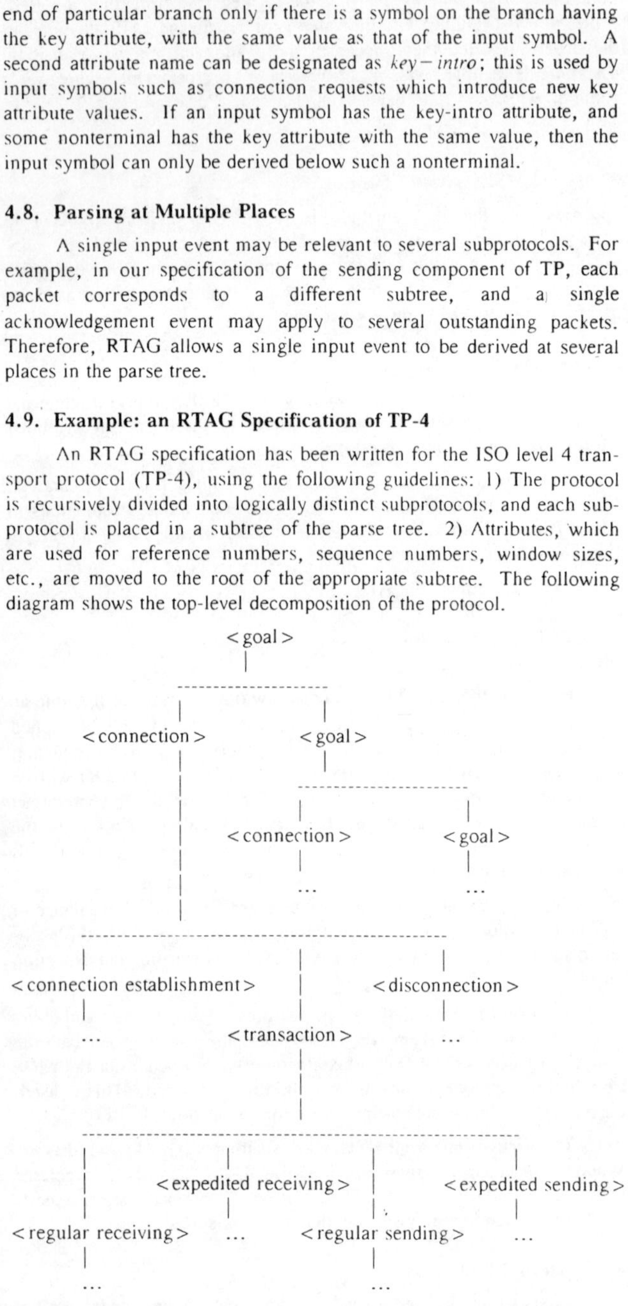

Appendix A contains an RTAG specification of the "regular sending" subprotocol.

5. AN RTAG PARSING ALGORITHM

This section outlines an algorithm for an RTAG parser, i.e., a real-time program which, given an RTAG specification, processes input messages, handles timers, and generates output events in such a way that the protocol defined by the grammar is obeyed. Our design emphasizes simplicity and portability rather than efficiency.

5.1. Static Data Structures

The parser uses the following static structures, which depend on the RTAG grammar being parsed, and are computed in advance·

Symbol Descriptors

Each symbol X is described by record giving its class (input, output, nonterminal, special) and the number of attributes.

If X is an input symbol, the descriptor also contains a list of nonterminals that derive X in the underlying CFG, a list of nonterminals that left-derive X in the underlying CFG, and for each nonterminal Y which left-derives X, a list of the production sequences by which X can be left-derived from Y in the underlying CFG. Because left recursion is not allowed in the underlying CFG, the length of such a production sequence is bounded by the number of symbols, and therefore the list is finite and easily computable.

If X is an output symbol, the descriptor contains the address of the corresponding event performance routine.

If X is a nonterminal, the descriptor contains a list of its immediate and timed productions.

Production Descriptors

Each production in the RTAG specification is described by a record containing: 1) the enabling condition (in the current version, expressions are encoded in an intermediate form that is interpreted by the parser); 2) a list of encoded attribute assignments; 3) pointers to the parent symbol's descriptor, and to those of each RHS symbol; and 4) information describing the concurrent grouping of the RHS symbols.

5.2. Dynamic Data Structures

The parser maintains the following dynamic (time-varying) data structures:

Symbol Instance Descriptors:

Each symbol instance is described by a record containing 1) pointers to its parent and children instances; 2) flags indicating whether the symbol is active, expanded, or fully expanded; 3) if expanded, a pointer to the production descriptor; 4) pointer to attribute descriptors for this symbol; 5) a pointer (possibly null) to a timer descriptor. The attributed parse tree is rooted by an instance (called *root*) of the goal symbol of the grammar.

Attribute Instance Descriptors:

Each attribute is described by a record containing old and new values, and a list of "immediate production links", each of which points to an active symbol instance having an immediate production whose enabling condition depends on this attribute.

Timer Descriptors

Each timer is described by a record containing pointers to a symbol instance, a timed production, and a delay. The parser maintains an incremental delay queue of active timers.

The following FIFO queues are used by the parser during processing:

(1) A queue of input events to be processed.

(2) A queue of timed-out timer descriptors waiting to be processed.

(3) *Newly_active_queue*, a queue of symbol instances that have just become active and are awaiting processing. A symbol is added to the tail of this queue when the left siblings of its concurrent-activation group are fully expanded.

(4) *Immed queue*, a queue of active nonterminal instances that may be eligible for an immediate production. A symbol is added to this list when an attribute value on which the enabling condition of the production depends is changed.

5.3. Processing Input Events

The first step in parsing an input event **[x]** is to find the "candidate set", namely the set of nonterminal leaves which left-derive **[x]** in the underlying CFG, and which are eligible under the "key attribute" restrictions. This can be done by a depth-first search of the parse tree. For each candidate **<a>**, there is a set of production sequences which left-derive **[x]** from **<a>**. These sequences are attempted in an arbitrary order.

When a sequence longer than one production is being attempted, its attribute assignments must be performed since later productions in the sequence may use the target attributes in their enabling conditions. On the other hand, if a production sequence fails (due to an enabling condition not being met) then all its assignments must be undone. Hence while a multi-production sequence is being attempted, its attribute assignments are "logged" (i.e., modified attributes are linked and their old values saved). If the production sequence fails, the changes are undone.

The immediate and newly active queues are processed when the candidate list is exhausted. This processing, which may add new entries to the queues, is continued until both queues are empty.

An element on the immediate queue is processed by testing the enabling conditions of its immediate productions, and applying the first production whose enabling condition is met.

The processing for a newly active symbol X is as follows: if X is a nonterminal, the conditions of its immediate productions are evaluated, and if one is true that production is applied. If none are true, links from the attributes on which the conditions depend are established, so that if one of the attributes is changed the conditions can be retested. For each timed production of X, the enabling condition is evaluated. If it is true, the expression assigned to the "interval" attribute of **/timer/** is evaluated, and a timer is started with that value. A link is set up so that if X is later removed from the tree, or is expanded by another production, the timer can be canceled.

If X is an output symbol then the corresponding external function is called; the attribute values of X are passed as arguments. If X is a special symbol such as **/remove/** or **/freedata/**, the appropriate internal action is taken. **/remove/** removes from the parse tree the descendants of the symbol instance pointed to by /remove/'s *where* attribute, and marks this symbol as fully expanded.

5.4. Processing Timeouts

A timeout is processed by performing the associated production on the associated symbol instance. As with input events, this can result in additions to the newly active queue and the immediate list, which then must be processed as described above.

5.5. Deallocation of Fully Expanded Subtrees

When all the symbols in the RHS of a production are fully expanded, the parser deallocates them and their attributes, and recursively flags and processes the parent. The other case where symbols are deallocated involves right-recursive productions. In a production of the form

<x> : S <x>.

when the symbols in S have all been fully expanded, and the two instances of **<x>** have identical attribute values, the symbols in S are deallocated and the two instances of **<x>** are merged (the child replaces the parent and the parent is deallocated). This is done to prevent right-recursive constructs from using unboundedly large amounts of memory.

6. RTAG SOFTWARE SYSTEM

In the section we describe software tools which, given an RTAG specification, produce an implementation of that part of the protocol which the grammar describes. Tools have also been developed to assist in user-level debugging of RTAG specifications and of interface routines. This software has been developed under 4.2 BSD UNIX, but could be built on other systems as well. There are two main components: the "grammar analyzer" and the "RTAG parser", which correspond roughly to the parser generator and the parser of a compiler generation system.

6.1. Grammar Analyzer

The grammar analyzer accepts a symbolic RTAG specification, and generates the static data structures of the RTAG parser (see Section 5.1). It assigns numbers to input symbols, and produces an "include" file of these assignments, for use in the user-supplied interface to the RTAG parser. It also produces a C source file containing declarations for the output event routines and external functions which the user must supply, as well as an address table through which these routines can be called by the RTAG parser.

The grammar analyzer was constructed using the UNIX Lex and Yacc utilities to facilitate changes to RTAG, and it uses the UNIX C pre-processor to handle comments, macro substitutions, and include files.

6.2. RTAG Parser

The parsing algorithm described in Section 6 has been implemented in C under UNIX. It can be run in user or kernel mode. It must be linked with appropriately-named output routines and external functions, which are called indirectly through the address table generated by the grammar analyzer.

The networking portion of the UNIX kernel includes the "mbuf" memory management system, in which variable-size blocks of data are represented by chains of "mbuf" records, and routines are available to manipulate mbuf chains in various ways. On UNIX, the RTAG *dataptr* attribute type is a pointer to an mbuf chain, and the RTAG parser performs concatenation and other operations by calling the appropriate utility routines.

6.3. Software Environments for the RTAG Parser

To obtain a kernel-mode implementation of an RTAG-specified protocol on UNIX, the RTAG parser is compiled into the kernel, along with the structure initializations generated by the grammar analyzer. Routines must be supplied to interface the RTAG parser to the UNIX socket system on the upper level, and to the UNIX IP implementation or directly to network interface modules on the lower level. We have also developed two user-level environments for protocol experimentation and development:

1) User-level Experimentation System

This consists of a simulated network layer based on the UNIX IPC facility, as well as routines which translate between output symbols and packets (correctly handling data pointed to by *dataptr* attributes). Also included is a routine for interactive perusal of the parse tree, and debugging options that allow logging the actions of the parser (productions, event occurrences, and attribute assignments) on a disk file. This system allows RTAG specifications to be tested with a minimum of programming.

2) User-level Kernel Simulation

In debugging kernel-level protocols it has been extremely helpful to simulate the kernel at the user level. This was done by compiling the relevant kernel code (such as the socket routines and the read/write system call routines) into the user program, together with a software simulation of the IP and network layers. The tree perusal and logging routines, as well as UNIX symbolic debuggers, can be used.

6.4. Example: an Implementation of TP-4 under 4.2 BSD UNIX

The RTAG specification for the TP-4 transport protocol has been combined with the kernel-mode RTAG parser to obtain a production version of TP-4 operating in the DoD Internet domain. The following interface routines and external functions were needed:

(1) Network-level event output routines for assembling packets and sending them to the UNIX IP (Internet Protocol) module.

(2) A network-level input routine to accept a packet from IP, verify the checksum, extract the data fields, and generate the appropriate input event.

(3) Upper-level event output routines for conveying information to the UNIX socket system. For example, the data output routine calls a UNIX kernel routine which appends the mbuf chain to the receive buffer of the appropriate socket, and wakes up any waiting user process.

(4) An upper-level input routine to handle "user requests" from the socket system. These, for the most part, translate directly into input events.

7. CONCLUSION

The RTAG methodology is rich enough to express a wide variety of protocol concepts. Furthermore, the RTAG methodology can be used to experiment with variations on protocols (by changing the grammar) and to quickly obtain experimental implementations of protocols. We have implemented an RTAG software system, have specified a RTAG for the TP-4 transport protocol, and have developed a grammar-based TP-4 implementation in the UNIX kernel.

Based on our experience with the current RTAG TP-4 implementation and with a number of other protocol implementation projects we have concluded that grammar-based formalisms are well suited to specifying complex protocols. We believe that RTAG has inherent advantages over protocol specification methodologies based on augmented FSA. In particular RTAG provides:

(1) Ease of Use and Understanding: An RTAG specification of a protocol compactly shows the legal event sequences which are often difficult to deduce from other specification formalisms. In addition, with RTAG one can isolate particular functions in well defined portions of the grammar and of the parse tree. The capability of RTAG to express concurrent activities makes it easier to express natural protocol interactions and dependencies, and for others to understand the mechanisms of the protocol.

(2) Range of Applicability: Distributed computing applications in databases and operating systems have introduced the possibility of protocols more complex than those found in data communication protocols. Such applications may benefit from RTAG's powerful underlying formalism.

We believe that development of the RTAG methodology can have a major impact on the way protocols for a wide range of applications are specified and implemented. In addition, we hope to develop RTAG-based tools to assist in the verification of both protocol specifications and implementations.

REFERENCES

Blu82

Blumer, T.P. and Tenney, R.L., A formal specification technique and implementation method for protocols, Computer Networks 6,3 (July 1982), 201-217.

Bri85

Brinksma, E. A tutorial on LOTOS. Proc. 5th IFIP Workshop on Protocol Specification, Testing, and Verification. June, 1985.

Har78

Harangozo, J. Protocol definitions with formal grammars, Proceedings Symposium on Computer Network Protocols, Liege, Belgium (Feb. 1978), F6.1-F6.10.

Mil80

Milner, R. A calculus of communicating systems. Springer Verlag, 1980.

Nas83

Nash, S.C. Automated implementation of SNA communication protocols. IEEE International Conference on Communication. Boston, MA., (June 19-22, 1983), 1316-1322.

Poz82

Pozefsky, D.P. and Smith, F.D. A meta-implementation for Systems Network Architecture, IEEE Trans. Commun., vol. COM-30 (June 1982), 1348-1355.

Sch80

Schultz, G.D., Rose, D.B., West, C.H and Gray, J.P. Executable Description and Validation of SNA, IEEE Trans. Commun., vol COM-28 (Apr. 1980), 661-677.

Sid83b

Sidhu, D.P. and Blumer, T.P. Verification of NBS class 4 transport protocol. SDC Report (Sept. 1983).

Sun81

Sunshine, C.A. Formal modeling of communication protocols. USC/ISI report ISI/RR-81-89.

Ten78

Teng, A.Y. and Liu, M.T., A formal approach to the design and implementation of network communication protocols, Proc. COMPSAC 78, Chicago (Nov. 1978), 722-727.

Yem83

Yemini, Y. and Nounou, N. CUPID: a protocol development environment. Proc. 3th IFIP Workshop on Protocol Specification, Testing, and Verification. May 1983.

APPENDIX I
RTAG SPECIFICATION FOR TP-4 PROTOCOL FRAGMENT

```
/* RTAG specification for regular send portion of TP-4 */

/* macro definitions for parameters */

#define RETRANS_TIME      5    /* retransmission interval */
#define GIVEUP_TIME       30   /* time before giving up */
#define RETRANS_COUNT     4    /* # of retransmissions */


/* TERMINAL SYMBOL DEFINITIONS */

output  [N-DT]/* data packet to network */
  int     src_ref, dst_ref, seqno
  boolean      eot
  dataptr data

input   [N-AK]/* acknowledgement packet from network */
  int     refno, credit, seqno

input   [U-GR]/* user initiated graceful close */
  int     refno

output  [U-GR]/* user graceful close indication or confirm */
  int     refno
```

```
input    [U→DT]/* data from user */
  int      refno
  dataptr data


/* NONTERMINAL SYMBOL DEFINITIONS */

nonterm <transact>          /* data transfer and graceful close */
  boolean start_send        /* set when connection established */
  boolean start_receive
  int      initial_credit
  int      nxoutstanding    /* # of unacked expedited packets */
  boolean GRarrived         /* true if have in-sequence GR. */
  boolean GRsent            /* true if have received GR from user */

nonterm <reg send>
  int      windowstart      /* first unacknowledged packet */
  int      nextseq          /* seqno of next packet */
  int      windowend        /* seqno just beyond end of send window */

nonterm <send message>
  boolean ready             /* true iff can start sending message */

nonterm <send message tail>
  boolean ready

nonterm <transmit GR>
  boolean ready

nonterm <transmit message>
  boolean ready
  dataptr data

nonterm <send packet>
  dataptr data
  int      seqno
  boolean eot

nonterm <send packet tail>
  boolean eot

nonterm <deliver DT>

nonterm <retransmit DT>
  int      count

nonterm <get GR>


/* DECLARATIONS OF EXTERNAL FUNCTIONS */

extern   boolean #between#(int,int,int)
         /* (i,j,k); true if i < j <= k in cyclic order */
extern   dataptr #extract#(dataptr,int)
         /* return copy of initial segment of given size */
extern   boolean #eot#(dataptr)
         /* true if empty */
extern   dataptr #copy#(dataptr)
         /* copy mbuf chain */
```

```
/* PRODUCTIONS */


<reg send>   :   <send message tail>.
        $0.nextseq = 0
        $0.windowstart = 0
        $0.windowend = <transact>.initial_credit
        $1.ready = true
;

<send message tail>   :   {<send message> <send message tail>}.
        $1.ready =* $0.ready
        $2.ready = false

        |
        if <transact>.GRsent
;

<send message>   :   [U→DT] <transmit message>.
        $2.data = $1.data
        $2.ready =* $0.ready

        |         [U→GR] {<transmit GR> <get GR>} [U←GR].
        if not <transact>.GRarrived
        $2.ready =* $0.ready
        $4.refno = <TC>.refno

        |         [U→GR] <transmit GR>.
        if <transact>.GRarrived
        $2.ready =* $0.ready
;

<get GR>   :   .
        if <transact>.GRarrived
;

/* productions of <transmit GR> omitted for brevity */

<transmit message>   :   <send packet tail>.
        if $0.ready
        $1.eot = #eot#($0.data)
;

<send packet tail>   :   {<send packet> <send packet tail>}.
        if not $0.eot
        $1.data = #extract#(<transmit message>.data,
            <TC>.tpdu_size)
        $1.eot = #eot#(<transmit message>.data)
        $1.seqno = <reg send>.nextseq
        <reg send>.nextseq = (<reg send>.nextseq + 1)
            mod <TC>.max_seqno
        $2.eot = $1.eot

        |         /freedata/.
        if $0.eot
        $1.data = <transmit message>.data
        <send message tail>/<send message tail>.ready = true
;
```

```
<send packet> :          [N-DT] <retransmit DT>.
        if <reg send>.nextseq != <reg send>.windowend
          && <transact>.nxoutstanding == 0
        $1.src_ref = <TC>.refno
        $1.dst_ref = <TC>.foreign_refno
        $1.eot = $0.eot
        $1.seqno = $0.seqno
        $1.data = #copy#($0.data)
        $2.count = RETRANS_COUNT
;

<retransmit DT>  :  /timer/ [N-DT] <retransmit DT>.
        if $0.count > 0
        $1.interval = RETRANS_TIME
        $2.src_ref = <TC>.refno
        $2.dst_ref = <TC>.foreign_refno
        $2.eot = <send packet>.eot
        $2.seqno = <send packet>.seqno
        $2.data = #copy#(<send packet>.data)
        $3.count = $0.count - 1

        |        [N-AK] /freedata/.
        if #between#(<send packet>.seqno, $1.seqno,
           <reg send>.nextseq)
        <reg send>.windowend = ($1.seqno + $1.credit)
          mod <TC>.max_seqno
        $2.data = <send packet>.data

        |        /timer/.
        if $0.count == 0
        $1.interval = GIVEUP_TIME
        <TC>.transerror = true
;
```

Panel 2

DARPA Interdomain Addressing

Chairperson
V. Cerf
MCI
Panelists
D. Clark, *MIT*
D. Comer, *Purdue University*
L. Peterson, *University of Arizona*
D. Terry, *Xerox PARC*

Domain Names: A Hierarchy In Need of Organization

Douglas E. Comer
Purdue University

The DARPA Domain Name project is an attempt to move from a centralized naming authority to a distributed one. In essence, the domain system computes a single binding from a character string known as a host *name* to another string known as the host *address* (a 32-bit integer). Besides the syntactic changes in names, the chief difference between the old-style names and domain names is that in the old system, a database containing the name-to-address bindings was distributed to each host where bindings were performed locally, while in the domain system computation of the name-to-address binding uses the internet to contact search the hierarchy of domains until a server is found that can perform the binding. Unfortunately, to make the system efficient, servers cache bindings they look up, and clients begin a search by contacting the nearest "leaf" server to avoid searching the hierarchy. If sufficient memory is available, the cache at a server corresponds exactly to old-style host tables.

The domain scheme raises other questions. In principle, the hierarchical structure of domain names permits easy delegation of naming authority because it allows names within independent subtrees to be assigned by independent organizations. In practice, however, the hierarchical scheme collects together organizations under fixed top-level domains. Because extremely small organizations cannot afford to maintain a domain name server, they must be supported by (and obey the authority of) a domain server for the subhierarchy under which they lie.

A hierarchical organization of names also imposes restrictions on the use of the system. For example, even though domain name servers contain (name, address) pairs, it is impossible to locate the desired binding given only the address.

DOMAIN NAMES: MORE QUESTIONS THAN ANSWERS

Larry L. Peterson
University of Arizona

The DARPA domain name project raises several interesting questions about the underlying assumptions and issues involved in providing a hierarchical naming scheme in an environment the size of the Internet. For example, the project was motivated by the growth of the Internet. However, because the proliferation of hosts took place **within** the sites belonging to the Internet (due in part to an increase in the number of workstations and other small machines), and not necessarily in the number of sites, a less general solution that maintained a centralized and fully distributed *site* table, with each site running a name server for its hosts, would have been sufficient.

In addition, the distributed implementation of the name to address database uncovered several fundamental naming and addressing issues that had been ignore by the original centralized approach. For example, name servers and resolvers must be able to select a reasonable Internet address for hosts connected to the Internet at multiple points. The decentralized solution also raises questions about authority over host names.

Panel 3

Operational Experience with Packet Switching Networks

Chairperson
K. Joseph
Bell Canada

Panelists
A. Dobson, G.D. White, and D. Jeannes, *Bell Canada*
R. Stubbs II, *BellSouth*
S. Poppe, *Tymnet*
D. Dingus, *AT&T Communications*

PANEL: "OPERATIONAL EXPERIENCE WITH PACKET SWITCHING NETWORKS

T.C. Buttle
Telecom Canada

Chairman's Remarks

ABSTRACT

From its experimental 1968 inception in ARPANET, followed by CCITT standardization of the X.25 protocol in 1976, packet switching now provides inexpensive worldwide data communications capabilities for its subscribers. Customer connections to public packet switched network services are increasing at a phenomenal rate, typically 30-40% year over year with no foreseen slowdown. This 2-3 year doubling effect presents numerous challenges to network operators in trying to effectively support and evolve their packet switched networks while simultaneously increasing customer service levels.

This Panel presents an overview of current customer packet switching applications and then focuses on how network operators are innovatively meeting the challenges presented by high growth on the operation, control, support, configuration and evolution of their packet switched networks.

PANEL: "OPERATIONAL EXPERIENCE WITH PACKET SWITCHING NETWORKS

A. Dobson
Assistant Director - Product Management (Datapac)
Bell Canada

Datapac

ABSTRACT

Packet Switched Networks are well accepted and provide cost-effective data communications for a wide range of applications. This paper will examine the growth of packet switching by profiling Telecom Canada's Packet Switched Network, Datapac. It will focus on the following:

- the emergence of applications by new users of data communications,
- development of new applications by existing users,
- migration to packet switching networks from dedicated leased line networks,
- the development and introduction of new service features to continue the phenomenal growth rates achieved in earlier years.

PANEL: "OPERATIONAL EXPERIENCE WITH PACKET SWITCHING NETWORKS"

G.D. White
Section Manager - Systems Operations (Data Networks)
Bell Canada

Line Operations Network Growth Issues

ABSTRACT

The task of keeping customer impact to a minimum while attempting to grow, modernize and improve the network in todays packet switching environment is becoming increasingly more difficult. As the network evolves and the customer demand for higher network availability increases, the need for line operations personnel to reduce customer impact has presented a unique challenge for innovative ideas to accomplish this objective.

Although the equipment supplier may provide excellent technical documentation for the provisioning, testing and maintenance of their products, they cannot provide the means to administer, manage and control a large network. This is the sole area where the network provider is responsible for deciding how his network will impact the customer.

This presentation will deal with some of the operational problems encountered in growing a network from 5 nodes to 55 nodes over a period of 9 years. Some of the subject areas that will be covered are: the maintenance window - is it required?; administrative documentation; utilization of network statistics; hardware retrofits; and software updates - major and minor. These are but a few of the problem areas challenging packet switching network providers today. Their resolution are limited only by the imagination of the personnel charged with the responsibility of resolving them.

PANEL: "OPERATIONAL EXPERIENCE WITH PACKET SWITCHING NETWORKS

D. Jeannes
Manager - SL-10 Systems Architecture
Bell Northern Research

Technology for Managing Large Packet Switching Networks

ABSTRACT

The evolution of packet switching network technology over the past ten years has been strongly influenced by the dramatic growth of these networks. The ability to control and manage large networks has required many innovations and has been most successful where the growing network itself has provided the mechanisms needed for network management.

This discussion will cover the increasing use of decentralization and parallel operation of functions required to monitor the network, to remotely test and maintain subscriber facilities, to collect billing and statistical data, and to distribute and install new software. Specific problems which require special attention in large network management will be covered; these include time synchronization, routing optimization, and network security.

The SL-10 packet switching system from Northern Telecom is referenced through this discussion as an example of a system which has built on the facilities and technology of the network switching equipment to provide the management function. Since all of the attributes of capacity, availability, reliable communication, configuration flexibility, and security are the same for both packet switching subscribers and network management, using a common base provides real advantages.

The discussion will show how the exclusive use of network terminal interfaces, virtual circuits, and X.25 host interfaces for all management data transfer functions provides the required function with virtually unlimited growth potential. Support of both packet switching and management functions on the same base of processor, storage devices, and software, further extends the benefits of commonality.

PANEL: "OPERATIONAL EXPERIENCE WITH PACKET SWITCHING NETWORKS"

R.D. Stubbs II, L. Dee Swymer, T.L. Quinn
Corporate Communications
Bell South Services

Operations Considerations in a Large Private Packet Switching Network

ABSTRACT

In early 1977, Southern Bell began to study ways to reduce terminal and leased line costs associated with internal computer communications. Because a large volume of these communications were of an asynchronous, character-mode nature, packet switching emerged as the most attractive solution to reducing these costs. In May, 1979, the first application went on-line with one switch, one host concentrator and three terminal concentrators. Operation of this network was simple, with operations and planning residing in the same work group.

Today, the network consists of twenty tandem switches, 185 local switches and 85 packet assembler/disassemblers, interconnected by 15,928 miles of trunk facilities. The network architecture has evolved to geographically distributed "principle switching points" interconnected through a single major switching center in Atlanta, Georgia. Each "principle switching point" serves an area of company interest by supporting a number of local switches and concentrators.

As the complexity of the network has increased, so has the complexity of network operation. Operations have been both functionally and geographically differentiated to keep up with the evolution of the network, and to stay consistent with changing packet switch architecture. This talk will cover the network operations and maintenance considerations associated with the operation of today's very large network.

PANEL: "OPERATIONAL EXPERIENCE WITH PACKET SWITCHING NETWORKS"

Stephen C. Poppe
Manager, Network Analysis and Design
Tymnet

Solving Growth Problems in a Rapidly Expanding PDN

ABSTRACT

Tymnet's public data network has been expanding in size and revenues at about a 35% per year rate since its inception. Maintaining this growth has put strain more on the organization than the technology. I will discuss several dimensions of the problem of managing prolonged rapid growth, including capacity management (bandwidth, switching, interfaces, and floor-space), topological complexity, job specialization, order processing, forecasting, and deploying new versions of code. The topic selection will favor issues that are likely to be common to other operations and the discussion will be biased toward solutions and pro-active management.

Session 3

Communication System Performance

Chairperson
D. Ferrari
University of California, Berkeley

WINDOW SELECTION IN FLOW CONTROLLED NETWORKS*

Mario Gerla, H.W. Chan
Computer Science Department
UCLA, Los Angeles, CA 90024

ABSTRACT

The end-to-end window scheme is a popular mechanism for flow and congestion control in packet switched networks. The window scheme may be implemented in the transport layer protocol (like in TCP), or in the source-to-destination protocol(like in ARPANET), or in the network layer protocol (like X.25). In many implementations, the window size is chosen at connection set up time.

In this paper, we provide guidelines for window selection. Specifically, we show that if the network becomes overloaded, that is, the offered load exceeds network capacity, then the selection of user windows has a critical impact on individual user throughputs. Thus, user windows should be chosen judiciously, so as to satisfy a well defined "fairness" criterion.

We formulate the optimal window assignment as a mathematical programming problem, and show that the exact solution is computationally impractical because of the combinatorial nature of the problem and the complexity of the underlying multiple chain, closed network of queue model. We then develop a heuristic approach which is computationally very efficient and provides nearly optimal solutions. Numerical results are provided to illustrate and validate the method.

1 INTRODUCTION

Packet switched networks are known to be prone to congestion when they are subject to heavy user demands. Congestion generally manifests itself in the form of buffer overflow at some internal node, with consequent traffic blockage which may propagate to neighbor nodes and, possibly, to the entire network. A symptom of network congestion is the fact that an increase in offered traffic leads to a decrease in effective throughput. Eventually, under sustained offered load, the network may become deadlocked [Gerl80].

*This research was partially supported by the Defense Advanced Research Projects Agency of the Department of Defense under Contract MDA 903-82-C-0064, by the Air Force Office of Scientific Research under Contract AFOSR 820304A and by the National Science Foundation under Contract NSF-INT 82-05252.

A common remedy to congestion is flow control. Flow control is a generic term used to refer to a class of protocol mechanisms which regulate the flow of packets in the network [Gerl80]. A popular form of flow control is end-to-end window flow control, or, window control in short. In window control, each packet issued by the source node is acknowledged by the destination node (source and destination here could be the sending and receiving Hosts, or the entry/exit nodal switch pair). The source is allowed to have up to N packets outstanding, for which the acknowledgements have not yet been received. N is the "window size" for this source to destination session. Congestion control is exercised in the following way: if the network becomes heavily loaded, then the queueing delay along the path builds up, and packet and acknowledgement travel time increases. This causes the effective offered rate to decrease. Thus, as congestion builds up, the offered load decreases, a desirable property in a congestion protection scheme.

End-to-end window mechanisms are found in most packet network implementations. In ARPANET, the flow of messages between Host pairs is controlled by a window of size 8 [McQu72]. In X.25 networks, a window of size varying from 1 to 8 can be enforced on each end-to-end virtual circuit using the D-bit option of the X.25 protocol [Rybc80]. In the TCP transport level protocol, a variable size window is used to regulate the traffic on each end-to-end session.

In most network implementations, the window can be negotiated at connection set up time (e.g. X.25). In some cases, it can be changed dynamically during the course of a session (e.g. TCP). The issue then arises of how to optimally choose window sizes for the various sessions present in the network. The goal is to optimize the throughput within given delay and fairness constraints. It should be clear, in fact, that as window sizes are progressively increased, throughput increases, but queueing delay also increases, since more packets are allowed in the network. The delay increase has a feedback effect on throughput via window flow control. This way, congestion is prevented. In this process, however the throughputs on different user sessions are not always evenly regulated, even if window sizes are the same for all sessions. In order to achieve a fair distribution of throughputs, and at the same time meet some specified delay constraints, the window sizes must be properly adjusted.

Flow control, fairness and, therefore, window optimization problems are particularly critical in a network with heavy file transfer requirements. In this case, we want to maximize network throughput, since this will minimize file transfer times. Yet, we must maintain the delay within given bounds, otherwise interactive and control traffic (here assumed to be a small fraction of the total traffic) will be adversely affected. Also, we must provide a fair share of the bandwidth to the various file transfers, since the users expect equal quality of service.

In this paper, we consider a file transfer environment, where several transfers are in progress at the same time. A virtual circuit transport scheme is assumed, where a virtual circuit is established between each source/destination pair, and data flows on this virtual circuit. For each file transfer we define a demand d, which is the rate at which packets could be transferred on the virtual circuit if no other (interfering) traffic was present in the network. Clearly, if other traffic is present, the actual throughput is less than d. We define a proper fairness measure for this traffic environment, which implicitly includes the notion of throughput efficiency. We then propose to optimize fairness such that a given average delay constraint is met.

In section 2, the fairness measure is defined and the optimization problem is formulated. In section 3, a heuristic approach (based on Mean Value Analysis and Lagrangian Decomposition) is introduced for the solution of the problem. An efficient algorithm is presented in Section 4, and is applied to some sample cases in Section 5. Section 6 discusses some possible extensions of this work.

2 PROBLEM FORMULATION

The packet network consists of M nodes linked by NL communication channels. R user sessions are present in the network. Each session is supported by a virtual circuit between a given source and destination. Traffic on each virtual circuit is controlled by a window mechanism which limits the number of packets (the window size) travelling on the circuit. The traffic demand of user i is d_i. Due to window flow control, the actual throughput of user i is $\lambda_i \le d_i$. Both λ_i and d_i are measured in units of information(e.g.,bits) per unit of time.

In our model, we assume single path routing i.e., all packets on a given virtual circuit flow along a fixed path instead of being distributed over multiple paths. The usual assumptions regarding input traffic and service statistics are made; namely, Poisson arrivals, exponential packet length distribution and Kleinrock's independence assumption [Klei76].

The following fairness measure was chosen for the optimization of window sizes: [Gerl84]

- *Minimize*: $P = \sum_{i=1}^{R} \frac{d_i - \lambda_i}{\lambda_i}$

The system is optimally fair if $P = 0$. Thus, the goal of the problem is to minimize P.

This fairness measure has two pleasing properties:

(a) the throughput of the individual user cannot be driven to zero otherwise the value of the objective function becomes infinite; and

(b) the actual throughput achieved by a user tends to be proportional to his traffic demand.

With the above objective in mind, we formulate our fair window flow control problem in large network as follows:

- *Minimize*: $$P = \sum_{i=1}^{R} \frac{d_i - \lambda_i}{\lambda_i} \qquad (1)$$

- *Over sets of window sizes*: $\mathbf{N} = \{N_1, N_2, \cdots, N_R\}$

- *Subject to the constraints*:

$$T = \frac{\sum_{i=1}^{R} \lambda_i T_i}{\sum_{i=1}^{R} \lambda_i} \le T_{\max} \qquad (2)$$

$$0 \le \lambda_i \le d_i, \qquad i = 1,\ldots R \qquad (3)$$

where

d_i - traffic demands of user i

λ_i - actual throughput of user i

T - average network delay

T_i - average delay of user i

$T_{\max}$ - maximum delay

N_i - window size of virtual circuit i

The quantities λ_i and T are functions of the set of window sizes and should be represented as $\lambda_i(\mathbf{N})$ and $T(\mathbf{N})$. However for simplicity of notation we express T and λ_i without explicitly showing the dependence on $\mathbf{N}$.

3 APPROACH

As it may be noticed from Eq. (1) and (2), the solution to our problem requires the evaluation of λ_i and T for a given window allocation **N**. This is accomplished by using a closed queueing network model with multiple classes of customers, where each class represents a different virtual circuit. The service centers are the communications channels [Reis79]. The exact solution is computationally too cumbersome, except for very small cases. Thus, approximations must be used. Two different types of approximations have been used in this work, namely, Linearizer [Chan82] and Schweitzer [Schw79] approximation.

After defining a procedure for the computation of λ_i and T for a given **N**, we must solve the problem of optimizing **N**. A simple inspection reveals that this problem is of formidable complexity, since the variable **N** is a vector of integers, and objective and constraints are nonlinear. One solution is to try all possible window combinations **N**. This, however, is feasible only for small networks, with a small number of virtual circuits and small window range. For situations of practical interest, a different approach must be used.

The proposed solution is a heuristic solution based on the marginal analysis of incremental improvement of fairness per increment of average delay for each chain, when window is increased by one. The chain which minimizes the ratio of fairness increment over delay increment is selected. More specifically, the procedure starts with unit window assignment to all chains, i.e. $\mathbf{N} = (1,1,1, \ldots ,1)$ and progressively increases windows step-by-step. At each step, the window is increased for the chain minimizing the ratio:

$$\frac{\Delta P}{\Delta T} = \frac{\text{fairness increment}}{\text{delay increment}}$$

The procedure stops when:

$$\lambda_i = d_i \qquad \forall i; \qquad or$$

$$N_i = M_i \qquad \forall i$$

where M_i is the maximum admissible window size for chain i.

This procedure will generate a curve $P = P(T)$ of the type shown in Figure 8. The desired solution is represented by the point (P^*, T^*) on this curve such that T^* is the largest value $\leq T_{max}$.

The heuristic is intuitively appealing since at each step we choose the direction which gives the best reduction in P for the same increase in delay T. We expect the method to generate a suboptimal set of solutions, for a range of values of T. From this set, we obtain the solution for the desired value T_{max}.

This heuristic is based on a mathematical programming principle known as Lagrangian Decomposition [Ever63]. According to this principle, the above mentioned incremental method yields all the nondominated solutions in the (P, T) plane. We recall that solution A dominates solution B iff:

$$P_A < P_B \text{ and } T_A \leq T_B$$

Geometrically, finding all the nondominated solutions is equivalent to find the lower envelope of the set of points in the (P, T) plane.

The Lagrangian Decomposition principle may be applied if the following conditions on P and T are met:

- P and T are separable, i.e. are expressed by sums of terms, where each term depends only on one variable. Namely:

$$P = \sum a_i \, P_i \, (N_i)$$

$$T = \sum b_i \, T_i \, (N_i)$$

- $P_i = P_i \, (T_i)$ is convex.

Unfortunately, these conditions on P and T are not verified in our problem except in the (not very interesting) case in which there is no interference among chains. In practical situations, each term P_i and T_i depends on *all* window sizes, and not only on N_i. (However, the strongest dependence is on N_i, i.e. a chain's delay and throughput is affected the most by a change in its own window size).

The application of the Lagrangian Decomposition method thus provides a set of suboptimal solutions. That is, not all the nondominated solutions are generated. The numerical results, however, show that the set of points so generated approximates very closely the lower envelope (see Figure 6).

In order to improve the computational efficiency of the algorithm in handling large networks, another approximation is introduced. We recall that, at each step of the algorithm the ratio $\Delta P/\Delta T$ must be evaluated for each chain. This implies recomputing λ_i and T for each window increment. The exact approach requires solving for the entire network again after each increment - a *very* costly proposition even if Schweitzer or Linearizer approximations are used. To simplify the computation, an approximation technique is used, as reported in Appendix A. This approximation allows us to compute the changes in λ_i and T due to a window increment directly from the current solution. More specifically, it allows us to increase window size by 1 in one of the chains, to estimate the throughput increase in that chain and the throughput decrease in the remaining chains.

4 ALGORITHM

This algorithm generates the set of nondominated (P,T) solutions.

1. Set the maximum number of iterations

2. Set the window size for each chain to be 1, i.e. $N_i = 1$ for $i = 1,2, \ldots , R$.

3. Calculate the throughput λ_i of each chain and the average delay of the network using MVA with this set of window sizes.

   Repeat steps 4 through 7 for all the chains $i = 1,2,\ldots,R$

4. Let $N_i = N_i + 1$, other window sizes unchanged.

5. Calculate the throughput of chain i using the formula [Reis80] :

$$\lambda_i(\mathbf{N} + \mathbf{e}_i) = \frac{N_i + 1}{\sum_{k=1}^{NL} \theta_{ik} x_{ik} \left[1 + L_k(\mathbf{N})\right]}$$

6. Calculate other chains $j(\neq i)$throughputs using the formula

$$\lambda_j(\mathbf{N} + \mathbf{e}_i) = \frac{N_j}{\frac{N_j}{\lambda_j(\mathbf{N})} + \sum_{k=1}^{NL} a_{jk} \frac{L_{ik}(\mathbf{N})}{N_i}}$$

   derived from Schweitzer's approximation as shown in Appendix A, Eq. (A2).

7. Compute the unit decrease in objective value per unit increase of average delay, that is

$$\eta = \frac{P - P'}{T - T'}$$

   where P is the new objective value and P' is the previous objective value;
   T is the new average delay and T' is the previous average delay.

   Note: T is always increasing for increasing window size. P however, may increase or decrease.

8. Let $N_i = N_i - 1$;

9. Find the chain m^* which minimizes η

10. Let $N_{m^*} = N_{m^*} + 1$

11. Go back to 3 until the maximum number of iterations, or the maximum window sizes have been reached.

The above algorithm identifies at each step the best chain on which the window should be increased for the given set of window sizes. Starting from $\mathbf{N} = (1,1, \ldots ,1)$ the algorithm finds the next window increment. This gives us a new set of objective values and average delays. The algorithm stops when all the windows have reached their maximum values or the maximum number of iterations has been reached. The result is a curve of fairness values versus network average delays.

The algorithm has several nice characteristics :

1. It is simple and requires no complete enumerations of all possible window combinations.

2. The step to find the chain candidate for window increase is fast because of the ease of estimating throughputs from the approximate formula.

3. There is an upper bound of $\sum_{i=1}^{R} M_i - R + 1$ iterations as compared to $\prod_{i=1}^{R}(M_i - 1)$ number of possible window selection in the exhaustive method, where M_i is the maximum window size of chain i.

4. The algorithm gives a range of objective values versus average delays, rather than one single solution. Thus, sensitivity studies can be performed at no extra cost.

The algorithm just presented is an approximate algorithm. The approximation in the algorithm comes from three steps. First, for large networks, exact MVA cannot be used. Therefore, an error is introduced by the approximate technique (e.g. Linearizer or Schweitzer's) used in step 3. Secondly, the incremental throughputs in step 6 are based on Schweitzer's approximation. Thirdly, the objective and the constraint are not separable functions, thus violating the Lagrangian Decomposition requirements for the generation of all nondominated solutions.

Due to the approximate nature of the algorithm, it is important to validate the results with several examples. This is done in the next section.

5 NUMERICAL RESULTS

In this section we present the results of the application of the algorithm to several network examples. First we use the four chain and eight queue example shown in Figure 1. The traffic requirements and the paths used by the chains are shown in Table I. Queues are assumed to be FIFO with exponential service time distributions. We limit the individual window size to be 8. This example is small enough to permit us to explore the set of all possible (4096) window

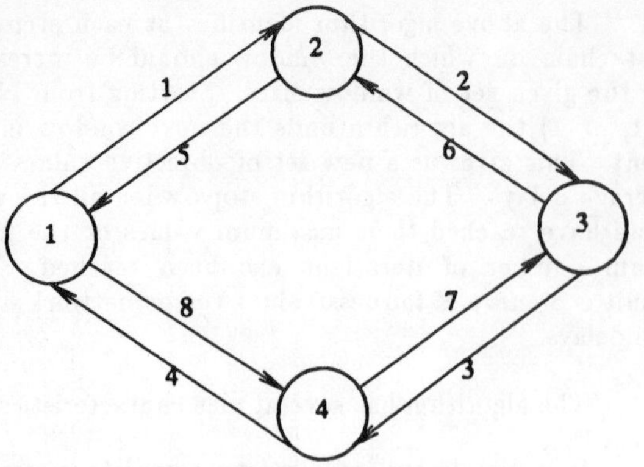

Fig. 1 Four node example.

Table I
Four node example - case 1.

chain	source	dest.	demand	route
1	1	4	2	1-2-3
2	1	2	2	8-7-6
3	3	1	2	6-5
4	4	2	2	4-1

size combinations. In Figure 2, we plot the objective values against network average delays for the lower 1000 objective value points. Our algorithm attempts to find the lower envelope points using the heuristic presented in the previous section. The connected points in the figure are the points generated by the algorithm using Mean Value Analysis (MVA) in step 3. They do represent most of the nondominated solutions and thus provide us with a suboptimal set of window sizes.

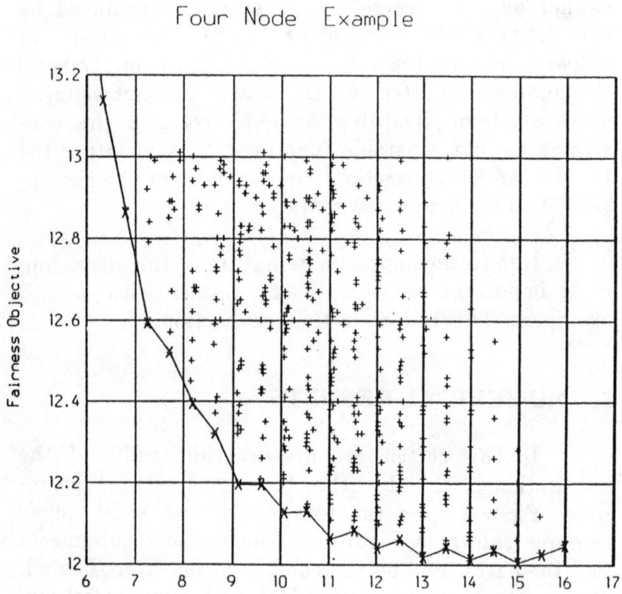

Fig. 2 Family of suboptimal solutions for the four node example in Table I.

The near optimality of the above results is mainly due to the fact that there is little interference among the chains. From Table I we see that chain 1 interferes with chain 4 only at link 1 and chain 2 interferes with chain 3 only at link 6. Our algorithm cannot sense this second order interference among the chains since it is based on Schweitzer's approach derived in Appendix A.

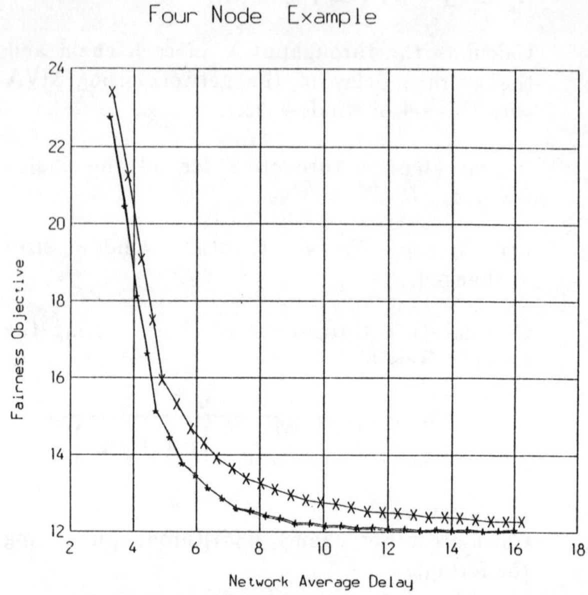

Fig. 3 Plot of lower envelopes from MVA(+), Linearizer(*) and Schweitzer(X) for the example in Table I.

In Figure 3, we plot the three sets of suboptimal solutions when different network of queue algorithms are used in step 3; namely MVA, Schweitzer [Schw79] and Linearizer [Chan82]. We see that Linearizer and MVA give comparable curves. Table II gives us the run time (in CPU seconds) using the three aforementioned approaches for our example. We see that MVA tops the list while Linearizer and Schweitzer have comparable run times. From this we see that Linearizer is probably the best choice for large network.

Table II
Run time for the four node example in Table I.

	MVA	Linearizer	Schweitzer
Run time	34.4s	6.6s	4.1s

Next we use the ten node example shown in Figure 4. The links are assumed to be full duplex. The demands and the routes used by the chains are shown in Table III. In the example we have three chains. Chain 2 has interaction at link 6 with chain 1 which in turn has interaction with chain 3 at link 10. Limiting maximum window size to 8, all 512 possible window size combinations can be plotted and compared with the set of window sizes obtained from our algorithm. Again Figure 5 shows that the results are near optimal in spite of the fact that more interference is experienced in this example. Table IV summarizes the set of window sizes generated by the algorithm. For a given T_{max} one can select the window set **N** that gives the best objective value.

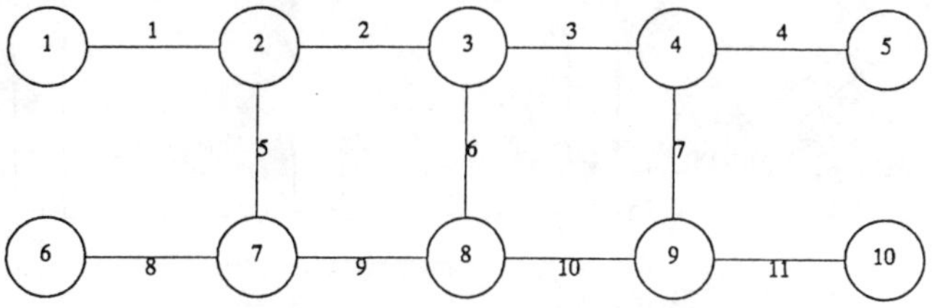

Fig. 4 Ten node example.

Table III
Ten node example.

chain	source	dest.	demand	route
1	1	10	4	1-2-6-10-11
2	5	6	4	4-3-6-9-8
3	2	4	4	5-9-10-7

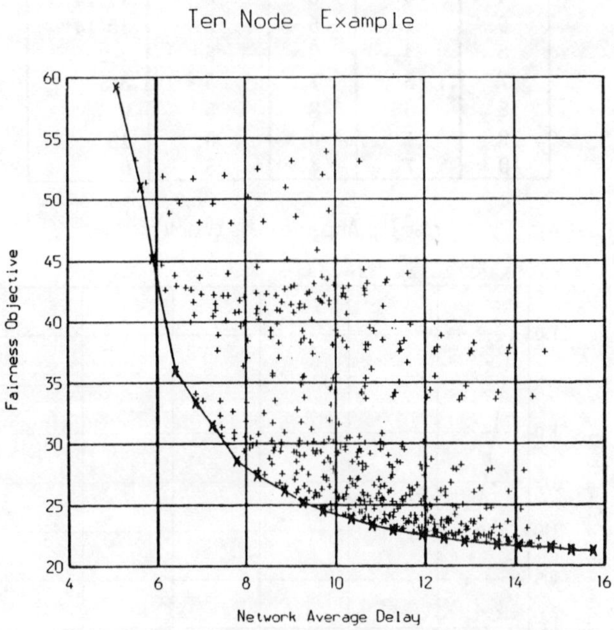

Fig. 5 Family of suboptimal solutions for the ten node example.

Next we consider again the same four node example in Figure 1, except that this time we introduce substantial interaction among the chains. The demands and the paths in use are shown in Table V. Chain 3 interacts with chain 4. Chain 4 interacts with chain 2 and chain 2 interacts with chain 1. Figure 6 gives the plot of fairness values versus network average delays. We observe that the set obtained from the algorithm is again very close to the optimal. Table VI tabulates the window sizes when different techniques are used in step 3 of the algorithm. There is only a slight difference between the set of windows found by Schweitzer's approximation and the set obtained using

Table IV
Delays and window sizes for the ten node example.

T	λ_1	λ_2	λ_3	**N**	P
5.1211	0.1762	0.1841	0.2255	(1,1,1)	59.1675
5.6534	0.1704	0.3113	0.2259	(1,2,1)	51.0343
5.9204	0.1637	0.3117	0.3691	(1,2,2)	45.1006
6.4210	0.2775	0.3020	0.3549	(2,2,2)	35.9286
6.8846	0.2689	0.3920	0.3559	(2,3,2)	33.3198
7.2557	0.2589	0.3930	0.4506	(2,3,3)	31.5022
7.8081	0.3371	0.3815	0.4341	(3,3,3)	28.5664
8.2591	0.3270	0.4478	0.4360	(3,4,3)	27.3393
8.8651	0.3854	0.4343	0.4211	(4,4,3)	26.0872
9.2774	0.3718	0.4370	0.4846	(4,4,4)	25.1643
9.7300	0.3611	0.4874	0.4876	(4,5,4)	24.4877
10.3702	0.4050	0.4733	0.4717	(5,5,4)	23.8071
10.8196	0.3914	0.4771	0.5178	(5,5,5)	23.3277
11.2776	0.3806	0.5163	0.5219	(5,6,5)	22.9229
11.9491	0.4149	0.5021	0.5057	(6,6,5)	22.5169
12.4224	0.4017	0.5068	0.5405	(6,6,6)	22.2512
12.8864	0.3911	0.5379	0.5455	(6,7,6)	21.9979
13.5846	0.4190	0.5238	0.5295	(7,7,6)	21.7380
14.0780	0.4077	0.5487	0.5352	(7,8,6)	21.5736
14.7993	0.4315	0.5341	0.5210	(8,8,6)	21.4372
15.2628	0.4196	0.5406	0.5467	(8,8,7)	21.2481
15.7605	0.4078	0.5469	0.5681	(8,8,8)	21.1641

MVA. The two sets of windows obtained using MVA and Linearizer coincide.

Table V
Four node example - case 2.

chain	source	dest.	demand	route
1	1	3	75	1-2
2	4	3	75	4-1-2
3	3	4	75	3
4	3	1	75	3-4

Lastly, the algorithm was tested on the network shown in Figure 7 corresponding to an early 8 node ARPANET configuration [Klei76]. Ten virtual circuits are present. Source, destination, demand and the path of each circuit are shown in Table VII. Figure 8 shows the curve of objective values versus network average delay. The slight increase of fairness values towards the end of the tail is due to the fact some of the window sizes have reached their maximum values 8. Apparently, fairness deteriorates by further increasing the windows on the remaining chains. By examining Figure 8, we observe that the set of suboptimal solutions is very dense, that is, the gaps between suboptimal solutions are very small. Also, the curve tends to be nearly convex. These two facts guarantee that the set of solutions is close to optimal. We expect this trend to be maintained as network size increases.

Table VI
Window sizes for the four node example in Table V.

Iter.	Window Sizes		
	MVA	Linearizer	Schweitzer
1	(1,1,1,1)	(1,1,1,1)	(1,1,1,1)
2	(1,2,1,1)	(1,2,1,1)	(1,2,1,1)
3	(2,2,1,1)	(2,2,1,1)	(2,2,1,1)
4	(2,2,1,2)	(2,2,1,2)	(2,2,1,2)
5	(2,2,2,2)	(2,2,2,2)	(2,2,2,2)
6	(2,3,2,2)	(2,3,2,2)	(2,3,2,2)
7	(3,3,2,2)	(3,3,2,2)	(3,3,2,2)
8	(3,3,2,3)	(3,3,2,3)	(3,3,2,3)
9	(3,4,2,3)	(3,4,2,3)	(3,4,2,3)
10	(4,4,2,3)	(4,4,2,3)	(3,4,3,3)
11	(4,5,2,3)	(4,5,2,3)	(4,4,3,3)
12	(5,5,2,3)	(5,5,2,3)	(4,4,3,4)
13	(5,6,2,3)	(5,6,2,3)	(4,5,3,4)
14	(6,6,2,3)	(6,6,2,3)	(5,5,3,4)
15	(6,7,2,3)	(6,7,2,3)	(5,6,3,4)
16	(6,7,2,4)	(6,7,2,4)	(6,6,3,4)
17	(6,7,3,4)	(6,7,3,4)	(6,7,3,4)
18	(7,7,3,4)	(7,7,3,4)	(6,7,3,5)
19	(7,8,3,4)	(7,8,3,4)	(6,7,4,5)
20	(7,8,3,5)	(7,8,3,5)	(6,7,4,6)
21	(8,8,3,5)	(8,8,3,5)	(6,7,5,6)
22	(8,8,4,5)	(8,8,4,5)	(7,7,5,6)
23	(8,8,4,6)	(8,8,4,6)	(7,8,5,6)
24	(8,8,5,6)	(8,8,5,6)	(7,8,5,7)
25	(8,8,5,7)	(8,8,5,7)	(7,8,6,7)
26	(8,8,6,7)	(8,8,6,7)	(7,8,6,8)
27	(8,8,6,8)	(8,8,6,8)	(7,8,7,8)
28	(8,8,7,8)	(8,8,7,8)	(8,8,7,8)
29	(8,8,8,8)	(8,8,8,8)	(8,8,8,8)

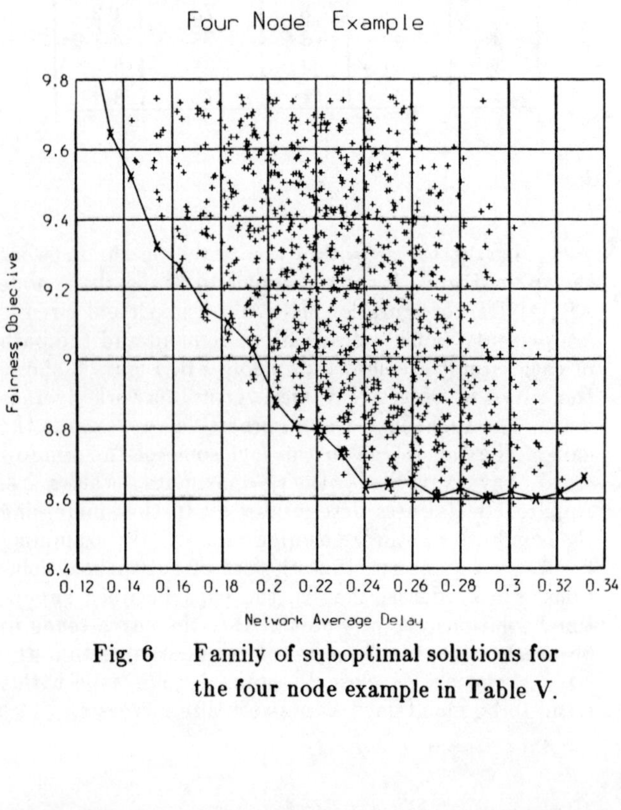

Fig. 6 Family of suboptimal solutions for the four node example in Table V.

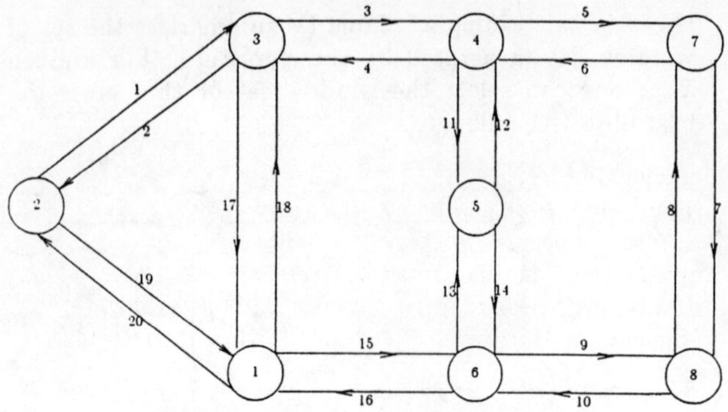

Fig. 7 Early ARPANET example.

Table VII
Early ARPANET example.

chain	source	dest.	demand	route
1	1	7	5	18-3-5
2	3	8	5	17-15-9
3	2	5	5	19-15-13
4	4	6	5	11-14
5	1	5	5	15-14
6	4	2	5	4-2
7	3	7	5	3-5
8	6	8	5	9
9	1	3	5	18
10	7	4	5	6

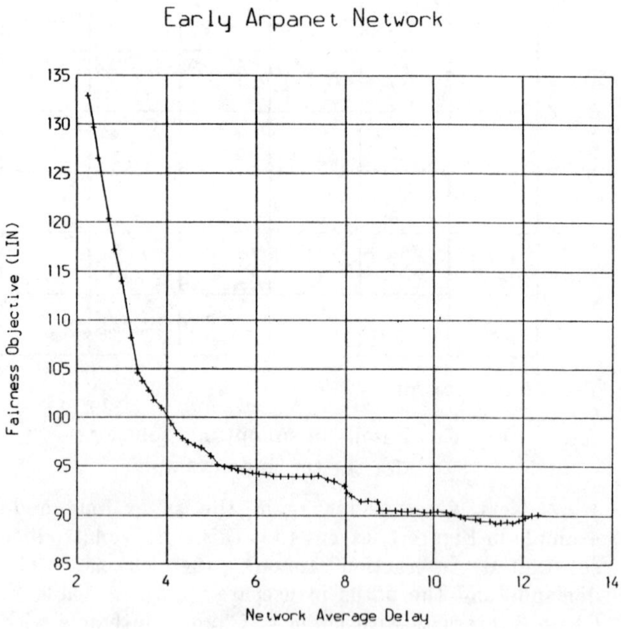

Fig. 8 Family of suboptimal solutions for early ARPANET example.

6 CONCLUSIONS

We have presented a suboptimal algorithm for the optimization of window sizes in a window flow controlled network. The algorithm is computationally efficient and yields a subset of the global set of optimal (nondominated) solutions corresponding to different average delay constraints. The algorithm may find applications in file transfer environments, where windows must be properly tuned to achieve throughput efficiency within given delay and fairness constraints.

The method may be extended to the optimization of windows in mixed traffic environments, where both file transfer and interactive traffic components are present. Another area in which the approach could be valuable is the development of dynamic window selection schemes, where windows are dynamically adjusted during network operation, to adapt to changing traffic patterns.

APPENDIX A

The goal is to express $\lambda_i(\mathbf{N} + \mathbf{e}_c)$ as a function of $\lambda_i(\mathbf{N})$, $\forall i$. We start by using Schweitzer's approach to approximate average queue length L_{ik} at center k for chain i, after the addition of one more customer in chain c, namely:

$$L_{ik}(\mathbf{N} + \mathbf{e}_c) = L_{ik}(\mathbf{N}) \qquad i \neq c$$

$$L_{ck}(\mathbf{N} + \mathbf{e}_c) = \frac{N_c + 1}{N_c} L_{ck}(\mathbf{N}) \qquad i = c$$

We can rewrite the above expressions for population $\mathbf{N} - \mathbf{e}_i$, i.e. one less customer in chain i as follows:

$$L_{ik}(\mathbf{N} + \mathbf{e}_c - \mathbf{e}_i) = L_{ik}(\mathbf{N} - \mathbf{e}_i) \qquad i \neq c$$

$$L_{ck}(\mathbf{N} + \mathbf{e}_c - \mathbf{e}_i) = \frac{N_c + 1}{N_c} L_{ck}(\mathbf{N} - \mathbf{e}_i) \qquad i = c$$

Therefore:

$$L_k(\mathbf{N} - \mathbf{e}_i + \mathbf{e}_c) = L_k(\mathbf{N} - \mathbf{e}_i) + \frac{L_{ck}(\mathbf{N} - \mathbf{e}_i)}{N_c}$$

We recall that from MVA principles [Reis80] and from Little's Result the waiting time W_{ik} can be expressed as:

$$W_{ik}(\mathbf{N}) = x_{ik}[1 + L_k(\mathbf{N} - \mathbf{e}_i)]$$

where x_{ik} is the mean service time of chain i at center k.

If one customer is added to chain c the expression becomes:

$$W_{ik}(\mathbf{N} + \mathbf{e}_c) = x_{ik}[1 + L_k(\mathbf{N} - \mathbf{e}_i + \mathbf{e}_c)]$$

$$= x_{ik}\left[1 + L_k(\mathbf{N} - \mathbf{e}_i) + \frac{L_{ck}(\mathbf{N} - \mathbf{e}_i)}{N_c}\right] \qquad \text{(A1)}$$

Using Schweitzer's approximation we get:

$$L_k(\mathbf{N} - \mathbf{e}_i) = L_k(\mathbf{N}) - \frac{L_{ik}(\mathbf{N})}{N_i}$$

and

$$L_{ck}(\mathbf{N} - \mathbf{e}_i) = L_{ck}(\mathbf{N}) \qquad i \neq k$$

Substituting in (A1) we get:

$$W_{ik}(\mathbf{N} + \mathbf{e}_c) = x_{ik}\left[1 + L_k(\mathbf{N}) - \frac{L_{ik}(\mathbf{N})}{N_i} + \frac{L_{ck}(\mathbf{N})}{N_c}\right]$$

$$= W_{ik}(\mathbf{N}) + x_{ik}\frac{L_{ck}(\mathbf{N})}{N_c}$$

This follows from the fact that

$$W_{ik}(\mathbf{N}) = x_{ik}\left[1 + L_k(\mathbf{N}) - \frac{L_{ik}(\mathbf{N})}{N_i}\right]$$

Hence, chain i mean throughput with one more customer in chain c can be expressed as:

$$\lambda_i(\mathbf{N} + \mathbf{e}_c) = \frac{N_i}{\sum_k \theta_{ik} W_{ik}(\mathbf{N}) + \sum_k \theta_{ik} x_{ik} \frac{L_{ck}(\mathbf{N})}{N_c}}$$

$$= \frac{N_i}{\frac{N_i}{\lambda_i(\mathbf{N})} + \sum_k a_{ik} \frac{L_{ck}(\mathbf{N})}{N_c}} \qquad \text{(A2)}$$

where θ_{ik} is chain i visit ratio at center k

Note that all the quantities on the RHS of Eq. (A2) are known since they are part of the current solution.

REFERENCES

[Chan82] Chandy, K.M. and D. Neuse, "Linearizer: A Heuristic Algorithm for Queueing Network Models of Computing Systems," *Communications of the ACM,* Vol. 25, No. 2, Feb. 1982, pp. 126-134.

[Ever63] Everett, H., "Generalized Langrange Multiplexers Method for Solving Problems of Optimal Allocation of Resources," *Operations Research,* Vol. 11, 1963, pp. 399-418.

[Gerl80] Gerla, M. and L. Kleinrock, "Flow Control: A Comparative Survey," *IEEE Transactions on Communications,* Vol. 28, No. 4, April 1980, pp. 553-574.

[Gerl84] Gerla, M., H.W. Chan, and J. Marca, "Routing, Flow Control and Fairness in Computer Networks," in *Proceedings ICC,* Amsterdam, Holland: May 1984, pp. 1272-1276.

[Klei76] Kleinrock, L., *Queueing Systems, Vol. 2: Computer Applications,* New York, New York: John Wiley & Sons, 1976.

[McQu72] McQuillan, J.M. and et al., "Improvements in the design and performance of the ARPA network," in *Proceedings Fall Joint Computer Conference,* 1972, pp. 741-754.

[Reis79] Reiser, M., "A Queueing Network Analysis of Computer Communication Networks with Window Flow Control," *IEEE Transactions on Communications,* Vol. 27, No. 8, Aug. 1979, pp. 1199-1209.

[Reis80] Reiser, M. and S. Lavenberg, "Mean-Value Analysis of Closed Multichain Queueing Networks," *Journal of the Association for Computing Machinery,* Vol. 27, No. 2, April 1980, pp. 313-322.

[Rybc80] Rybczynski, A., "X.25 Interface and End-to-End Virtual Circuit Service Characteristics," *IEEE Transactions on Communications,* Vol. 28, No. 4, April 1980, pp. 500-510.

[Schw79] Schweitzer, P., "Approximate analysis of multiclass closed networks of queues," in *Proceedings Int'l Stochastic Control and Optimization,* Amsterdam, Holland: 1979.

STRATEGIES FOR OPTIMAL CAPACITY ALLOCATIONS IN DAMA SATELLITE COMMUNICATION SYSTEMS*

by

Kevin Wong** and Nicolas D. Georganas
Department of Electrical Engineering
University of Ottawa
Ottawa, Ontario
Canada K1N 6N5

I INTRODUCTION

In a circuit-switched Demand-Assignment Multiple-Access (DAMA) Satellite Communication System many users, each with a different service time requirement, may request communication capacity allocation.

In view of the finite total capacity available (eg. TDMA slots), some users will be rejected and they will have to try again.

Mathematically, the problem may be formulated as follows: there are N user groups, each characterized by a finite or infinite Poisson source of calls generated with rate λ_i and capacity demand of K_i units (eg. TDMA slots), i = 1, 2,...M. The satellite server has N capacity units available in total and services in FIFO order with exponential service time per call.

If a new call arrival cannot be admitted because of lack of capacity available, the call is rejected. Several schemes applying to this problem have been proposed in the past under the generic name of "buffer management" strategies. Examples: Complete Sharing (CS) of the resource, Complete Partitioning (CP) to each user group, Sharing with Minimum Allocation (SMA) of resource units reserved for each group and Sharing with Maximum Capacity (SMC) limits imposed on each user group.

All of these strategies possess the so-called "Coordinate Convexity" property and the resulting queueing problem has a "product form" solution. Performance measures such as mean system throughput, mean transit delay, utilization and blocking probability can then be obtained.

The question then arises: which of these schemes can optimize one or more of the performance measures and for what parameters?

It has been recently proven that for the case of M = 3, the SMC scheme is optimal.

In this paper, we conjecture that this holds for arbitrary M and we use a "pattern search" approach for optimizing the weighted performance measures. This procedure is described in the paper. The sensitivity of the optimal parameters obtained is also studied.

II DAMA CIRCUIT-SWITCHED SERVICE

2.1 MODEL

The system is modelled as a pool of N units of capacity shared by M groups of users on a Demand Assignment Multiple Access (DAMA) basis. In this system each user group i, i = 1,2,3,...M, is characterized by the number of traffic sources R_i, mean arrival rate λ_i and average call length $1/\mu_i$. Assuming that the system operates in a blocked calls-cleared circuit switched mode with incoming traffic of Poisson arrivals and exponentially distributed service time, its statistical model leads to a closed product form solution [1], [2].

The state of the system can be defined by the vector n, where $n = (n_1, n_2, \ldots n_M)$; each n_i is the number of ith group calls in progress. The steady-state probability, P(n), of the system being in state n is given in equation (2.1).

* This work was supported in part by the National Sciences and Engineering Research Council of Canada under operating Grant A-8450.

** Mr. K. Wong is currently with Canadian Astronautics Limited, Ottawa, Canada

$$P(\bar{n}) = P(\bar{0}) \prod_{i=1}^{M} V_i\,(n_i), \quad \bar{n} \in A \qquad (2.1)$$

where $V_i(n_i) = \dfrac{\rho_i^{\,n_i}}{n_i!}$ for an infinite population in group i

or $V_i(n_i) = \binom{R_i}{n_i}\rho_i^{\,n_i}$ for a finite population in group i

and $\rho_i = \dfrac{\lambda_i}{\mu_i}$

A = set of feasible states which depends on the access discipline, system capacity N and number of user groups M.

$P(\bar{0})$ = probability of the state (0,...,0) occurring

$P(\bar{0})$ is also the normalization constant so that

$$\sum_{\bar{n} \in A} P(\bar{n}) = 1$$

$$P(\bar{0}) = \left\{\sum_{\bar{n} \in A}\left[\prod_{i=1}^{M} V_i(n_i)\right]\right\}^{-1} \qquad (2.2)$$

Hence $$P(\bar{n}) = \left\{\sum_{\bar{n} \in A}\left[\prod_{i=1}^{M} V_i(n_i)\right]\right\}^{-1}\left\{\prod_{j=1}^{M} V_j(n_j)\right\} \qquad (2.3)$$

2.2 SHARING SCHEMES

For a system with a total of N units of capacity and each user from group i using K_i units of capacity, the total number of capacity units engaged at any instant is $\sum_{i=1}^{M} K_i n_i < N$.

If the capacity needed to serve a new request were to make the summation to exceed N, then that request will be rejected. Additional constraints can be imposed depending on the access strategy, giving rise to a set of feasible states, A.

For each sharing strategy, its corresponding A-set can be represented by a M-dimensional geometrical object residing on the positive coordinate axes (Figure 2.2). The outer surfaces of the object represent the states of the A-set at which new arrivals will be blocked. Thus, the blocking probabilities of this system can be evaluated by summing up all the probabilities of being in the states represented by the outer surfaces.

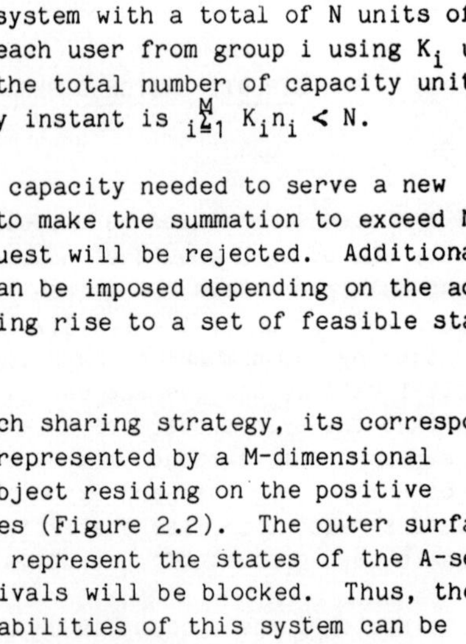

Figure 2.2a Complete Sharing, M = 3

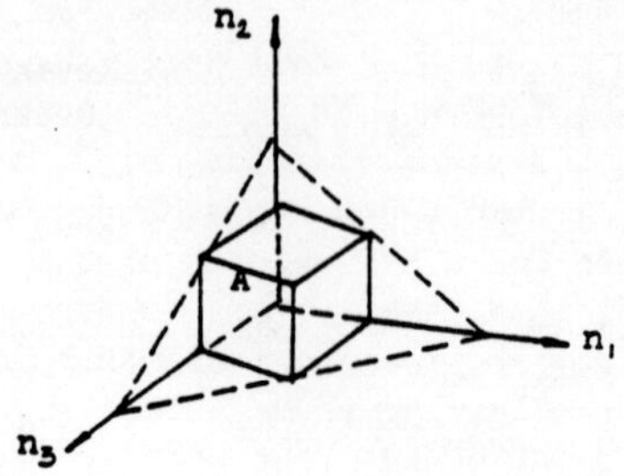

Figure 2.2b Complete Partitioning, M = 3

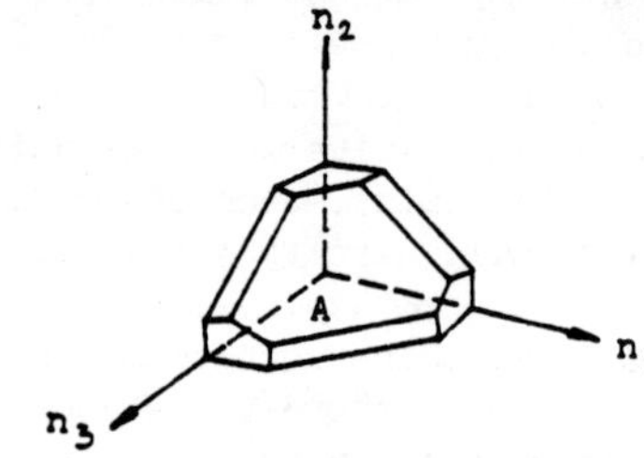

Figure 2.2c Sharing with Maximum Allowed, M = 3

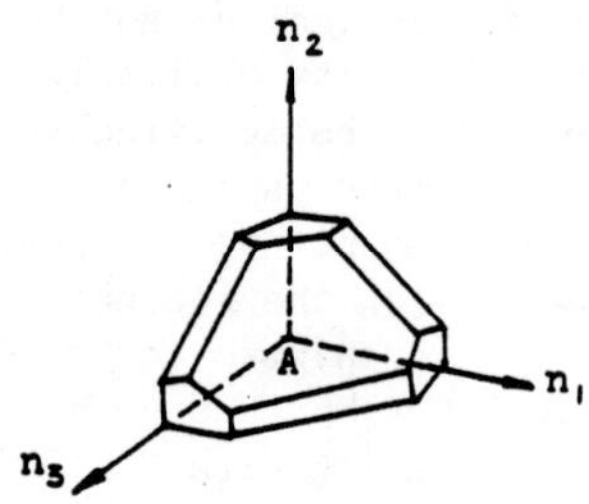

Figure 2.2d Sharing with Minimum Allocated, M = 3

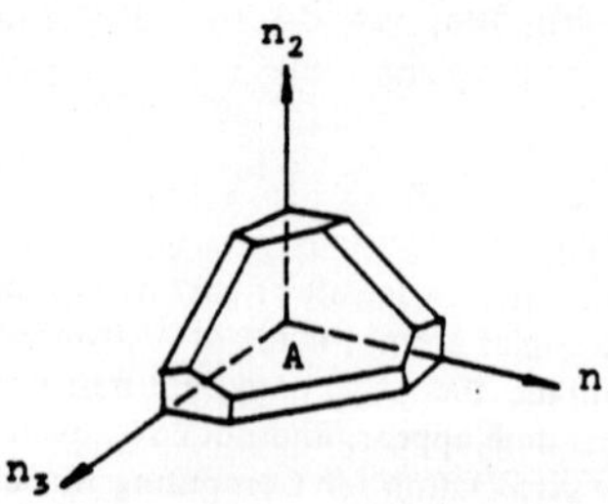

Figure 2.2e Intermediate Sharing, M = 3

III SOLUTION ALGORITHM

3.1 ONE-DIMENSIONAL & TWO-DIMENSIONAL ALGORITHMS

From the product form analytical solution (2.3), the performance of the system in terms of throughput, utilization, delay, etc. can be evaluated [2]. However, direct application of this equation even for a system with a small number of capacity units, N, and a small number of user groups, M, may require an intolerable amount of computation. Clearly, an efficient algorithm is required for the design of realistic systems.

Kaufman, treating the total number of capacity units being occupied as a random variable [4], derived a one-dimensional recursive method which is only applicable for a system that adopts the complete sharing policy. Barberis and Brignlol produce a more general two-dimensional recursive method which is applicable for all five sharing policies [3]. These algorithms allow a DAMA system's various performance parameters to be obtained without resorting to the state probabilities of (2.3) while keeping the amount of computation manageable.

3.2 NUMERICAL EXAMPLES

As an example, a system with 50 capacity units (N = 50) shared by 5 classes of users (M = 5) was considered. The traffic intensities of the 5 different user classes were arbitrarily chosen at $1.5\beta:3\beta:6\beta:12\beta:15\beta$ where β is a constant.

From the point of view of the system owner, utilization is the main factor in a design problem. Figure 3.1 compares the system utilization for the cases of Complete Sharing, Sharing with Maximum Allowed (for $N_5 = 25$) and Complete Partitioning. In Complete Partitioning, the system was divided into 5 subsystems, the capacity ratio of these systems was made to be the same as that of the traffic intensities while keeping the sum of capacity N to be 50. Obviously, this strategy gives a poorer utilization compared with that of the Complete Sharing and Sharing with Maximum Allowed disciplines.

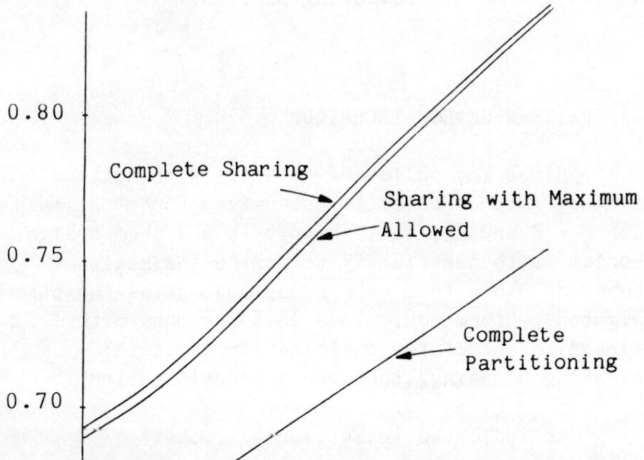

Fig 3.1 System Utilization vs β

IV SYSTEM OPTIMIZATION

4.1 OPTIMIZATION CRITERIA

In the preceding sections, analytical expressions as well as efficient algorithms for evaluating the DAMA system performance were presented. In a system design problem, however, the question posed to the designer is: given a DAMA system with N units of capacity to be shared by M groups of users, how should the resources be shared in order to fulfill a set of optimization criteria? These criteria can be the maximizations of throughput, utilization or minimization of blocking probability. Moreover, in order to represent the relative importance of different user groups, weighting factors can be introduced so that optimization can be done with respect to weighted blocking, weighted utilization or weighted throughput. An example is that the system owner charges different user groups according to different fee schedules.

1. Weighted Blocking

   The blocking probability of the ith groups users is B_i; their weighted sum can be used to find the blocking in the system, i.e.

   $$\sum_{i=1}^{M} B_i W_i$$

   The weighting factor, W_i, serves to emphasize the blocking of each type differently. To optimize the system with respect to weighted blocking, the above sum should be minimized.

2. Weighted Throughput

   Througputs can be expressed in terms of blocking, B_i, in the following way:

   $$T_i = \lambda_i(1-B_i).$$

   The system weighted throughput is:

   $$\sum_{i=1}^{M} W_i T_i = \sum_{i=1}^{M} \lambda_i W_i (1-B_i).$$

   To maximize the system weighted throughput corresponds to minimizing the sum:

$\sum_{i=1}^{M} \lambda_i W_i B_i$.

3. Weighted Utilization

The mean number of users from group i, $E\{n_i\}$ is $E\{n_i\} = \rho_i(1-B_i)$.

The mean number of capacity units used by group i users is $K_i E\{n_i\} = K_i \rho_i (1-B_i)$.

Therefore, the total weighted utilization of the system is

$$U = \frac{1}{N} \sum_{i=1}^{M} W_i K_i \rho_i (1-B_i)$$

where N is the total number of units available in the system. To maximize the weighted utilization, the sum $\sum_{i=1}^{M} W_i K_i \rho_i B_i$ is to be minimized.

In summary, the optimization can be done by using any one of the following criteria:

Minimize weighted blocking:
minimize $\sum_{i=1}^{M} W_i B_i$

Maximize weighted throughput:
minimize $\sum_{i=1}^{M} W_i \lambda_i B_i$

Maximize weighted utilization:
minimize $\sum_{i=1}^{M} K_i \rho_i W_i B_i$

All three procedures can be reduced to the minimization of a sum in the form of $\Sigma\, C_i B_i$ where C_i are weighting factors.

The optimal policy will be dependent on the weights W_i. Also, for each set of W_is, optimizing the throughput may not give the optimal blocking nor optimal utilization. However, it can be proven [5] that in the case of M = 2 and M = 3, the structure of the optimal policies, using any one of the three criteria is the same. In particular, the optimal policy is that of "Sharing with Maximum Allowed", with the maximum number of the ith group users, N_i, depending on the weights W_i and the optimization criterion. Therefore, any optimal policy may be implemented by limiting the number of each group to N_i. The proofs for the two-dimensional and three-dimensional cases are beyond the scope of this presentation (see Ref. 5). Figures 4.1 and 4.2 show the optimal policy for the two-dimensional and three-dimensional cases respectively.

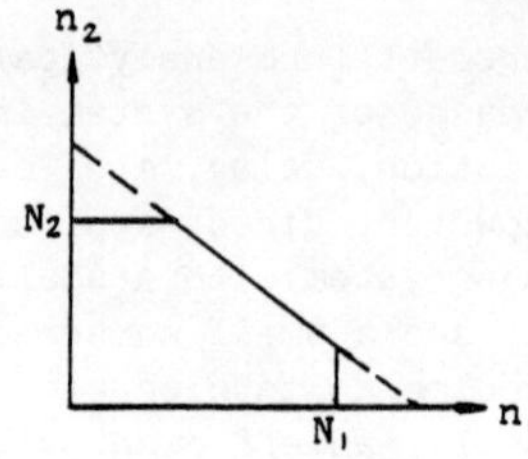

Figure 4.1 Optimal Policy for a Two-Dimensional System

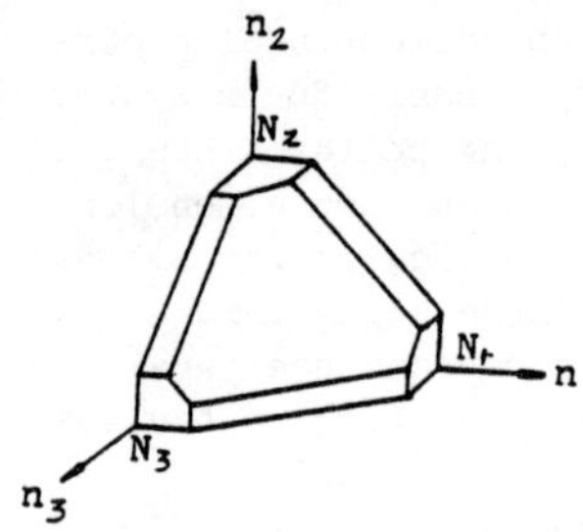

Figure 4.2 Optimal Policy for a Three-Dimensional System

With the policy of "Sharing with Maximum Allowed" proven to be optimal for M = 2 and 3 amongst the five sharing policies described in Section 2, the first task of our system design is accomplished. What remains as the second and last task is the numerical evaluation of the set of optimal system parameters, i.e. the maximum number of ith group users, N_i. This is the subject of the following section.

4.2 PATTERN SEARCH TECHNIQUE

Since the policy of "Sharing with Maximum Allowed" is theoretically determined to be optimal (for M = 2 and 3), what is left in a system design problem is to numerically determine the maximum number of users from each group that gives the best weighted performance. This is to be done with respect to one of the optimization criteria: weighting blocking, throughput or utilization.

As indicated in Section 4.1, the optimization involves minimizing a sum in the form of $\sum_{i=1}^{M} C_i B_i$; C_i being a weighting factor for group i. In general, the goal is to minimize the objective function $f(N_1, N_2, \ldots, N_M) = \sum_{i=1}^{M} C_i B_i$ (B_i is function of N_is) and determine the corresponding set $(N_1, N_2, \ldots, N_M)$.

The pattern search technique [6] was chosen for this task for the following reasons:

a) it is well adapted to a computer since our task tends to use repeated identical arithmetic operations;

b) it is logically simple - which simplifies computer programming;

c) it permits integral increments to be taken - suitable for this problem since all N_is must be non-negative integers.

Pattern search is a direct search routine for minimizing a function $f(N_1, N_2,...,N_M)$. The arguments N_i are varied one by one until the minimum f is obtained. The pattern search routine determines the sequence of values for N_is. An independent routine computes the functional values of f. An exact description of the pattern search routine is given in [6].

4.3.1 OPTIMIZATION RESULTS

Two numerical examples are considered here; both are four-dimensional (M = 4) with N = 200. The first one has balanced traffic. By balanced traffic we refer to the fact that the products $K_i\rho_i$ for all i are identical. The second one has non-balanced traffic, i.e. $K_i\rho_i$ are not equal for all i.

For each of the two examples, the optimal vector $(N_1,...,N_4)$ and the corresponding system performance is determined (with respect to each of the three criteria).

The results are summarized as follows.

Example 1: N=200, M=4
Input

i	1	2	3	4
K_i	40	20	8	4
ρ_i	1.0	2.0	5.0	10.0
W_i	0.1	0.2	0.3	0.5

Output

Criterion	f	N_1	N_2	N_3	N_4
Weighted Blocking	0.084	2	9	19	31
Weighted Throughput	0.201	1	4	21	49
Weighted Utilization	4.930	2	8	18	30

f = objective function

It is clear from the results that optimizing the system in terms of weighted blocking does not give optimal results as far as weighted throughput or weighted utilization is concerned. To optimize weighted blocking, the $\{N_i\}$ vector should be (2,9,19,31) while if optimizing the weighted throughput is desired, the $\{N_i\}$ vector should be (1,4,21,49).

Example 2: N=200, M=4
Input

i	1	2	3	4
K_i	40	16	16	4
ρ_i	0.1	1.0	5.0	10.0
W_i	0.9	0.05	0.01	0.04

Output

Criterion	f	N_1	N_2	N_3	N_4
Weighted Blocking	0.0107	4	6	1	24
Weighted Throughput	0.022	4	9	5	40
Weighted Utilization	31.5	4	11	4	27

f = objective function

Again, the same observation made in Example 1 applies here. Therefore in a practical design problem, when more than one criterion is deemed to be significant, a compromise is required. Judgement of this nature can only be passed on a per system basis.

Obviously, cross comparisons between Examples 1 and 2 should not be made since they are two entirely different systems. As for computational efforts, for these particular examples, the package developed for this work used approximately 60 iterations for each optimization. The requirement increases with M and N.

4.3.2 SENSITIVITY STUDIES RESULTS

To conclude this numerical exercise one question remains to be answered: how much tolerance in the resultant optimal N_is do we need? This is an important question to ask since it determines how long the pattern search should be continued before a solution is declared acceptable.

Sensitivity studies of this type were conducted on both of the examples here based on two parameters; namely, the objective function f with respect to the N_i and the objective function with respect to ρ_i:

1. At each of the optimal points $[N_i]$, find the required change in each of the coordinates in order to produce, say, a 5% change in the objective function, eg. take the weighted blocking case in Example 1, the optimal $[N_i]$ is (2,9,19,31). To determine how sensitive f is to N_4, n_4 is varied from 31 until f is increased by 5%. It was found that N_4 can be changed by as much as 45% before the objective function increases by 5%. This is repeated for all i's. The results are summarized below.

Example 1: Balanced Traffic
ΔN_i = required percentage change in N_i to give 5% change in f.

Criterion	ΔN_1	ΔN_2	ΔN_3	ΔN_4
Weighted Blocking	50%	56%	47%	45%
Weighted Throughput	100%	25%	48%	60%
Weighted Utilization	50%	75%	67%	47%

The results show that the allowed changes in N_i in this particular example range from 25 to 100% before the objective function increases by 5%. The indication here is that f is not very sensitive to $[N_i]$. Therefore no significant advantage is gained by allowing the pattern search procedure to proceed too long once little improvement is observed in f, i.e. instead of using 60 iterations (as indicated in Section 4.3.1), the search could have been terminated earlier without much degradation in the final results. This is significant when a large system is under consideration.

A similar observation can be made on the Example 2 as indicated by the following results.

Example 2: Unbalanced Traffic

Criterion	ΔN_1	ΔN_2	ΔN_3	ΔN_4
Weighted Blocking	25%	33%	100%	33%
Weighted Throughput	50%	67%	20%	53%
Weighted Utilization	75%	97%	75%	63%

2. The same procedure as that in 1 is repeated but instead of changing N_i, the traffic intensity ρ_i is changed so that at the optimal point $[N_i]$, the change in each of the ρ_i required in order to increase the f by 5% is determined. Again, the choice of 5% is arbitrary. In most systems, the traffic pattern is assumed to be known. This study shows how sensitive the system is in case of variation of this information.

Example 1: Balanced Traffic
$\Delta\rho_i$ = required percentage changes in ρ_i to give a 5% change in f

Criterion	$\Delta\rho_1$	$\Delta\rho_2$	$\Delta\rho_3$	$\Delta\rho_4$
Weighted Blocking	8%	6%	6%	6%
Weighted Throughput	9%	4%	4%	4%
Weighted Utilization	20%	40%	20%	12%

Example 2: Unbalanced Traffic
$\Delta\rho_i$ = required percentage change in ρ_i to give a 5% change in f

Criterion	$\Delta\rho_1$	$\Delta\rho_2$	$\Delta\rho_3$	$\Delta\rho_4$
Weighted Blocking	10%	10%	40%	5%
Weighted Throughput	10%	10%	2%	5%
Weighted Utilization	10%	30%	94%	20%

The results indicated that the allowed change in ρ_is is again of relatively large amount although it is not as much as that of N_is. One can conclude that the systems under consideration in these two examples are considerably more sensitive to ρ_i than N_i. However, there is no indication that a general conclusion of this nature can be made on all DAMA system using the policy of sharing with Maximum Allowed.

V CONCLUSIONS

A Demand Assignment Multiple Access (DAMA) system with a pool of resources shared by several user groups was considered here. Sharing policies of such resources that are simple to implement and require no detailed measurement of arriving traffic (i.e. accuracy of statistical parameters such as ρ is not important) are of considerable practical interest. Five of these policies, namely, Complete Sharing, Complete Partitioning, Sharing with Maximum Allowed, Sharing with Minimum Allocated and Intermediate Sharing, were studied in detail. All of these policies possess the property of coordinate convexity, giving rise to the well-known product form solution for the steady-state probabilities.

To allow use of these theoretical results in systems of realistic size, efficient algorithms were presented. From that, commonly considered system performance measures can be determined with a reasonable amount of effort. Moreover, the contribution of the two-dimensional algorithm was shown to be more significant when it was adapted in system designs.

In a system design problem, the designer is required to choose a sharing policy and then find the optimal system parameters numerically. For such optimization, three of the commonly used criteria, namely, weighted blocking, weighted throughput and weighted utilization, were considered. Theoretical results have been presented in the literature indicating that for two- and three-dimensional systems, the optimal policy for any of the three criteria is that of Sharing with Maximum Allowed. For the determination of the maximum number of users allowed, the pattern search technique was presented. By adopting the pattern search procedure, in conjunction with the two-dimensional recursive algorithm, an optimization computer package was developed.

Finally, two numerical examples were given. In each case, the optimization was repeated for each of the three criteria discussed. Sensitivity studies were also carried out to determine how sensitive the objective function was with respect to both the variation of N_i from its optimal value and to the variation of ρ_i. It was concluded that for these two examples, they are quite insensitive.

REFERENCES

1. J. Aein. "A Multi-user-class, Blocked-calls-cleared Demand Access Model", Trans. on Comm., Vol. Com-26, No. 3, March 1978, pp. 378-385.

2. J. Aein, O. Kosovych. "Satellite Capacity Allocation", Proceedings of the IEEE, Vol. 65, No. 3, March 1977, pp. 332-342.

3. G. Barberis and R. Brignlol. "Capacity Allocation in a DAMA Satellite System", IEEE Trans. Comm., Vol. Com-30, No. 7, July 1982, pp. 1750-1757.

4. J.S. Kaufman. "Blocking in a Shared Resource Environment", IEEE Trans. Comm., Vol. Com-29, No. 10, October 1980, pp. 1474-1481.

5. G.J. Foschini, B. Gopinath. "Sharing Memory Optimally", IEEE Trans, Comm., Vol. Com-31, No. 3, March 1983, pp. 352-360.

6. R. Hooke, T.A. Jeeves. "Direct Search Solution of Numerical and Statistical Problems", J. Assn. Comp., Vol. 8, April 1961, pp. 212-229.

INTEGRATING SATELLITE LINKS INTO A LAND-BASED PACKET NETWORK

M. Chu, D. Drynan, L.R. Benning

BNR, Ottawa, Canada

Abstract

It is becoming apparent that communication via satellite links will play an increasing role in data networks. The challenge is to integrate this new communications option into an existing land-based network such that users can realize its full benefits. This paper discusses that solutions for the SL-10* Packet Network. The design criteria and performance objectives are also examined.

Introduction

Packet switching technology has been commercially available for over a decade now. During this time, the communications between nodes or switches relied mainly on terrestrial facilities. This is changing rapidly as satellite communication technologies mature and move into the data communications market. Satellite facilities can be leased as dedicated links between locations or as private satellite networks of earth stations. Satellite facilities can provide flexibilities that are unmatched by the terrestrial ones. In geographically remote locations, terrestrial links can be very difficult and expensive to obtain. On the other hand, satellite links can reach into every corner of the earth.

As the strategic and economic advantages of satellite communications technology become more obvious, there is strong pressure to integrate satellite links into existing land-based packet networks. One fundamentally different property between satellite and terrestrial link is the propagation delay. Communication satellites orbit at a distance of 35,000 kilometers above the earth in a geosynchronous orbit. A single round trip on a satellite link requires at least 540 milliseconds without including queuing and transmission delays in the network. This is more than 20 times longer than that of a terrestrial link. The long delay cannot be masked from the network users. The challenge is to integrate this different communications option into existing terrestrial networks in such a way that the advantages can be shared by the network users and operators.

This paper presents both the planned and implemented solutions on the SL-10 Packet Switching System for integrating satellite links into its subnet. This paper does not include user access employing satellite links into the packet network or internetworking with X.75 gateways via satellite links, althought they are under development by Bell-Northern Research.

Design Criteria and Performance Objectives

1. Integrating satellite links into the network should not cause erratic delays for packet arrivals. Network delay as perceived by the user should be relatively consistent. To achieve this the number of terrestrial and satellite hops traversed by a call should be maintained over the duration of the call. Otherwise the significantly longer satellite link delay would cause packets to arrive out-of-order at the destination switch which would have to resequence the packets by delaying their delivery. Frequent resequencing of packets would result in an uneven quality of service that would be especially unacceptable to interactive users.

2. The internodal trunk protocol should be able to adapt to the long delay of a satellite facility to allow maximum throughput on the link. Otherwise the benefits of satellite links cannot be fully realized.

3. Users of the packet network should be able to specify whether or not their calls can go over satellite links. This is especially important because some user applications are delay-sensitive and should be routed over terrestrial links only.

4. The availability and reliability of the existing subnet should not be

----
* Trademark of Northern Telecom Limited

compromised. The transmission quality of the satellite link can deteriorate significantly under inclement weather conditions which can last for hours. The throughput on these links under such conditions can be drastically reduced. Thus transmission quality should be monitored and, when appropriate, the trunk should be disabled to allow rerouting of the traffic over other available facilities. However, the disabling criteria should be controlled by adjustable thresholds to minimize unnecessary network topology changes.

5. The introduction of satellite trunks into an existing network should be accomplished in an easy and flexible manner. Network growth and management should be assisted by network engineering tools.

6. Existing network features should still be able to work over the satellite links.

Solutions

Class of Service

Employing satellite links in the subnet has the potential to affect all services offered by the network. The long propagation delay cannot be masked from the end users. Some applications that use external higher level protocols with short time-outs may not be able to tolerate this delay. Without changing the application, the only solution is to restrict these calls to terrestrial paths.

In general, most interactive users can accept the delay incurred in one satellite hop without noticeable effect. Other factors like processor delays, however, may cause these calls to be restricted to terrestrial trunks.

Non-interactive batch traffic is usually high volume and requires high throughput. Although not a direct factor, propagation delay can result in restricted throughput if it is not managed properly. This traffic, however, is suitable for transmission over satellite trunks, freeing the more costly terrestrial links for interactive traffic which requires fast response time. Therefore, from the user's point of view, users should be able to select whether or not their traffic can go over satellite link. This objective can by met be offering two classes of service for user traffic:

1. Delay Sensitive (DS) Traffic - traffic that requires short round-trip delay, such as interactive traffic and service-implemented protocols with short time-outs.

2. Delay Insensitive (DI) Traffic - traffic that does not require short round-trip delay, such as high volume batch traffic and file transfers.

The user should be able to specify the class of service at call setup time.

User Access

As currently planned, the user will be able to signal to the network at call setup time the class of service of the call. On an X.25 service, the Transit Delay Indication facility can be used for this purpose. On a PAD (Packet Assembler Disassembler) service like X.3, the X.28 Select command may have an optional class selector field. Alternatively, users can select their default class of service at service subscription time and all their calls will be restricted to the class subscribed. In this way, the DTEs (Data Terminal Equipment) do not have to implement different procedures or software. However, a user can always signal the class of service at call setup time, even if one is defined by subscription. The signaled class of service will override the subscription option. Direct calls, Permanent Virtual Circuits (PVCs) and Fast Select calls can all make use of the ability to specify class of service.

Class Mapping

The SL-10 will always try to set up the call as specified. The routing system will determine the best route for a certain class. However, the trunk facilities that handle a particular class may become unavailable for a period of time due to trunk outage or temporary network topology changes. The SL-10 may redirect the call over the facilities serving the other class, if these facilities have enough capacity. This is referred to as Class Mapping. When the network topology is restored and the trunk facilities serving the particular class become available again, then the traffic will be mapped back to the original class. The mapping applies to both calls in progress and new calls. Mapped calls will continue to send data, possibly with degraded performance, on the assumption that full performance will be restored within a short time.

The network administration will retain complete control over all aspects of class mapping via the use of service data. On a switch-by-switch basis, class mapping can be totally or selectively prevented from DS to DI or DI to DS. Complete blocking may be appropriate if the network is not engineered to handle these possible changes in traffic pattern. All affected calls would be cleared with a message indicating the reason for clearing the call. The planned DS class of traffic may be blocked completely from satellite trunks to guarantee minimum network delay. Conversely, the high-volume DI traffic can flood the terrestrial trunks if mapping is allowed.

The network will always try to route every data packet over the facilities serving the requested class. The user could assume that if the call is set up, it will generally use the facilities requested. The user, therefore, is not informed of any call mapping. Network accounting statistics will record the class of service specified in the call request, even if the call is mapped to a different class. Since mapping is only temporary, accounting will always assume that calls are established at the class requested. If the network administration decides this is not appropriate, class mapping can be blocked. Call requests which specify an unavailable class of service will be blocked.

Routing

The SL-10 Routing System employs a fixed path algorithm which reacts dynamically to changes in network topology. Except under heavy load conditions, once a set of network paths is computed it remains in effect until another change in topology occurs. The most common cause for a topology change is a temporary trunk outage due to the failure or fading of a facility. Routing decisions are made independently at each switch. The routing process builds its view of the network by creating routing tables and exchanging them with neighboring switches. The tables contain the best routes, determined by the metric, as seen from each switch to all other switches in the network. The end-to-end metric is the summation of all the individual metrics for each hop on the path. A single hop metric, calculated between directly connected switches, is a function of the total bandwidth, number of trunks, and assumed trunk utilization. The propagation delay of the trunk path is not incorporated into the metric.

The routing system also allows the network administration to specify a preferred path between any two switches. To prevent accidental engineering of permanent routing loops, only minimum metric paths are eligible. Following the same rules, the administration may also select an alternate minimum metric path for load sharing under heavy traffic conditions. Both features are optional on a switch-by-switch basis and are effected using service data. In general, service data can be updated without taking any switch out of service. A network engineering tool assists the network administration to analyze and make choices regarding preferred path or secondary path assignment.

To support the planned class of service feature outlined in the previous section, the routing system should be able to segregate satellite and terrestrial trunks in such a way that different types of traffic can be sent over the appropriate trunk. If both types of trunks are allowed to form a path in parallel, the packets belonging to a call would frequently arrive out of sequence and this would result in erratic delays to users. Network efficiency would also suffer because of long buffer holding times. The solution is to provide two separate routing systems, each responsible for one class of service. These two routing systems are not mutually exclusive but are complementary.

One routing system will handle the DS class of traffic and the other the DI class. In essence, two different sets of routing tables containing the appropriate metric values would be used. Packets belonging to each class of traffic will be routed based on the information in the corresponding routing table. Every trunk belongs to both routing systems and will be assigned two metric estimates (instead of one) for the two classes of service.

To set up paths that favor the terrestrial trunks for the DS class, a penalty value will be added to the DS metric for all satellite trunks. For the DI class, the penalty will be added to the DI metric of terrestrials. Thus if there are two paths with differing facility types but identical aggregate bandwidths, the satellite path would have a lower metric and be chosen for DI traffic. The penalty values will be assigned by network administration and may be tailored to meet individual network requirements. However, a single penalty value for each class will be used network-wide.

The previous section discussed various ways to route and map one class into the other class of service. These would be under the control of network administration on a routing class and switch-by-switch basis. If one class, say the DI traffic, were allowed into the DS terrestrial system, the DS traffic would maintain priority over the DI traffic. The quality of service for the specified class, DS in this case, would not be degraded. Because both routing systems maintain the fundamental strategy of always choosing the minimum metric path, the proven stability of the routing system is guaranteed and all current routing features retained without changes. Switches may be connected by both terrestrial and satellite trunks, each carrying the appropriate class of traffic. Singly connected switches, regardless of the trunk type, may serve both DS and DI class of traffic.

Link Level Trunk

In order to meet the requirements of the different properties of the satellite link, a new trunk processor has been developed by Bell-Northern Research. This trunk processor is designed to meet the current and projected requirements for both terrestrial and satellite facilities. The new trunk processor employs a store-and-forward strategy and is equipped with sufficient memory to buffer all packets for outgoing transmission and in flight. The trunk processors resides on each switch and connect via standard interfaces to physical facilities. A trunk consists of the physical link that connects two switches and the trunk processor at each end. Each trunk processor is responsible for managing the link and ensuring error-free packet delivery. This new trunk processor employs a HDLC based, layer two protocol which was also developed by Bell-Northern Research. A higher layer protocol handles the network and transport layer functions like routing decisions and end-to-end virtual circuits.

The new trunk processor has been successfully system-tested to operate with terrestrial and satellite links at speeds up to 128 kbit/s at maximum throughput regardless of packet size. Higher speed can also be handled. For a two-hop satellite facility, a maximum round-trip delay of 1.5 seconds is allowed. During link initialization time, the trunk processors actually measure the round-trip delay of the link and this information is used by the higher layer processes to determine whether the link is satellite or terrestrial.

The link-level protocol has been designed to achieve high throughput on long delay links. It is based on the High-level Data Link Control (HDLC) procedure since it is flexible and commonly available in fast silicon. The standard HDLC flags and the Cyclical Redundancy Check (CRC) field enclose the link-level protocol. In order to achieve maximum efficiency, the protocol allows multiple frames to be in error. Only the incorrect frames are retransmitted and this does not affect the other frames in the pipe. This is important on a long delay link as the number of frames in flight can be very big and the probability of frames being corrupted is high. The conventional selective repeat error control procedure can greatly reduce the throughput on a high error rate link.

To further improve the efficiency of the link, short user packets are concatenated together to form a larger frame for transmission at high link speed under high traffic conditions. This is totally transparent to the high layer protocols and the end users.

The frame sequence number field width is determined by the round-trip delay and the link speed. This minimizes the overhead carried by each frame because the difference between a low- speed, short terrestrial link and a high- speed, two-hop satellite link is large. Although the existing implementation has been successfully tested for trunk speeds up to 128 kbit/s, the variable frame sequence number field is designed to accommodate speeds of up to 44 Mbit/s.

Network Operations

To augment the access services and the routing system, a comprehensive set of network control functions for satellite trunks is provided. All the trunks in the network are continually monitored from a network control center. The display is updated every minute to show the link utilization, alarms outstanding, and switches connected by trunks.

A satellite trunk may became unavailable occasionally due to facility problems. Before a satellite trunk is allowed into service, it is thoroughly tested to ensure that both the hardware and the facility can provide the required quality of service. The trunk must pass both loopback and far-end packet tests before it is returned automatically into service.

Satellite links can have long periods of high error rate when there is heavy fog or rain. The link throughput can be drastically reduced under such weather conditions, and it is desirable to reroute traffic away from this link to maintain the quality of service. Link quality is determined by the received CRC errors and the number of retransmissions. The error rate on each satellite link is monitored continually. Every minute, the trunk collects statistics and send them for performance analysis. The errors are checked against a preset threshold. The threshold, adjustable by operator command, gives the bit error rate tolerance. If the link error rate exceeds the threshold for a defined interval, the satellite trunk will be automatically disabled, allowing traffic to be rerouted through other available links. If a switch is singly connected to the network via a satellite link, it would be allowed to operate under this high error situation. The disabled satellite link is periodically tested and as soon as the error rate becomes acceptable again the link is allowed back into service.

To manage the subnet effectively and to create the complex service data required for these routing features, a set of network engineering and administrative tools is provided. Trunk statistics and traffic profiles are collected for all trunks and sent to the network administration system for off-line analysis. Trends in network behavior can be predicted and potential problems corrected before they become serious. The data collected are also used by network engineering tools to help the administration analyze network topologies, predict the level of trunk utilization, and simulate trunk failure scenarios.

Conclusion

The integration of satellite trunk facilities into a packet network can offer significant benefits to network operators and cost savings to users not adversely affected by the longer delay. General performance criteria should be adhered to if application services are to use this new communications option effectively. Accommodating both terrestrial and satellite trunks in a single packet network poses a complex but workable challenge, especially for trunk processors and the network routing system. Each packet network uses a different method according to the subnet strategy employed. A specific approach subject to the stated performance criteria has been presented in this paper for the SL-10 networks.

Panel 4

Directory Services/IBM Versus CCITT

Chairperson
R. Perlman
Digital Equipment Corporation

Panelists
E. Stecher, *Software Research*
P. Karp, *Skyline Associates*

Session 4

Packet Switching Network Management

Chairperson
A. Salazar
Infinet

AN INTEGRATED TEST CENTER FOR SL-10* PACKET NETWORKS

M.W.A. Hornbeek
BNR, Ottawa, Canada

Summary

The sheer scale and complexity of large data networks makes testing them a daunting task. System commissioning, release acceptance, network troubleshooting, performance testing, host conformance testing, and certification are all operational activities that involve testing. Packet switching systems typically provide builtin features to help with hardware level test operations such as modem loopback commands, system failure alarms and system selftests. However, testing system and protocol level functions still required the use of separate tool boxes. Typical data product test tools are designed to test single data lines, not multiline distributed networks (1,2,3). Considering all of the above, the SL-10 test tools group set out to develop network tools (4). Tools have been implemented which are well-integrated into the SL-10 product. The same tools can be used alone, independent of the SL-10 network. This tool set makes the job of network testing more manageable for developer, manufacturer and network operator alike. Terminals and host computers, as well as total networks, can now be tested for conformance and performance automatically, instead of manually. As a result, major productivity improvements and quality gains have been observed since the tools were introduced.

The tool set consists of four tools: an interactive protocol tester (IPT) for checking protocol implementations against specifications; the network load test system (NLTS) which generates simulated traffic and measures the network's performance; the network process monitor (NPM) which examines operating software at the module level; and the network test system (NTS) Driver which is an automatic or interactive test sequencer for coordinating and controlling distributed network testing. Each tool can be operated individually using a standard data terminal. The IPT, NLTS and NPM tools run on SL-10 RAPID* network processors and can be operated remotely via the network or the NTS Driver. Using these tools in combination, a fully integrated and automatic operational test center can be realized, and the need for separate tool boxes can be eliminated.

The design and capabilities of each of these tools are described. This paper describes how these tools meet the special needs of network test operations.

* SL-10 is a trademark of Northern Telecom Limited

Need For Tools

Data communications systems let two or more computers and computer terminals exchange information effectively and efficiently. When complex systems – data networks, for example – pass information, in the interest of accuracy and order, they must abide by commonly agreed rules or protocols. Northern Telecom's SL-10 packet switching system implements more than ten distinct protocols and three protocol layers. These protocol implementations are designed to interact with other suppliers' versions of the same protocols so that SL-10 will work harmoniously with other products and systems.

Protocol specifications can be imperfect, leaving aspects of communication ambiguous. For example, protocol implementations for data circuit-terminating equipment (DCE) can be incompatible with those for data terminal equipment (DTE), and must be debugged before the two can work with each other. In such cases, conformance test tools are helpful when preparing new protocol implementations for certification and eventual interaction. Before each major software release for SL-10, BNR has to execute thousands of test cases. On receiving new network software, operating companies need tools to check that new features work to their satisfaction. In addition, they must certify each host and terminal implementation to ensure they interact correctly with their active networks.

Testing how well a networking system performs is important – both to ensure it will work reliably under all conditions, and to measure performance. To pinpoint the source of problems found during such system-level conformance or performance testing, a reliable means of debugging is necessary. Therefore conformance testing, network loading and performance evaluation are all areas where test tools are needed. Problems identified in this way can then be reproduced in a controlled manner and rectified with the assistance of the debugging tools.

Packet switching network development teams and, more recently, international associations responsible for setting standards – such as the International Telegraph and Telephone Consultative Committee (CCITT) and International Organization for Standardization (ISO) – have recognized the need for tools to carry out this kind of testing. ISO is currently coordinating efforts to develop standards for conformance testing for the seven protocol layers needed for open systems interconnection (OSI) (5). Most of this effort is concentrated on a methodology for testing the transport and higher layers. Little work has been done yet on developing standards for testing the layers most relevant to networks (network layer and below). A set of tools for testing conformance for all existing and new protocols as well as performance and debugging would both advance the state of the art and give the company which developed it a business advantage.

In the BNR laboratory environment, a single-pass test on SL-10 used to take many work weeks. Even performing regression tests on existing product features proved very time-consuming. New test tools with the capacity to automate thousands of complex test cases for protocols and applications would obviously be helpful. These new tools should be designed to accommodate new protocols as well as those already in existence – for example, those evolving from OSI.

Architectural Approach

With such a broad range of requirements, it was clear very early in the development stage that an overall architectural approach would be best.

Since many protocols needed testing and many more would have to be added later, a protocol interface whose hardware and software could be modified easily was needed. Other prerequisites for a successful tool kit were:

- an interactive test capability to assist protocol debugging;
- a simple user interface to speed up automatic test case development for the hundreds of test cases for each protocol;
- a good report writer to reduce the mass of data generated while the test cases are running;
- a means of testing typical network configurations involving multiple data links and protocols;
- a means of distributing test interfaces for invoking, controlling, and observing remote systems under test from a central location;
- a system for generating data traffic;
- a means of testing throughput and delay parameters, as well as reliability and availability characteristics;
- a system which provides both expandable fan out and very precise timers; and
- the ability to debug components at remote sites.

These requirements indicated the need for tools with the processing power, flexibility and kind of peripheral equipment support found in a general purpose minicomputer, as well as the real-time data processing and timing capabilities of a data communications processor. In addition, it had to be possible to distribute the real-time protocol interfaces as well as parts of the debugging tool to the site of the system under test. A multilink, high capacity, general purpose test driver implemented on a minicomputer is capable of controlling separate test interfaces for specific applications. It can also distribute these interfaces to network, host or terminal systems under test. Separate applications for conformance, performance and debugging tests can be developed and optimized for specific protocols or systems.

The SL-10 RAPID system (RAPID stands for remote access protocol interface device), developed at BNR for Northern Telecom, proved to be the ideal vehicle for providing test application interfaces (6). (Figure 1 shows the SL10 RAPID configured with test software; Figure 2 shows the SL-10 RAPID hardware architecture.) It is a multi-microprocessor system in which each processor can perform various functions, depending on what software has been loaded into it, and what peripheral interface (PI) circuit cards have been connected to it. One processor functions as a system load interface processor when connected to a disk interface. The same processor functions as an SL-10 network link interface when loaded with the integrated access method (IAM) network communications software. Other processors can be loaded with test tool or service software for other functions. It is possible to configure up to 14 processors in one circuit pack shelf. Since the tools can be packaged separately in an SL-10 RAPID, they can also be used independent of the SL-10 network.

Interactive Protocol Testers

For conformance test applications a set of single-line protocol handlers, called interactive protocol testers (IPT), was developed. The IPT tool software runs on the SL-10 RAPID. These IPT tools can emulate protocol implementations for both data terminal equipment (DTE) and data circuit-terminating equipment (DCE). Each IPT has the ability to control one interface under test; send test protocol data units (PDU) to the system under test; receive, decode, and display PDUs from the system under test; and insert and check error conditions. The IPT provides physical level (modem) monitoring and control. Various data traces can be selected to display lower protocol layer activity while upper layer tests are in progress. PDUs can be time-stamped, and automatically acknowledged. Since the IPT is implemented on SL-10 RAPID, line speeds up to 72 kb/s are easily supported, and the cost per line is very low compared to conventional single-line emulators. The system is made user-friendly through the use of on-line help and English-like commmands and responses.

A typical system (Figure 3) consists of user interface, encoder-decoder, and lower level service. One processor with one PI can support up to two IPT applications. A second PI and a peripheral bus extender (PBE) can be added to support two more IPTs. The IPT software module can be downloaded through the network, as well as accessed, controlled and observed remotely. For protocol testing, the IPT is switched into a data link connected to a DTE or DCE. Since the IPT has a resident editor and macro facility, it is possible to create test scenarios and save them on disk.

Network Load Test System

Performance tests in development and manufacturing environments, as well as those in operational networks, need powerful data generation and measurement capabilities. The network load test system (NLTS) was developed to provide these capabilities, and to test the performance of the various SL-10 protocol network implementations. NLTS was implemented on SL-10 RAPID because it has the required throughput, line fan-out, and delay timing accuracy necessary for these kinds of tests. Another unique benefit of such an implementation is the ability of the tool to coexist with network services. NLTS can generate traffic load on the network, (see Figure 4) then measure call and call-clear rates, transit delay, and throughput. Like IPT, NLTS can be connected to the DCE to carry out a specific function – in this case network performance testing. The NLTS modules can be downloaded, and accessed, controlled and observed remotely in the same way as the IPT module.

NLTS is a multilink, multiprotocol network data generator and measurement tool implemented as a multiprocessor system. It is composed of two module types: common control (CCM) and supervisor (SUP). The CCM provides the user interface and controls the test session. It is also capable of producing summary reports and storing test configuration data. The supervisor module can control up to sixteen individual links simultaneously. Each link has a protocol encoder/decoder (PED) handler. Since one CCM can control up to thirteen SUP modules, fifty lines can be tested simultaneously. Because the PED handlers are coded in assembler language it is possible to achieve a data speed of 72 kb/s on each link, and timing accuracy within 10 milliseconds.

Each PED performs a range of functions. It:

- implements a common essential subset of the protocol;
- sets up calls, exchanges data, clears calls, and recovers from faults;
- can be programmed with speed and protocol-specific parameters;
- provides default test parameters which can be changed or restored;
- acts as a data source or sink (is set up to generate or receive data);
- selects appropriate tests for measuring call rates, transfer delay, throughput, and reliability; and
- records statistics such as errors, restarts and number of packets of a given type.

All of these functions can be carried out automatically from a standard data terminal in accordance with the users' own parameters.

NLTS is designed to be robust, having recovery mechanisms that will restart a test automatically should failures occur. It collects performance statistics such as counts of message types and rates, and failure occurrences. A macro/editor capability lets users create their own tests, save them to disk, and recall or modify them.

Testing with NLTS is a simple four-step process. One, the hardware is connected. Two, the user logs in using the appropriate password, which is under administration's control. Three, commands are issued to select the appropriate NLTS SUP processors to assign the source and sink PEDs according to the desired protocol options, and to run the tests. Commands can be saved in files and edited later if necessary. Complex configurations can thus be run by issuing the command file name only. (Default options are provided.) Four, NLTS reports the results to the user automatically.

Using NLTS, a broad range of network performance parameters can be measured. For example, all of the parameters of the American National Standard Institute draft X3S35/135 and CCITT X.135/136 can be determined (7).

Network Process Monitor

SL-10 and SL-10 RAPID have been designed with a unique built-in test feature. Each processor in the network has a resident processor interface processor (PIP) that can examine and modify internal transformations of any executable process. This eliminates the need to insert special test software when debugging network components. The tool known as network process monitor (NPM) implements a single user interface to all PIPs in the network. NPM runs on an SL-10 RAPID processor.

The NPM module provides several useful capabilities which include a simple user interface, a macro command language, logon password security, the ability to build macros from various commands, a file subsystem, an editor, and a datagram interface to remote PIPs. Test macros can be developed using the resident editor, saved to disk, retrieved from disk, then executed. A file list feature helps users build their own file directories should they wish to do so, but the system is delivered complete with some predefined macro functions included.

The special purpose NPM language, expressly designed for SL-10 and SL-10 RAPID, includes variables, literals, operators and expressions appropriate for use with this kind of system. Special features of the language include

- creating and running processes,
- reading and writing to memory,
- executing, editing, and storing command files,
- storing file lists (subdirectories),
- acquiring free message and process blocks,
- inserting delays,
- querying parameters,
- GOTO branches,
- protecting and unprotecting memory,
- executing operator commands,
- intercepting messages,
- subroutines,
- terminal input/output,
- processing break commands, and
- accessing remote processors.

Each PIP module implements a datagram slave interface to NPM as well as a set of primitives for performing local operations whenever the NPM instructs it to do so. Reading and writing to memory, process interception, memory protection and unprotection, getting free memory blocks, releasing blocks, and running processes are examples of these primitives. Although a number of standard functions are included in the NPM tool, the NPM language lets users develop their own macro functions – for example, running an audit of internal memory usage.

Network Test System Driver

The network test system driver (NTS Driver) was devised in response to the very real need for a means of coordinating, controlling and automating distributed network testing operations. The NTS Driver serves a variety of purposes: for example, protocol development testing, protocol and network verification, and automatic regression testing. It can also be used to build host and terminal certification and conformance tests for any network protocol when combined with a suitable IPT, or to coordinate performance test operations and results analysis when combined with NLTS. The NTS Driver is a multilink test sequencer which can be used either as a programmable multilink test controller or an interactive test system (see Figure 6). It can be used on its own or in combination with other tools. It can coordinate the operation of other tools (such as IPT, NLTS and NPM) through asynchronous data links, as well as collect, display and check results.

Since NTS is programmable, users can elect to develop their own test cases instead of using those already provided. If desired, an entire testing operation can be automated across a distributed network. There are six features which make this possible. First, the interactive mode when used with the IPT allows network (DCE), host or terminal (DTE), or gateway to be tested. Second, development aids help users create test cases either interactively on-line or off-line. Third, a rich programming language facilitates a wide range of applications. Fourth, simple control commands and data structures make the task of running tests automatically a simple one. Fifth, there is a variety of report types for recording and tracking test results. Last, the database capability helps users manage thousands of test cases.

The NTS driver is independent of any particular protocol. It implements a user interface, programmable test sequencer, and multiple link controller, as well as a central controller for distributed tool interfaces. The driver connects to various test interfaces by means of asynchronous RS232 data lines.

Under program control the driver can send ASCII commands to an IPT which, in its turn, sends protocol data units to the system under test. The IPT decodes response protocol data units received from the system under test and returns the decoded ASCII equivalent command response to the driver for evaluation. Such simple command and response facilities in addition to the driver's multilink control capability, the powerful test language, and a special test case development aid make it easy to develop and manage network test suites involving thousands of interactions over many links and protocols. Since the IPTs are distributable, NTS can be used to test the distributed network. Test cases can be built in a similar fashion to control NLTS and NPM as well as line monitors, or to test equipment that can be controlled via standard data terminals.

The driver hardware is based on Digital Equipment Corporation's configurations. The same company's VAX VMS operating system software proved useful because of its ability to support multiple users on each processor. The VAX family of processors proved the best choice for three main reasons. First, it has the requisite data processing and and filing capacity. Second, it is available in many compatible configurations ranging from small single user to very large multi-user systems. Third, asynchronous terminal interfaces provide standard ASCII RS232 protocol, making them ideal for test control links. The test driver can therefore be used with any system capable of being controlled and observed through RS232 ASCII, character mode protocol.

BNR's NTS Driver software package consists of a test language interpreter, a user interface, a report writer, a switch, and an asynchronous interface (ASIM). After logging into the system the user can select from three optional user interfaces. The first – the NTS command line interpreter – permits simple command-response interactions at the user's terminal. The second provides a paged screen which separates report output from command responses. The third, still under development, acts as a session controller for users who are inexperienced or who only wish to run existing test suites, rather than develop new ones.

The specialized test language (TLAN) is an interpretive interactive language designed in BNR expressly to test data protocols. Because it provides such features as powerful data communication test sequence compilation, test cases may be written in TLAN for all known protocols and run independent of the system under test.

A line in a program written in TLAN can be of four types: comment, send data, receive data, or program statements. The last type, program statements, can be of one of the following types: assignments, conditionals, loops, arithmetic and logical operators, or predefined functions. These functions include string manipulation, ASIM programming, hex number base conversion, terminal output and input, fail and note reporting, timing, and test checking. Commands can either be written in a file and run, or entered line by line while interacting with the system under test.

User interface: A feature specially designed to assist test case development lets the user record interactions line by line in a log or journal file. The contents of this file can be run automatically at a later point. Being able to do this eases test case development greatly by making a machine, rather than a human being, do much of the tedious work. Time spent in debugging is also greatly reduced because less human intervention is required. Responses from the system under test need not be programmed in advance because they are recorded automatically while the test is running. An example following this section illustrates this.

Report writer: Another specialized module presents several attractive features. The report writer can accept data from the interpreter, switch, and ASIM modules; it can write up to fifteen reports simultaneously; it allows a user to select from various predefined report formats and specify the destination for each. For example, it is possible to specify a full report with trace, including interpreter commands, data exchanged with the ASIM, the result of each test case, and totals of test cases passed, failed, and omitted. A summary report would only have included results. If the EXCEPTION qualifier is specified, only failed test cases are reported for any of the report formats. Reports can be sent to terminals, printers, or files.

Switch: This module allows NTS to control up to fifteen ASIMs at the same time by acting as a selector between the interpreter and the ASIMs. It checks the ASIMs' level of activity by means of a timer.

Asynchronous interface: Each ASIM interfaces to one asynchronous ASCII RS232 character mode data line. The ASIM can be programmed to support various link speeds, parity, block size, buffer size, terminator sequences, as well as protocols for sending and receiving data. Automatic responses (specified by the user) to particular data sequences (also specified by the user) can be programmed into the ASIM. A number of timer features implemented by the ASIM check for line integrity.

The following example illustrates one of the ways the NTS Driver tool can be used in combination with other tools to perform network level testing.

Conformance testing with NTS

The unique capability of the network test system (NTS) to develop automatic test cases interactively improves test case productivity dramatically. All test configurations defined for OSI conformance testing (5) can be realized with NTS when combined with IPTs. The example shown here (Figure 7) illustrates how a typical network test case for the X.25 and ITI (X3/X38) protocols as implemented on SL-10 are developed. There are hundreds of such cases, each of which must be executed many times during the testing period.

First, the X.25 and ITI ports on the SL-10 are enabled (put into active service) by entering network operator commands at a terminal. The ITI operator makes a data call to the X.25 host port. An X.25 interactive protocol tester (IPT) receives the incoming call and the X.25 operator accepts it. The ITI operator can now send data and the X.25 operator can accept it. When the X.25 operator clears the call, the ITI operator observes the call clear signal. This is the minimum number of actions required to execute one simple test case manually. The sequence takes approximately ten minutes to complete. To carry out a full protocol check would have meant running hundreds of similar tests. There are more than 10 000 test cases for the entire SL-10 system, 3500 of which have to be executed for each major software release.

The obvious solution to such a time-consuming problem is to perform these test sequences automatically. NTS allows this to be done.

First NTS connects the operator to ITI and the X.25 IPT, forming a single switchable user interface. The operator can at this point turn on a report option (LOG report) to record all subsequent interactions (Figure 8). Program syntax is added automatically in comment form. The resultant file can be edited easily to produce a test case which can then be rerun automatically (Figure 9). This same case checks whether or not the system under test always responds in accordance with the original test case. Because the NTS operator need never enter responses during the manual test session (they are captured and added by NTS), the incidence of error is greatly reduced and the process speeded up. A test case that takes ten minutes to execute manually can now be run automatically in less than one minute. Hundreds of these tests can be run in a batch completely unattended. When the test run is complete, the user simply reviews the summary report produced automatically by the system.

Results

The NTS driver and IPTs proved successful in developing over 3500 automatic test cases, including tests on ten protocols. All these cases have proven to be very effective when testing new SL-10 software generics, as well as the new SL-10 RAPID product where using NTS resulted in productivity gains of 500 percent. Regression testing can be carried out much more thoroughly than before using NTS (8). Features for aiding test case development have contributed to the very high completion rate of up to 50 debugged test cases per work week. Test case development for NTS continues to advance and is quickly becoming the main test tool for the SL-10 development and verification teams.

Both NLTS and NPM proved useful during the development and evaluation phases of the SL-10 RAPID product. NLTS, so much easier to set up than previous tools, was used extensively to evaluate dynamic performance. The results were easy to interpret. Manufacturing uses NLTS to test the system hardware for each SL-10 RAPID before it is released. This tool has shown itself to be an easy-to-use, high capacity load and performance tester. NPM is also in high demand for debugging where its macro language, capable of running either on the SL-10 RAPID simulator or on the product itself, proved very effective.

Developing a comprehensive tool strategy has paid off and the base systems for a full range of test tool types have been firmly established. We have made many productivity gains while at the same time improving product quality. Network operating companies have only recently been introduced to this tool kit, but already their interest has been encouraging. We expect most of these tools to be in use in various operational environments in 1985.

Because the IPT, NLTS, and NPM tools can run on the SL-10 RAPID we have been able to add unique value to SL-10 networks. No supplier of packet switching systems, other than Northern Telecom, is known to have this kind of integrated capability. As a bonus, these tools are eminently suitable for testing other data products.

Adding new applications and enhancing existing capabilities as the need arises will be relatively simple. We look forward to feedback from users, both in the laboratory and in the field, to point out future directions for this exciting area.

References

[1] Weir, D.F., Proter, W.E., Dam, X.N., "X.25 Test Facilities on DATAPAC", ICCC, 1978.

[2] Melici, J.A., "The BX.25 Certification Facility", Data Communications, December 1982.

[3] Rayner, D., "A System for Testing Protocol Implementations", Data Communications, December 1982.

[4] Lee, C.H., Hornbeek, M.W.A., Naujokas, T., "A Network Test Environment For Packet Switching Networks", P.570, IEEC&E proceedings 1983.

[5] Rayner, D., "Conformance Testing With OSI", 2nd Int'l Conference on Introduction of OSI Standards, proceedings P.43, Ottawa, May 1984.

[6] Bunner, B., "Integrated Packet Network Access for Remote Users", ISS ' 84, 41B3, May 1984.

[7] ANSI Committee X3, Subcommittee X3S3, Task Group X3S35, "Data Communication Systems And Services User Oriented Performance Measurement Methods", draft Standard X3S35/135, June 1984.

[8] Lee, C.H., Hornbeek, M.W., Duong, C., "Network Software Engineering Capability And Its Delivery", ISS'84 Florence, proceedings 14C3, May 1984.

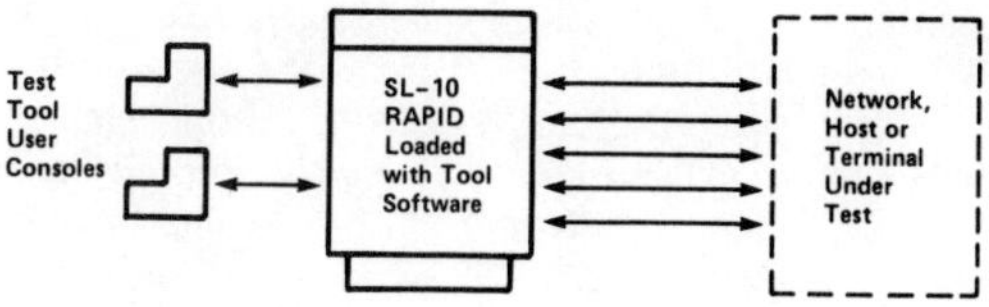

Figure 1 The SL-10 RAPID Based Test Tools

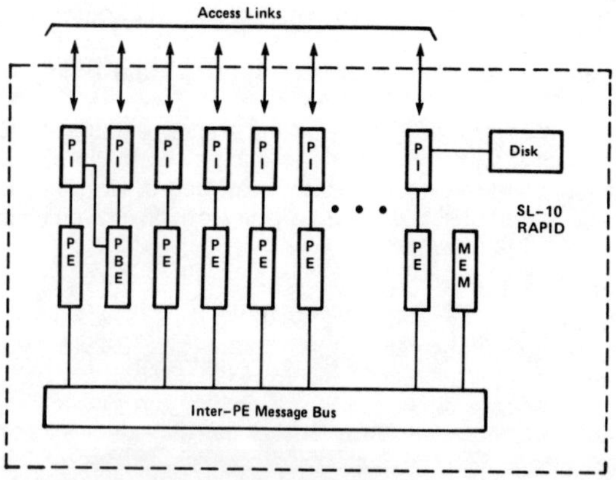

Key: PE = processor element (up to 14 per shelf) (one has to run the office and disk loader)
PI = peripheral interface (up to four per PE. PIs two to four each require a PBE)
MEM = memory
PBE = P-bus extender card

Figure 2 The SL-10 RAPID Internal Architecture

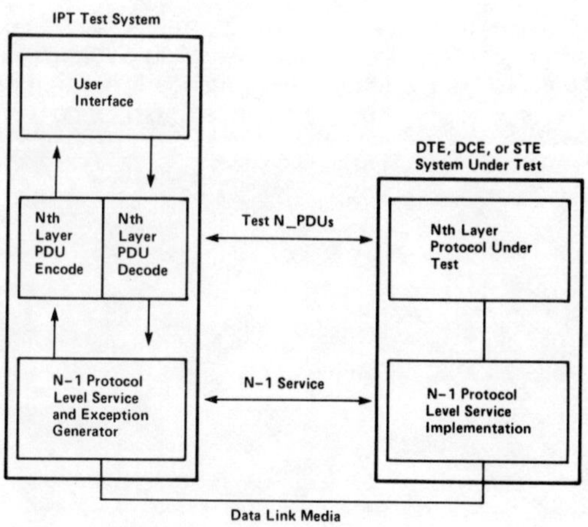

Key: IPT = interactive protocol tester
PDU = protocol data unit
DTE = data terminal equipment (host or terminal)
STE = signalling terminal equipment

Figure 3 The Interactive Protocol Tester (IPT)

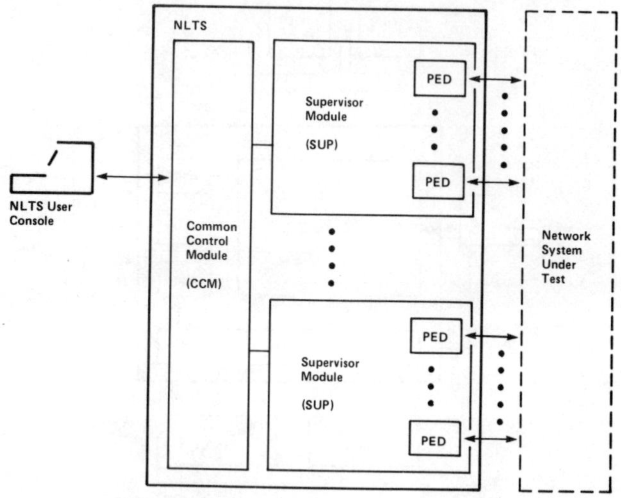

Key: PED = protocol encoder/decoder
PE = processor element
CCM = common control module
SUP = supervisor module
PI = peripheral interface

Figure 4 The Network Load Test System (NLTS)

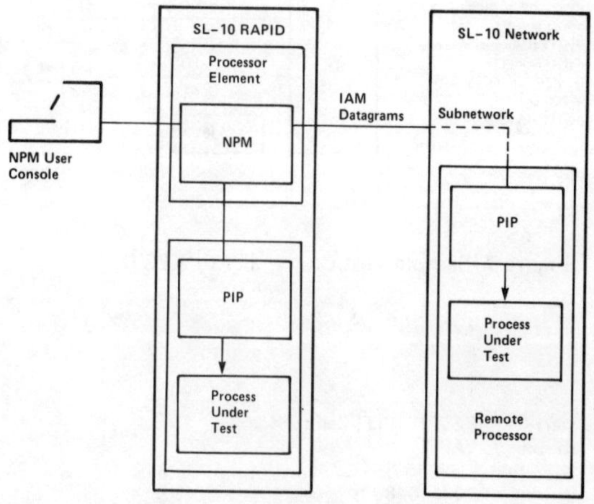

Figure 5 The Network Process Monitor (NPM)

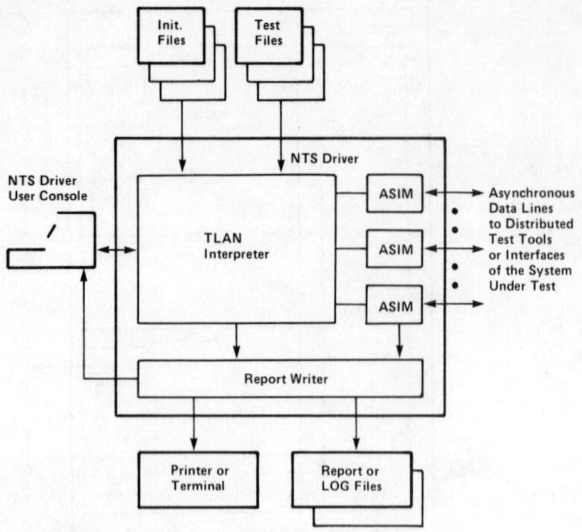

Figure 6 Network Test System Driver Internal Architecture

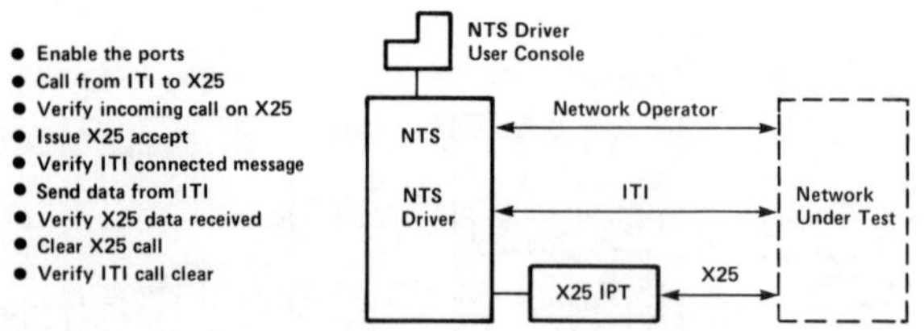

Figure 7 Sample Test Case — ITI to X25 Call

```
  start-test ("2.2.1", "ITI Call to X.25");
  iti-line = "A 5";
  X25-line = "B 48";
  iti-dna = "02110480";
  X25-dna = "02100050";
S (operator) EN 'iti-line'
S (operator) EN 'X25-line'
S (iti)      'X25-dna'
S (X25)      A,1001
S (iti)      VARIABLE LENGTH DATA STRING
S (X25)      CLEAR, 1001, 00
  end-test ( );
```

Figure 8 Lines Typed During Interactive NTS Session

A NEW METHOD FOR TOPOLOGICAL DESIGN IN LARGE, TRAFFIC LADEN PACKET SWITCHED NETWORKS

Kim-Joan Chen Jerrold F. Stach

Tsong-Ho Wu

UNINET, INC.
10951 Lakeview Avenue
Lenexa, Kansas 66219

ABSTRACT

The AN/1 network architecture employs compact LANs as integrated switching nodes. The multibussed nodal architecture imposes a natural hierarchy to the network interconnection problem. A nodal interconnection method is introduced as a generalization of hierarchical topology design techniques. Based on traffic and distance, a flat nodal topology is decomposed into several level, overlapped subnetworks. Each subordinate subnetwork communicates with its ordinate subnetwork via two formal gates. The traffic matrix for each subnetwork can be obtained. Linear programming techniques are used to determine bandwidth requirements. Perturbation procedures are applied to determine the appropriate groupings of LANs into subnetworks and the appropriate hierarchy. A flat topology may be returned if it is optimal.

1. INTRODUCTION

Research in the design of packet network systems has been devoted to the development of efficient methods and operational procedures. Treatments are generally limited to small or moderate sized topologies with non-congestive traffic loads. As the size of the topology and traffic load increase, computational complexity precludes the use of traditional, iterative algorithms which treat the topology as a flat gragh.

For networks with a large number of switching elements, present procedures fail because of the computational complexity of the suboptimization. Kleinrock and Karmoun have shown that hierarchical design can reduce this complexity by employing flat gragh treatments to each hierarchical level [12].

AN/1 is an experimental data network architecture employing diverse types and speeds of transmission media, to connect switching sites which are formal Local Area Networks based on a modified ETHERNET CSMA/CD protocol.

Asynchronous multiprocessor clusters are attached to a 30 feet, 10 Mbps ETHER via one or more processor modules. Processor modules and/or line modules are tightly coupled into Backbone, Access Network or Application clusters via an 8 Mbps multimaster bus. Clusters common to the ETHER constitute a node. Nodes locally connected via an Internodal Gateway define a swithing site.

This paper considers the efficient interconnection of LAN's. The configuration of the LAN's can be found in [13]. The interconnection of the LAN's will proceed from the following procedures which are a generalization of hierarchical topology design. The design problem can be desribed as follows :

Given :	The switching sites (LAN's) Channel capacity options and costs
Minimize :	The line cost
Subject to:	Delay constraint, reliability constraint, Inter-switching site traffic requirements

Hierarchical structures have been proposed [2,6,12] to substantially reduce the computational complexity and the storage requirement involved in the optimization. Based upon the AN/1 architecture and the LAN's natural hierarchy, these techniques have been extended in this paper. The system architecture is introduced in section 2.

In section 3, we proceed a top-down procedure which satisfies the AN/1 LAN configuration. Using the linear function of traffic and distance, a flat nodal topology is decomposed into several level, overlapped subnetworks. Each subordinate subnetwork communicates with its ordinate subnetwork via two formal gates. The traffic matrix for each subnetwork can be obtained from the distribution of traffic between the disjoint subnetworks. The decomposition should consequently lead to the design of the interconnection of the LAN's within each subnetwork for which the total flow and total bandwidth required can be obtained from present design strategies.

In section 4, linear programming techniques are used to determine bandwidth requirements for the interconnection of subnetworks based upon link delay goals. Optimized perturbation procedures are applied after the class of topology algorithm determines the appropriate groupings of LAN's into subnetworks and the appropriate hierarchy. A flat topology may be returned if it is optimal. Optimal search techniques are employed to achieve grouping and levels.

The numerical results show that the algorithm can reduce design cost and improve the computational complexity.

2. SYSTEM ARCHITECTURE

Figure 1 depicts the AN/1 nodal architecture. In the AN/1 network, the cluster is the basic building block of the AN/1 switching node. Each cluster consists of combinations of several modules. A number of clusters can be bussed together on a single Ether-like bus to form a node, which provides full interconnectivity among all the clusters on the node.

A number of AN/1 nodes can be combined together to form an AN/1 switching site. An AN/1 switching site is a Local Area network which is a (n+2) level subnetwork, providing the interface between the local access network and the distributed long-haul system. Given the locations of the AN/1 switching sites and an assignment of DTE to the sites, the total network design problem is decomposed into separate design problems for the local access network and for the backbone network.

A number of the switching sites can be treated as a (n+1) level subnetwork. Each (n+1) level subnetwork has two switching sites as gates that provide the interface between the (n+1) level subnets and the n-th level subnet. The collection of (n+1) level gates forms the n level subnet which also provides the interface between two different level subnets. Two (n+1) level subnets are adjacent, if they share at least one gate. Otherwise they are disjoint. The traffic between two different (n+1) level subnets will pass through the n level subnet. Figure 2 depicts an example of a 3-level topology.

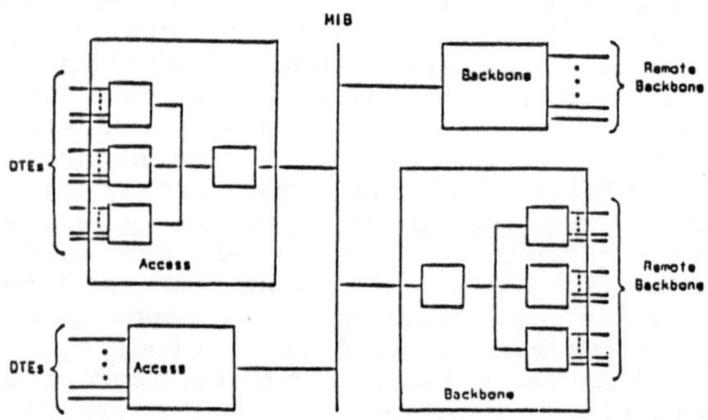

Figure 1. AN/1 NODAL ARCHITECTURE

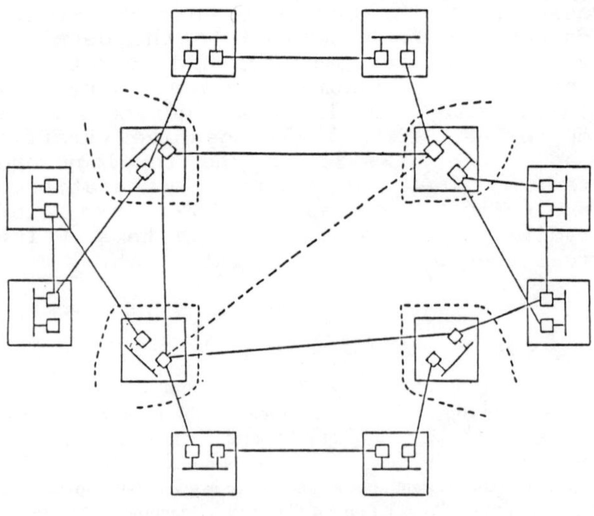

Figure 2. AN/1 ARCHITECTURE

3. DESIGN METHODOLOGY

The main idea behind this design is to impose a decomposable structure on the design problem which results in a set of tractable subproblems, that contribute delay and reliability constrained by network wide goal. Using the formulation of the traditional topological design problem [14] as a basis, this algorithm determines whether the topology should be flat or hierarchical. For a hierarchical topology, the correct number and location of the switching hierarchies is returned.

The traditional delay constraint in topological modeling provides a default bound on the number of levels induced into the switching hierarchy. In our design, three level structures are used. The first level subnetwork will be formed by the gates. Several LAN's will be grouped into the second level subnetwork, and the LAN itself is the third level subnetwork. In this paper, we emphasize the inter-connection of LAN's. Configuration of the third level subnet (level n+2) can be found in [13]. The treatment of three levels is artificial.

Several natural grouping and level techniques have been proposed in literatures [12] to create a hierarchical topology. These techniques use a nearness measure which take into account traffic, cost and delay. We impose the linear function of traffic and distance for each switching site such that the grouping and level can be further optimized by the search.

3.1 OVERLAPPED SUBNETWORK DEFINITION

Based on a linear function, the weighting factor for each switching site can be set, which will determine the gates for all subnetworks at all levels. The assumption underlying the overlapped subnetwork is as follows :

1. The switching site whose weighting is greater than a bound will be assigned to the first level.
2. The switching site, not in the first level, will pick two first level switching sites as gates based on the weight for the switching site to all the first level switching sites.
3. Any switching site, not in the first level, that shares the same gates, will be considered in the same group.

I. ASSIGNING LANs TO THE FIRST LEVEL SUBNETWORK (Level n)

(a) Let $X = \{ X_j \mid X_j$ means j-th switching site$\}$

be the set of all the switching sites, and let $A = (a_{ij})$ be the traffic matrix;

a_{ij} is the traffic from X_i to X_j. Let $D = (d_{ij})$ be the distance matrix; d_{ij} is the distance in latitude and longitude coordinate from X_i to X_j.

(b) Let $R = (r_{ij})$ be the weight matrix;

$r_{ii} = 0$ and $r_{ij} = f_{ij}(a_{ij}, d_{ij})$, for $i \neq j$.

$$\text{Let } f(a_{ij}, d_{ij}) = w\, a_{ij} + (1 - w) \frac{CT}{|d - d_{ij}|},$$

where w is the parameter to decide the weights for the traffic load and the distance, and the proper value of w will be obtained by search, where

$$d = \frac{\sum_{i=1}^{n} \sum_{j=1}^{j \neq i} d_{ij}}{n(n-1)}$$

, and CT is a constant which normalizes the traffic load and the distance to the same units.

(c) Let $R_i = \sum r_{ij} + r_{ji}$ be the weighting factor;

R_i is the factor to determine which X_i (i-th switching site) is in the first level.

(d) Let RBOUND be the lower bound of the weighting factor to be qualified in the first level. The proper value of RBOUND will be determined by the search.

(e) Let $B = \{ X_i$, while $R_i \geq$ RBOUND$\}$, and B forms the first level subnet.

II. ASSIGNING LANs TO THE SECOND LEVEL SUBNET (Level n+1)

(a) From the previous assignment, we have the set $B = \{ X_i$, while $R_i \geq RBOUND \}$.

Let I be the index set such that if i belongs to I, then X_i belongs to B. Let I' be the index set which consists of $\{ i' \mid X_{i'} \notin B \}$.

(b) Let $e_{ij} = g_{ij}(a_{ij}, a_{ji}, d_{ij})$ for $i \in I$ and $j \in I'$, where $g_{ij}(a_{ij}, a_{ji}, d_{ij}) =$

$$v(a_{ij} + a_{ji}) + (1 - v)\frac{CTT}{d_{ij}},$$

where v is the parameter to decide the weights for the traffic and the distance, and the proper value of v will be obtained by search. CTT is a constant which normalizes the traffic load and the distance to the same units. CTT can be defined as the ratio of units.

(c) For $j \in I'$, find k and m such that e_{kj} and e_{mj} are the first two maximum values of $\{ e_{ij}, i \in I \}$

(d) Determine X_k, X_m and X_j, then X_j, X_k and X_m belong to the same group.

(e) Let $G_{(k,m)}$ be the second level subnet which consists of $U_j = \{ X_j$, X_k and X_m belong to the same group, for $j \in I' \}$.

Once the overlapped subnetworks are defined and the gates selected, then the subnetwork relationships are established.

The following defintions are used to represent this property:

3.2 SUBNETWORK ADJACENCY DEFINITION

Any two subnetworks at the same level will have the following characteristics.

(a) Two subnetworks at the same level that share one switching site as their gate are defined to be adjacent.

(b) Two subnetworks at the same level that have no internetwork traffic are defined to be disjoint.

(c) The intersection between the first level and second level subnetworks are the gates.

From the above definition, the following assumptions are made:

1. The flow between two different second level subnets will pass through the first level switching sites.

2. The traffic between two switching sites in different subnets will be assigned to the shortest length path that passes all the necessary gates.

3.3 SUBNETWORK TRAFFIC ASSIGNMENT

The subnetwork traffic, therefore, can be obtained from the distribution of traffic between the disjoint subnetworks.

(a) Let I_i be the index set which consists of $\{ j \mid X_j \in S_i \}$. Let X_{k_i} and X_{m_i} be the intersection of the first level subnet and the i-th second level subnet.

(b) Let $Gate(i) = \{ k_i, m_i \}$ be the subset of I_i, and let a^i_{jp} be the traffic matrix for the i-th second level subnet, and let A_{km} be the traffic matrix for the first level subnet ;

(c) Let $a1_{(ij)}$, $a2_{(km)}$ and $a3_{(np)}$ be the transit traffics;

The traffic matrix for each subnet can be computed as follows:

(i) if i and j belong to the same second level subnet, G_p, then $a^p_{ij} = a_{ij}$;

(ii) If i and j belong to the first level subnet, S, then we have $A_{ij} = a_{ij}$;

(iii) If $i, j \in I'$ and i and j do not belong to the same second level subnet, then let $i \in I_p$ and $j \in I_q$. Determine p_i and q_j such that $[\, i, p_i, q_j, j \,]$ is the shortest path from i to j passing Gate(p) and Gate(q), where $p_i \in$ Gate (p), $q_j \in$ Gate(q) .

Let $a1_{i\,Gate(p)} = a_{ij}$ and $a^{p}_{ip_i} = a^{p}_{ip_i} + a1_{i\,Gate(p)}$;

$a3_{Gate(p)\,Gate(q)} = a_{ij}$ and $A_{p_i q_j} = A_{p_i q_j} + a3_{Gate(p)\,Gate(q)}$;

$a2_{Gate(q)\,j} = a_{ij}$ and $a^{q}_{q_j j} = a^{q}_{q_j j} + a2_{Gate(q)\,j}$;

(iv) If $i \in I$, $j \in I'$ and i and j do not belong to the same second level subnet, then let $j \in I_q$. Determine q_j such that $[\, i, q_j, j \,]$ is the shortest path from i to j passing Gate(q) where $q_j \in$ Gate(q) ;

Let $a3_{i\,Gate(q)} = a_{ij}$ and $A_{i q_j} = A_{i q_j} + a3_{i\,Gate(q)}$;

$a2_{Gate(q)\,j} = a_{ij}$ and $a^{q}_{q_j j} = a^{q}_{q_j j} + a2_{Gate(q)\,j}$;

(v) If $i \in I'$, $j \in I$ and i and j do not belong to the same second level subnet, then let $i \in I_p$. Determine p_i such that $[\, i, p_i, j \,]$ is the shortest path from i to j passing Gate(p), where $p_i \in$ Gate(p) ;

Let $a_{Gate(p)\,j} = a_{ij}$ and $A_{p_i j} = A_{p_i j} + a_{Gate(p)\,j}$;

$a_{i\,Gate(p)} = a_{ij}$ and $a^{p}_{i p_i} = a^{p}_{i p_i} + a_{i\,Gate(p)}$;

4. CONSTRAINTED OPTIMAL GROUPING AND LEVELS

Once the overlapped subnetworks are created and the traffic matrices computed, traditional network design techniques may be used to design the overlapped subnetworks independently.

The link capacity assignment for each subnetwork should satisfy the delay goal between two connected LAN's. Perturbation will be applied after the overlapped subnetworks are optimized.

The direct application of the decompositon techniques may lead to various subnetwork sizes. Kleinrock and Kamoun have introduced some techniques to optimize flat structures [12], hence the parameters which decide the grouping can be obtained. In this paper, we employ the binary search technique to determine grouping and levels. A flat topology may be returned if it is optimal.

4.1 LINK CAPACITY COMPUTATION

The interconnection of the LAN's within each subnetwork can be determined from the subnetwork design. The link capacity assignment will be based on the link delay goal for the interconnection of the LAN's which can be provided from the routing procedure. It is then important to determine the bandwidth requirements which minimize the line cost and also satisfy the delay goal.

To achieve the link capacity assignment for the interconnection of the LAN's, we define exclusive addition as follows :

$$\frac{b}{a} \oplus \frac{c}{d} = \frac{b+c}{a+d}$$

Let the link delay goal for the interconnection of the LAN's within the subnetwork be $F_i / (C_i - F_i)$, where the F_i and C_i are the link flow and the link capacity. Let the set of channels and their expected flows be c_{ij}, f_{ij}. It is easy to see that the exclusive addition of the set of channel delays is exactly equal to the link delay. Moreover if every channel delay is less than a predetermined value, then the the exclusive addition is also less than this value. Hence the linear programming techniques can be applied to determine bandwidth requirements for interconnection of LAN's. It can be described as follows:

Minimize the line cost (linear cost) subject to

$$\sum_{j=1}^{k} c_{ij} = C_i$$

$$\sum_{j=1}^{k} f_{ij} = F_i$$

$$f_{ij} < c_{ij} \quad , j = 1,2,\ldots,k$$

$$f_{ij} \lesssim T\,(c_{ij} - f_{ij}), \; j = 1,2,\ldots,k$$

$$c_{ij} \leq \text{channel capacity}$$

A class of topology for the overlapped subnetwork may be obtained from the subnetwork design. The combination of the overlapped subnetwork will form the topology for the network, which satisfies the delay and reliability constraint. Further perturbation is applied to optimize the flow between the n level and (n+1) level.

4.2 PERTURBATION

Once the subnetworks are defined and the traffic matrices obtained, the topology for each subnet can be obtained. The perturbation procedure is applied after the overlapped subnetworks are optimized, from which the topology for the network will be obtained.

(a) Let S be the optimized first level subnet, and let $S \cap S_i = \{ X_{k_i} , X_{m_i} \}$, where S_i are the second level subnets. From the flow between X_{k_i} and X_{m_i}, and the traffic matrix for each second level subnet, the topology for each second level subnet can be obtained.

(b) The topologies for the second level subnets generate the new flow between X_{k_i} and X_{m_i}. Replace the old flow by the new flow in the first level subnet.

(c) If the difference between the old flow and the new flow is less than the pre-determined value, stop thealgorithm.

(d) Perturb the first level subnet to get the new topology.

(e) Perturb the second level subnets with the new flow obtained from (d) and substract from the second level subnet. Go to (b).

4.3 GROUPING AND LEVEL SEARCH

With the different parameter values for the linear function, the grouping and levels may be different. Suitable values may be provided by some literatures which are based on the weighted graph technique. The search technique we use is the binary search. It is based upon the property of the quadratic form which we assume the optimal values of the parameters lie between two end values that converge to the optimal value.

Therefore the parameter values for the linear function may be found from the search. The RBOUND parameter constrains the level of hiearchy. If RBOUND is very large, then the structure will be flat. Finally, a flat topolgy may be obtained if it is optimal.

5. SUMMARY

In this paper we studied the major aspects related to the design of a large packet switched network. The focus was primarily on the determination of an overlapped structure. The algorithms determine whether the topology should be flat or hierarchical. For a hierarchical topology, the correct number and location of the switching hierarchies is returned.

The general problem of interconnection of LAN's can be transformed to a hierarchical design problem. Based on the AN/1 delay and reliability constraints, the one gate hierarchical topology design was generalized. Three level structures were studied in this paper. A general formulation may require further research.

REFERENCES

[1] R. Boorstyn, and H. Frank, "Large-Scale Network Topological Optimization," IEEE Trans. Commun., 25(1), 29-47(1977).

[2] G.H. Dysart, and N.D. Georganas, "NEWCLUST: An Algorithm for the Topological Design of Two-Level Multidrop Teleprocessing Networks," IEEE Trans. Commun., 26(1), 55-62(1978).

[3] S. Even, " An Alogrithm for Determining Whether the Connectivity of a Graph is at Least k," SIAM J. Comput., 4, 393-396(1975).

[4] H. Frank, I.T. Frisch, and W. Chou, "Topological Considerations in the Design of the ARPA Computer Network," AFIPS Nat. Comput. Conf. Exp. Conf. Proc., 36, 551-567(1970).

[5] H. Frank, and W. Chou, "Topological Optimization of Computer Networks," Proc. IEEE, 60(11), 1385-1397(1972).

[6] H. Frank, M. Gerla, and W. Chou, "Issues in the Design of Large Distributed Computer Communication Networks," Proc. of the Nat. Telecom. Conf. Atlanta, GA, November 1973, IEEE, New York, pp. 37A-1-37A-8.

[7] M. Gerla, H. Frank, and J. Eckl, "A Cut Saturation Algorithm for Topological Design of Packet Switched Communication Networks," Proc. NTC, 1074-1085(1974).

[8] M. Gerla, and L. Kleinrock, "Topological Design of Distributed Computer Networks," IEEE Trans. Commun., 25, 48-60(1977).

[9] M. Hoshi, "Local Network Area Size Optimization," IEEE Trans. Commun., 33(3), 199-202(1985).

[10] B.W. Kernighan, and S. Lin, " An Efficient Heuristic Procedure for Partitioning Graphs," Bell Syst. Tech. J., 49, 291-307(1970).

[11] A. Kershenbaum, and R.R. Boorstn, " Centralized Teleprocessing Network Design," Networks, 13, 279-293(1983).

[12] L. Kleinrock, and F. Kamoun, "Optimal Clustering Structures for Hierarchical Topological Design of Large Computer Networks," Networks, 10, 221-248(1980).

[13] T. Wu, J. Stach, and K. Chen, "Constrained Configuration of Local Area Networks for the AN/1 Switching Architecture", to be presented at the 10th Local Computer Network Conference, Minneapolis, October 1985.

[14] A.S. Tanenbaum, Computer Networks, Prentice-Hall, Inc., New Jersey, 1981.

Session 5

Prototype Networks

Chairperson

I. Rubin

UCLA

Design of an Integrated Services *Packet* Network

Jonathan S. Turner

Washington University
St. Louis, Missouri 63130
jst@wucs.UUCP

ABSTRACT

The Integrated Services Digital Network (ISDN) has been proposed as a way of providing integrated voice and data communications services on a universal or near-universal basis. In this paper, I argue that the evolutionary approach inherent in current ISDN proposals is unlikely to provide an effective long term solution and advocate a more revolutionary approach, based on the use of advanced packet switching technology. The bulk of this paper is devoted to a detailed description of an Integrated Services *Packet* Network (ISPN), which I offer as an alternative to current ISDN proposals.

Introduction

An integrated voice and data packet communications system has several advantages over existing methods.

- It uses a common set of switching and transmission facilities for both voice and data communication. This is less costly than current systems that use separate mechanisms.
- It allows voice communication to be done using less than 25 percent of the bandwidth currently needed, without sacrificing quality. This allows major savings in long distance transmission costs. It also allows customers to carry on two or three simultaneous voice conversations along with a substantial amount of data traffic over a standard copper loop.
- It provides much higher performance data communication and at lower cost than current systems. This is largely due to the integration of data communication with voice, which allows one to take advantage of the economies of scale possible in the large systems needed for a national telephone network.

The system is based on high performance packet switches which are large and fast enough to effectively support both voice and data communication on a large scale. Each packet switch can have a raw throughput of up to 1.5 Gbs, allowing it to support as many as 50,000 simultaneous voice conversations using a 32 Kbs voice encoding scheme. The one-way cross-network delay in a worst-case connection in a national network in the U.S. can be limited to about 150 ms. The key elements of the design are

- The use of high speed digital transmission facilities (1.5 Mbs) with excellent error performance.
- Simple link level protocols. In particular, there is no flow control or error correction done at the link level, which eliminates the need for state information in the link level protocol processors.
- A predominantly connection-oriented service. This allows the routing of most packets to be handled by a very simple method and facilitates bandwidth allocation and overload control. A connectionless (datagram) service can also be supported.
- Hardware implementation of basic switching and protocol functions. Switching is done using a large self-routing network containing roughly 1300 custom VLSI chips. All per packet protocol functions are handled by protocol processors (one for each link) consisting of one custom controller chip plus one large memory chip.
- Implementation of higher level protocol functions (including error correction and flow control) on an end-to-end and application-dependent basis.

The governing philosophy behind the design is that the communications network should provide transport of information at the highest possible level of performance, but nothing else. The network achieves generality, not by providing every service a user might conceivably require, but by providing only those services that every user requires. The system can be implemented with currently available technology at a cost that is comparable to that of conventional telephone switching systems.

The Trouble With ISDN

The Integrated Services Digital Network has been heralded as the mechanism that will usher in the Information Age. ISDN, it is said, will facilitate the development of new communications services, including a wide range of data services and maybe even video. It will allow such services to be implemented on a large scale at a cost most customers can afford, and spur the transformation to a 'post-industrial society.'

While I share the long range goals of ISDN proponents, I don't believe that current ISDN plans can take us very far. The fundamental problem is that current plans (see [1], for example) implicitly assume a network model based on circuit switched voice and a combination of circuit and packet switched data. This assumption is evident in current standardization efforts which focus on a transmission format for digital subscriber lines that provides two 64 Kbs circuit switched channels plus a 16 Kbs packet switched channel. While this plan can provide a limited data communications capability (it's certainly a vast improvement over the current situation), it is too inflexible to satisfy long term needs. What happens, for example to an application that requires a 75 Kbs channel? Do we implement it using the two 64 Kbs channels or one of them and the packet switched channel. Neither option is particularly attractive.

The source of the trouble is the reliance on circuit switching, which requires that the available bandwidth be divided up into fixed size channels. The channel size is a permanent feature which can't be changed easily, if at all. There's no reason to think that 64 Kbs is a particularly useful channel size. The only reason for picking it is that current telephone switching systems are based on that size. In fact, the driving force behind current ISDN plans is to preserve the investment in existing equipment while 'evolving' to a more flexible network.

Unfortunately, it won't work, because it's inherently a hybrid approach. The only thing integrated about ISDN is its name. Two (or more) switching networks are required to support ISDN as it's currently envisioned. Manufacturers may talk about their integrated architectures for ISDN, but what they mean is that they plan to put a packet switch and a circuit switch in the same box and call it an integrated system. This is not their fault—what else can they do? Packet switching and circuit switching are very different communications methods. They require different switching and transmission facilities and all the proposed schemes for combining them are little more than packaging.

What to do then? Is a hybrid network really necessary or is there an integrated solution that can satisfy the needs of both voice and data and remain flexible enough to meet satisfy new requirments as they arise? I claim there is such a solution, but it requires abandoning circuit switching and moving to new network designs based on packet switching.

Why Packet Switching?

Here are three reasons that make packet switching an attractive method for providing integrated voice and data services.

- *Adaptability to changing traffic.* Packet switching naturally provides the user with exactly the bandwidth required. As new services are developed with different bandwidth requirements, packet switching systems can adapt to the changing conditions easily. Circuit switching systems cannot.
- *Integrated internal architecture.* As outlined above, current ISDN plans require separate switching networks for different types of information. Packet switching can provide both an integrated customer interface and a single network solution for a wide range of communications needs, leading to substantial cost savings in switching systems and system administration.
- *Transmission efficiency.* Many data services are characterized by bursty communications patterns which make poor use of conventional circuit switched facilities. For example, interactive data users typically use only a few percent of the bandwidth available to them. Although it is less widely recognized, voice is also bursty. In the average telephone conversation, less than 40% of the available bandwidth is actually used. Packet switching can exploit this burstiness to double the number of conversations that can share a single transmission facility.

Given all these advantages, why has packet switching not been used extensively for voice? To answer this, we must take a closer look at the technical specifications and costs of conventional circuit switches and packet switches and see how they compare.

Circuit switching has been the technology of choice in the telephone network, since its origin about one hundred years ago. During that period, circuit switches have evolved from manually operated switchboards to small but automatic step-by-step switches, to larger panel switches to the computer-controlled electronic switching systems that currently dominate the scene. The current systems are very large—local switches can provide service to over 100,000 customers, large toll switches support as many as 50,000 simultaneous voice calls corresponding to a raw bandwidth of 6 gigabits per second (Gbs). The network as a whole is also very large. There are approximately 10^8 telephones in the United States at present and over 10,000 local switching offices. The performance of the switching systems is quite impressive—information passing through a modern digital switch is typically delayed no more than a few milliseconds, and new connections can be established in a fraction of a second. Equally impressive is the cost—while there is considerable variation, per line equipment costs for modern digital telephone systems is typically in the $100–200 range.

Just as circuit switching has long dominated the telephone network, so has packet switching dominated the data communications scene, principally because of the advantages cited earlier. Packet switching is exemplified by the Arpanet, which was the first major example of a large scale data communications network. In the Arpanet, the endpoints of the communication are typically large timeshared computers, called hosts. Communication is provided by packet switches, each of which typically connects to a few hosts and several other packet switches. There are currently about 300 hosts in the Arpanet and 100 packet switches. The packet switches are implemented using general purpose computers (usually minicomputers), although more recent versions have been supplemented with frontend communications processors to reduce the load on the main processor. The transmission facilities used by the Arpanet include low speed modem connections (1–10 Kbs) and higher speed digital facilities (56 Kbs). The throughput of the packet switches is generally under 1 Mbs and delays can be substantial (50–100 ms per switch). The cost of packet switching as provided by the Arpanet is quite high, since each packet switch supports a small number of hosts. Commercial data networks do better, but there remains a large gap between per-host costs in data networks and per-line costs in the telephone network.

This comparison explains the conventional wisdom that packet switching is poorly suited to the needs of telephony and the resulting conclusion that an ISDN implementation must include a circuit switching component to support voice. The fallacy in the conventional wisdom is that the disadvantages attributed to packet switching by this comparison are not due to any inherent properties, but are side-effects of the conventional implementations. The requirements and design constraints that shaped the development of the Arpanet and commercial data networks were completely different from those that shaped the telephone network. The scale, the performance requirements and the cost sensitivities are all very different. It's because the *needs* differ that the resulting systems differ so in their technical specifications. In the remainder of this paper, I will attempt to correct the widely held, but erroneous view that packet switching is poorly suited to the needs of voice by describing a system capable of supporting voice and data communication on a large scale and at a cost that is competitive with conventional telephone systems.

ISPN Architecture

Let's begin by deciding what general properties a packet switching system must have if it is to be a suitable vehicle for providing voice and data communication on a large scale. First and most obviously, it must be big—comparable in raw bandwidth to conventional telephone systems. Second, it must be fast—long end-to-end delays are annoying in voice connections and hence unacceptable. While opinions differ on the exact amount of delay that can be tolerated, most experts would agree that 100 ms is acceptable, while more than 500 ms is not. Third, it must be inexpensive. Any system that seeks to replace circuit switching must be able to provide voice services at a competitive cost. It is not enough to offer a better product at a higher cost—for a system to succeed on a large scale, it cannot significantly increase costs for basic services.

How can we achieve the requisite scale, performance and economy in a packet switching network? Well, the telephone network is one place to look for the answer. One of the first things one notices is that the telephone network makes extensive use of high bandwidth digital transmission facilities. Modern digital switching systems are designed to interface directly to 1.5 Mbs transmission facilities carrying 24 voice channels in a 64 Kbs format. Interfacing costs are cheap, since they are shared by the 24 channels and the bit error rates are excellent. Conventional packet switching systems, on the other hand interface to much smaller channels, and must be prepared to cope with the much higher error rates present in modem connections. A second thing one notices about the telephone network is the amount of special purpose hardware used to provide the switching function. While telephone systems contain general purpose computers, their function is primarily connection establishment. The computers don't move the bits. That function is handled by large specialized networks. In contrast, most conventional packet switches include a general purpose computer that must do some processing on every packet and can quickly become a bottleneck.

These observations suggest that a high performance packet switch should (1) interface directly to high speed digital transmission facilities and (2) use special purpose hardware to perform all per packet processing. Now, those familiar with link level protocols used in conventional packet networks may recognize this as a challenging task. Typical protocols are quite complex, requiring extensive state information in the protocol processors and complicated error recovery procedures. This complexity almost demands a programmable processor (a hardware implementation would never be completely debugged), but a microprocessor-based implementation is unlikely to be fast enough to keep up with a 1.5 Mbs link, and also may be too expensive. Fortunately, there is a way out—simplify the protocol. Current packet switching protocols were designed for hostile environments—the low speeds and high error rates typical of modem connections. In a network composed entirely of high speed digital facilities, much of the complexity of typical protocols can be eliminated from the lower protocol layers. In particular, error correction and flow control can be taken out of the link layer and provided on an end-to-end basis rather than a link-by-link basis. This eliminates the need for extensive state information at the ends of each link and the synchronization and recovery procedures required to maintain it and in turn makes possible the construction of inexpensive, high performance protocol processors. In addition, if these functions are provided on an end-to-end basis, they can be pro-

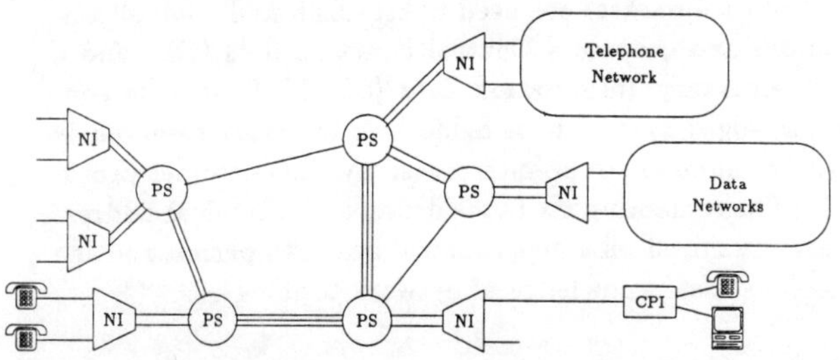

Figure 1: Network Architecture

vided selectively. Hence, delay-sensitive applications like voice can avoid the performance penalites they can cause.

Returning to the question of scale, how large must our packet switches be? In order to compete with conventional telephone switches, they should be capable of supporting at least 50,000 simultaneous voice conversations. How many conversations can a 1.5 Mbs link carry? When operated in a circuit switching mode using the 64 Kbs digital encoding method currently employed, the answer is 24. When operated in a packet switching mode that number jumps to about 50. One can obtain another factor of two improvement by using newer voice encoding methods. 32 Kbs adaptive differential pulse code modulation (ADPCM) is a good choice, since it provides comparable quality to the method currently used, but requires only half the bandwidth. This leads to a figure of 100 voice channels per 1.5 Mbs link, implying that our packet switch must terminate 1000 full-duplex links if it is to support 50,000 simultaneous voice conversations. This in turn, means that our switch must include some mechanism capable of receiving over two million packets per second, and sending each one out on the appropriate outgoing link.

Network Overview

Based on the discussion in the previous section, we can begin to develop an architecture for the ISPN. The major components are shown in Figure 1. The Packet Switches (PS) each terminate up to 1000 High Speed Links (HSL). Using 32 Kbs ADPCM coding for voice, they can support over 50,000 simultaneous voice conversations. Residential customers connect to the network over Medium Speed Links (MSL), which operate at 100 Kbs. High speed access can be provided to business customers. The Customer Premises Interface (CPI) can take a variety of forms depending on the kind of service required. At the low end of the spectrum would be a simple controller providing service for a single telephone and implemented using a single chip eight bit microcomputer. Customers wanting several phones and data communication would require a more complex controller. Businesses would typically have an interface to a Private Branch Exchange (PBX) or Local Area Network (LAN). The Network Interfaces (NI) provide concentration, accounting data collection and network protec-

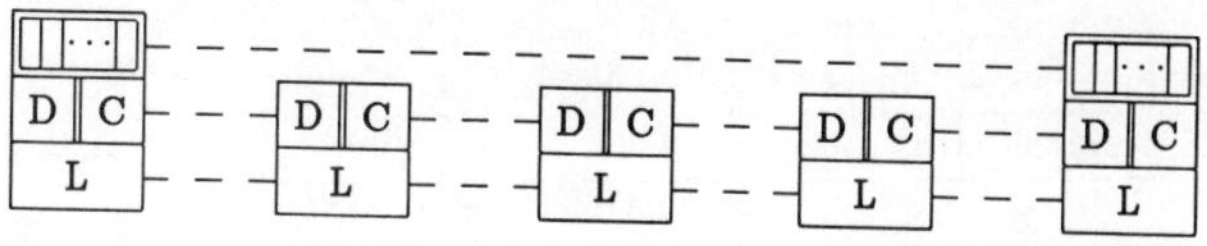

Figure 2: Protocol Structure

tion. A configuration designed for residential customers would support about 500 customers and would be connected to a local packet switch by four HSLs. Thus, one PS could have as many as 125 NIs and support over 60,000 customers.

NIs can also be designed to provide a conventional line interface as supported by current telephone systems. Even higher concentration ratios can be supported in this case—up to 2000 customers could be supported on a single NI connected to its host PS by four HSLs. While this configuration doesn't exploit the advanced capabilities of the system, it does provide a mechanism for easing the transition from a circuit-oriented network to a packet-oriented one. In a similar fashion, NIs can be designed to interface to the current telephone network. In this case, the NI would interface directly to up to 960 digital trunks.

The network provides two communications services.

- *Point-to-point channels.* These are two-way channels joining pairs of customers. The customer establishes a channel by sending a connection request message to the network specifying the destination and the average bandwidth needed. If the network accepts the connection, it in effect guarantees that the customer can expect to have the requested bandwidth available. The network may refuse to accept the connection if there are not adequate resources available. The connections do not provide perfectly reliable information transport. In particular, the network does not provide mechanisms for error correction and flow control. The network can provide connections at any speed up to 1.5 Mbs (although larger connections are more likely to be blocked). The bandwidth requirement may be asymmetric—that is, it may depend on the direction of transmission. No distinction is made between voice and data connections. A voice connection is simply one with an average bandwidth of about 12 Kbs.
- *Datagrams.* These are individually addressed packets, not associated with a pre-established connection. The network makes an effort to deliver them, but does not guarantee delivery.

The communications protocols are divided into three levels—the *link level*, the *network level* and the *customer level.* The interrelationships among the different levels are indicated in Figure 2. There is a single link level protocol, two network level protocols (one for connection-based communication and one for datagrams) and a variety of customer level protocols. The customer level protocols are not discussed in detail here, but the intention is that these protocols would be application-dependent and built on top

F	FTYP	PRI	TS	PTYP	LCN	I	FC

(a) Data Transfer Packet Format

F	FTYP	PRI	TS	PTYP	LCN	CF	SI	FC

(b) Connection Control Packet Format

F	FTYP	PRI	TS	ADR	I	FC

(c) Datagram Packet Format

F:	Flag (1 byte)	CF:	Control function (1 byte)
FTYP:	Frame type (4 bits)	SI:	Supplementary information
PRI:	Priority (4 bits)	ADR:	Address (8 bytes)
TS:	Time stamp (1 byte)	I:	Information field
PTYP:	Packet type (4 bits)	FC	Frame check sequence (2 bytes)
LCN:	Logical channel number (12 bits)		

Figure 3: Packet Formats

of one of the two network level protocols. One of these would be a telephony protocol. Another might be an internet protocol such as the DARPA IP, to facilitate communication among different data networks. Still another might be a connection-oriented internet protocol.

There are essentially three packet formats of interest to the network—*connection control packets, data transfer packets* and *datagram packets.* These packet formats are shown in Figure 3.

Link Level Protocol. The link level protocol provides frame delimiting, link transparency, error detection, packet timing, and congestion control, but not error correction. There are four fields used by the link level protocol, the Frame Type field (FTYP), the Priority field (PRI), the Time Stamp field (TS) and the Frame Check field (FC). The FTYP field identifies the frame as a test frame, a datagram or a frame belonging to a connection. The Priority field (PRI) contains a customer-specified priority. The network preferentially discards low priority frames to alleviate short-term overload conditions in the network. The network uses the Time Stamp field (TS) to record the delay encountered by the packet as it crosses the network. More precisely, it records the number of milliseconds the packet spent in each PS and NI it passed through (see [11]). This is important for applications such as voice that are sensitive to delay variations and need a mechanism to remove the timing 'jitter' that packet networks can introduce. It can also be used in distributed programming applications for clock synchronization. The Frame Check field (FC) uses a sixteen bit cyclic redundancy code to detect errors in the frame. Frames with errors are simply discarded.

Network Level Protocols. There are two protocols at the network level, the Simple Datagram protocol (SDP) and the Simple Connection protocol (SCP). SCP packets contain a four bit Packet Type field (PTYP) and a twelve bit Logical Channel Number field (LCN). The Packet Type field (PTYP) identifies each packet as either a data packet or a control packet and contains a Congestion Control subfield, used to inform the NIs and customers of internal network congestion. SCP data packets also contain an Information field, which can have any length up to 144 bytes. SCP control packets are used to establish and control connections and contain a Control Function field (CF) and a Supplementary Information field (SI). SDP packets contain an eight byte address field. Two bytes are reserved as a socket number, to be interpreted by the customer protocols. The remaining six bytes define a hierarchical address space organized on a geographical basis, to permit routing based on local knowledge of network topology.

Network Performance

Given the network architecture described above, we can make some approximate calculations of network performance. To do this we hypothesize a worst-case reference connection in which two customers access the network using MSLs and communicate over a 4000 mile connection passing through two NIs and six PSs. We assume that the access links each carry two voice calls and that the seven internal HSLs are all operating at an occupancy of 85%.

The delay experienced by a packet passing across the reference connection contains several components.

- *Speed-of-light delay.* This is the constant delay caused by the finite speed of light. It is approximately 40 ms for a 4000 mile connection.
- *Packetizing delay.* This is the delay involved in producing enough data to put in a packet. Using 32 Kbs ADPCM voice encoding, information is generated at the source at the rate of 32 bits per millisecond. 512 bits are placed in each packet, giving a packetizing delay of 16 ms.
- *Queueing delay.* This is the delay encountered by the packet as it is waiting to be transmitted over each of the seven HSLs and two MSLs in the reference connection. We will analyze this component in detail below.
- *Switching delay.* This is the delay encountered by a packet as it passes through a PS or NI, excluding queueing delays. We will see in the next section, this can be kept below 0.5 ms per switch, hence we allow 4 ms of switching delay in the reference connection.

We use an M/M/1 queueing model to approximate the queueing delay on the seven HSLs in the reference connection. The formula for the average delay through n tandem queues is

$$q = \frac{bn}{s(1-\rho)}$$

where b is the packet length in bits, s is the bit rate of the transmission lines (1.544 Mbs) and ρ is the link occupancy (0.85). We use an average packet length of 650 bits to model a mixture of voice packets with 1000 bit data packets. For seven queues we calculate $q \approx 20$ ms. Of course, this is just the average delay. For voice, it is important to have an upper bound on the delay experienced by *most* packets. We can determine this by using an Erlang distribution with a rank of seven and a mean of 20 ms. With this technique we estimate that 99% of all packets in the

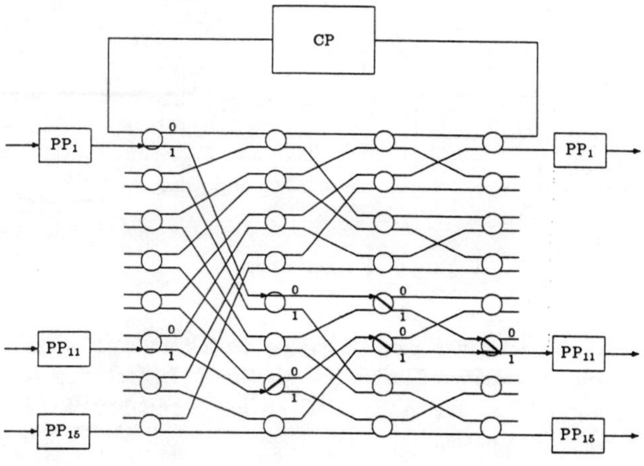

Figure 4: Packet Switch Structure

reference connection experience a delay of no more than 42 ms on the internal network links.

We still need to determine the queueing delays experienced on the access links. Here the M/M/1 model is not really applicable. On the access link going into the network, the delay is bounded by 12 ms, since we are assuming two simultaneous voice calls generating packets at 16 ms intervals. The worst thing that can happen is that packets are generated simultaneously from both voice encoders meaning that one packet must wait for the other to be transmitted. The delay of 12 ms results from the observation that voice packets are less than 600 bits long, and with an access line speed of 100 Kbs can be transmitted in 6 ms—hence it takes 12 ms to transmit a packet for each call. Packets experience a similar delay as they leave the network. Of course, the actual delays experienced at the access links is highly dependent on the nature of the services the customer is using at the time. One way to ensure good performance for voice at the access links is to use the packet priorities to give preference to voice packets. With this mechanism, we can add data to the traffic carried by the access link and be guaranteed a maximum delay of 22 ms on each end (assuming a non-preemptive priority mechanism). Based on these considerations, we assume 50 ms of access delay.

Summing the various delay components yields an estimate of about 150 ms for 99% of all packets.

Packet Switch Design

The network is built using large high performance packet switching systems, each terminating up to 1023 HSLs. The structure of such a packet switch is illustrated in Figure 4, which shows a small version with 15 HSLs.

The system is controlled by a Control Processor (CP) which performs all connection control functions, plus administrative and maintenance functions. The CP is a large general-purpose computer. Its role is analagous to that of the Control Processor in large telephone switching systems such as the No. 4 ESS [2].

Each HSL is terminated by a Packet Processor (PP), which performs the link level protocol for all packets and the network level protocol for data transfer packets. It also forwards connection control packets to the CP and datagram packets to the Datagram Routers (DR).

The heart of the switch is the Switch Fabric (SF) which consists of a large binary routing network. The important property of such networks is that the path each packet takes through the network is determined by successive bits of its destination address. The figure shows paths from two different SF input ports to output port 1011. Note that at the first stage the packets are routed out the lower port of the nodes (corresponding to the first '1' bit of the destination address), at the second stage they are routed out the upper port (corresponding to the '0' bit) and in the third and fourth stages they are routed out the lower ports. The self-routing property is shared by a variety of interconnection patterns, including the so-called delta, shuffle-exchange and banyan networks (see [5]). The PS uses a ten stage binary routing network with 1024 ports.

The Datagram Routers (not shown) are special-purpose devices used to route datagrams. The number of DRs can be engineered to suit the traffic. Each occupies a port on the SF, replacing one PP. Since most of the traffic is expected to be connection-oriented, the number of DRs required should be modest.

Duplication is required to provide the highly reliable operation expected in a large scale public communications network. The scheme used here (but not shown in the figure) is to duplicate the SF, but not the PPs. In normal operation, the duplicate SFs are operated in a load-sharing fashion, with each carrying half the traffic. If one side fails, all traffic is shifted to the other side.

Packet Processing

When a packet is received by a PP, it is placed in a buffer with several additional header fields added. The Destination field (D) identifies the destination port on the SF. The Source field (S) identifies the port where the packet arrived. The Length field (LNG) gives the packet length in bytes. The Switch Packet Type field (SPTYP) is used to identify various packet types within the PS. The Arrival Time field (AT) gives the time at which the packet arrived at the PS (this is used for processing the TS field).

For data transfer packets, the destination port is determined by the PP using the packet's LCN field and the PP's Logical Channel Translation Table (LCXT). Each entry in the LCXT contains an outgoing port number and a new LCN. The outgoing port is placed in the D field of the packet and the new LCN goes in the LCN field. The

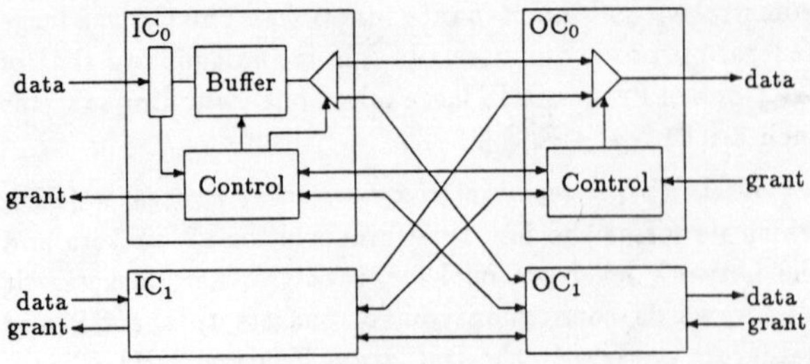

Figure 5: Switch Node

packet is then sent on to the switch fabric, which uses the D field to route the packet to the proper outgoing port as described earlier. When a packet arrives at the outgoing PP, it is buffered and then transmitted on the HSL with the extra header information stripped off.

The contents of the LCXTs is controlled by the CP, which can read and write the LCXT using special control packets sent to the PPs through the SF. Thus, the connection establishment process includes the sending of messages from the CP to the two PPs selected for the connection, updating their LCXTs appropriately.

Switch Fabric

The basic operation of the switch fabric has already been described. The SF's highly regular and parallel structure allows the construction of very large switches, without the bottlenecks that can arise in bus or ring based interconnection networks.

The nodes of the SF operate as miniature packet switches. Each node has a buffer at each input port capable of holding one maximum length packet. The data paths joining the nodes are bit serial and operate at 12 Mbs. This gives the SF an 8:1 speed advantage over the external HSLs. Thus, if all the external links are operated at an occupancy of 85%, the internal links will have an average occupancy of less than 11%. (This is assuming just one switch plane active. When both are active, the average occupancy is less than 6%.) There is also an upstream control lead joining each pair of adjacent nodes. This is used to implement a simple hardware flow control mechanism, which prevents buffer overflows within the SF.

A more detailed look at the switch node appears in Figure 5. It consists of two Input Controllers (IC) and two Output Controllers (OC). The ICs contain a buffer large enough to hold one packet and a controller implemented as a state machine. The OCs are simple state machines that arbitrate requests for their ports.

When a packet is received, the IC determines the proper outgoing port by examining the appropriate bit of the packet's destination field, then requests permission to use that port. If the desired port is immediately available, the packet is sent to it directly, bypassing the buffer.

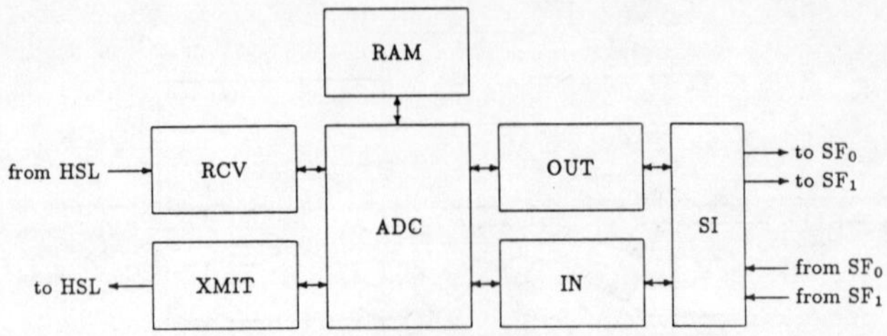

Figure 6: Packet Processor

Hence, a packet can pass through a switch node after experiencing a delay of just a few bit times. In fact, this is the normal case due to the relatively low occupancy of the internal data paths.

If the desired port is not available, the packet is shifted into the buffer. As soon as the desired port becomes available, the packet is sent out—even if not all of the packet has been received. If the port is still not available when the end of the packet is received, the IC holds its *grant* lead low to prevent the arrival of new packets. The grant lead is re-asserted as soon as the desired link becomes available, allowing a new packet to enter the buffer, while another is leaving.

Packet Processors

The structure of the Packet Processors (PP) is shown in Figure 6. It is organized around a 16 Kbyte RAM, which contains four packet buffers and the Logical Channel Translation Table (LCXT). The principal buffers are the Receive Buffer (RCB) used for packets received from the HSL on their way to the switch, and the Transmit Buffer (XMB) for packets going from the switch to the HSL. The Link Test Buffer (LTB) and the Switch Test Buffer (STB) are small buffers that provide loop-back paths for testing the HSL and switch respectively.

Access to the memory is provided through the Address Controller (ADC), which contains read and write pointers for the buffers and arbitrates memory access.

The Receive circuit (RCV) receives incoming packets from the HSL, removes the flag field, discards packets with errors, adds the extra header fields, initializes the length (LNG) and arrival time (AT) fields, converts from bit serial to eight bit parallel format and writes the packet to the RCB through the ADC.

The Output circuit (OUT) takes packets from the RCB, performs the logical channel translation described above and sends the packets onto the switch in bit serial format. The Input circuit (IN) takes packets from the switch and writes them to the XMB.

The Transmit circuit (XMIT) takes packets from the XMB, performs the time stamp calculation, strips the extra header information, adds the flag field and transmits the packets on the HSL.

The Switch Interface (SI) connects to the duplicated switch planes, normally routing packets to and receiving packets from the active plane. The SI can also send and receive packets from the standby plane—this is used for testing.

The PP can be implemented using two chips—one custom controller chip and one memory chip. The simplicity of the protocols makes this possible. There is no need to buffer unacknowledged frames that may need to be retransmitted as in conventional link level protocols that perform error correction and flow control. Similarly, there is no need for the recovery and synchronization procedures that are required by such protocols.

Performance of the Switch Fabric

Switching networks similar to the one considered here have been analyzed by several authors to determine their performance characteristics (see [4] and [8], for example). Unfortunately the earlier analyses don't quite suit the network considered here, since they assume packets are buffered at each stage. This is not the case for the network considered here, which permits packets to pass through a stage without being buffered at all. This buffering technique is called *virtual cut-through* by Kermani and Kleinrock [10], who analyze its performance in a somewhat different context. Their analysis does not apply to our situation either.

The analysis given by Jenq in [8] can be modified to suit our needs. Jenq provides an algorithm for calculating the delay through a binary routing network preceded by packet processors containing infinite buffers. The main difference between his model and the network described here is that it assumes that packets are buffered at each stage. This means that in his model, packets encounter a delay of at least nb/s, where n is the number of stages in the network, b is the packet length and s is the speed of the internal data paths. This is true even at very low occupancies. In our network, on the other hand, the delay experienced at low occupancies is $\approx 10n/s$, since each packet is delayed about 10 bit times per stage. Hence, at low occupancies the delay through the network can be obtained by subtracting $(b-10)n/s$ from the estimate obtained using Jenq's method. Since the network is normally operated at occupancies below 20%, this approach can be expected to yield satisfactory estimates.

Two other assumptions of Jenq's method need to be mentioned. First, it assumes that all packets have the same length—we will use it to estimate the delay under the assumption that all packets are of maximum length ($b = 1264$), which is a worst-case assumption for this network. Second, Jenq's analysis assumes that traffic is uniformly distributed throughout the network—more precisely, the destination field of each arriving packet is assumed to be selected at random from the integers 0–1023. This is called the *uniform traffic assumption* and while it appears reasonable, it ignores the effect of the network's connection-orientation on traffic patterns. This assumption is examined more closely in a later section and justified.

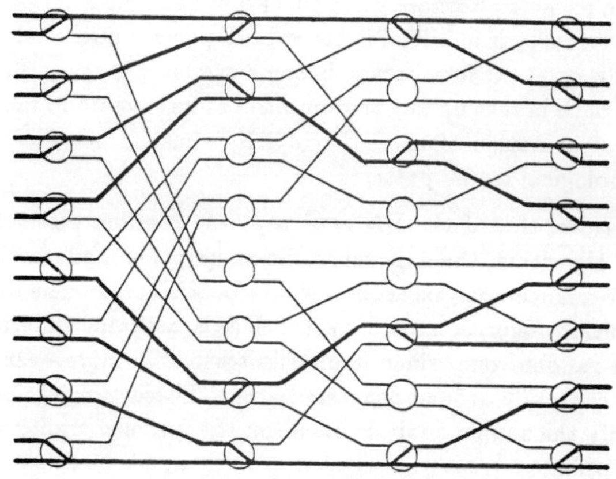

Figure 7: Congestion in Binary Routing Networks

Let $d_J(\rho, n)$ be the average delay through an n stage binary routing network for a given occupancy ρ, using the method described in section V of Jenq's paper [8]. We find, $d_J(.05, 10) = 1070\ \mu s$, $d_J(.1, 10) = 1090\ \mu s$, $d_J(.15, 10) = 1113\ \mu s$ and $d_J(.2, 10) = 1141\ \mu s$, Subtracting $1045 = (b - 10)n/s$, as described above gives corresponding estimates of 35, 45, 68 and 96 μs for our network. When all the HSLs coming into the PS are operated at occupancies of 80%, the occupancy of the internal links is about 11%, leading to an average switch fabric delay of about 50 μs.

Switch Fabric Revisited

One problem with binary routing networks is that they can become congested in the presence of certain traffic patterns. This is illustrated in Figure 7, which shows a traffic pattern corresponding to several *communities of interest.* In this pattern, all traffic entering the first four inputs is destined for the first four outputs, all traffic entering the second group of four inputs is destined for the second group of four outputs, and so forth. Note that with this pattern, only one fourth of the links joining the second and third stages are carrying traffic. Thus, if the inputs are heavily loaded the internal links will be hopelessly overloaded and traffic will back up. In a 1024 × 1024 network, there are ten stages and the links between the fifth and sixth stages can in the worst-case be carrying all the traffic on just 32 of the 1024 links.

This problem can be solved by using two networks instead of one. One of these is called the Routing Network (RN) and is a standard binary routing network. The other network sits in front of the RN and is called the Distribution Network (DN). It has the same structure as the RN, but instead of routing packets based on their destination address, it attempts to distribute packets evenly across all its output ports. This is done by having each switch node route packets alternately out its two ports. The alternate-port strategy is modified if one or both ports is unavailable. In this case, the first port to become available is used. This approach breaks up any communities of interest and makes the combination of the DN and RN robust in the face of pathological traffic patterns.

Note, that if the DN used a purely random distribution algorithm, the uniform traffic assumption made in the performance analysis section would be validated. The distribution algorithm used by the DN is actually superior to a random one, since it distributes traffic more evenly and can route around congested areas. These observations justify the simple analysis based on the uniform traffic assumption.

One drawback of the DN is that it doubles the number of stages in the SF, thus roughly doubling the packet delay and the circuit complexity. This loss can be recovered by using larger switch nodes for the RN and DN. Instead of 2×2, we can use 4×4 nodes. With the larger nodes we can construct a network with half the number of stages required with 2×2 nodes. This also halves the delay and the circuit complexity.

Other Issues

The circuit complexity of the switch nodes is dominated by the buffer, which is basically a 1280 bit shift register. Thus a 4×4 node contains about 5000 bits of memory and will fit comfortably on a single chip with 20 pins. These chips are combined onto circuit boards, each board containing either 32 or 48 chips. A single RN will contain 32 of these boards.

The PPs can be implemented using one custom controller chip and one $16K \times 9$ memory chip. Sixteen PPs are packaged together on a single board, meaning that 64 of these boards are required for the entire PS. The Datagram Routers, which haven't been described in detail can be implemented in a similar fashion to the PPs—they're actually a bit simpler.

The packaging of the PS can be remarkably compact. A single equipment frame, 3 feet wide by 6 feet high by 1.5 feet deep is all that's required for the duplicated switch fabric and all the packet processors.

Summary

This paper has described the design of a high performance packet switching network capable of supporting both voice and data communication on a large scale and at a cost to the user, comparable to current telephone service. The key features of the design are the use of high speed digital transmission facilities and simple link level protocols, a predominantly connection-oriented service, hardware implementation of all per packet functions and implementation of higher level protocol functions on an end-to-end and application-dependent basis. I claim that the advanced packet switching technology described here is inherently better suited for the provision of advanced communications services than the circuit switching technology which underlies current ISDN proposals.

Many aspects of the packet switching system described here have been patented by AT&T Bell Laboratories. See references [14]–[20] for further details.

Acknowledgements. The work described here was carried out between 1981 and 1983 when I was a Member of Technical Staff at Bell Laboratories in Naperville, Illinois. Many individuals contributed ideas. I particularly want to thank Harold Andrews, Danny Creed, Bill Hoberecht, Warren Montgomery and Len Wyatt who helped turn the original idea into a workable design. I also want to thank Bob Cardwell and Eric Nussbaum for recognizing the potential and providing the backing to develop it. Special thanks go to Maurizio Dècina, whose refusal to accept half-baked ideas forced me to make sure that mine were always well-cooked.

REFERENCES

[1] Aoki, D. J., D. H. Florin, S. D. McKenna, G. R Welsh and P. E. White. "Digital Feature Evolution in the 5ESS™ Digital Switch," *Proceedings of Globecom 83*, 11/83, 601–605.

[2] *Bell System Technical Journal*, vol. 56, no. 7, 9/77 (Special issue devoted to the No. 4 ESS).

[3] Dècina, Maurizio. "Performance Requirements for Integrated Voice/Data Networks," *IEEE Transactions on Communications*, vol. COM-30, no. 9, 9/82, 2117–2130.

[4] Dias, Daniel M. and J. Robert Jump. "Packet Switching Interconnection Networks for Modular Systems," *Computer*, vol. 14, no. 12, 12/83, 43–54.

[5] Feng, Tse-yun. "A Survey of Interconnection Networks," *Computer*, vol. 14, no. 12, 12/83, 12–30.

[6] Gruber, John G. "Performance Requirements for Integrated Voice/Data Networks," *IEEE Journal on Selected Areas in Communications*, vol. SAC-1, no. 6, 12/83, 981–1005.

[7] Hoberecht William L. "A Layered Network Protocol for Packet Voice and Data Integration," *IEEE Journal on Selected Areas in Communications*, vol. SAC-1, no. 6, 12/83, 1006–1013.

[8] Jenq, Yih-Chyun. "Performance Analysis of a Packet Switch Based on a Single-Buffered Banyan Network," *IEEE Journal on Selected Areas in Communications*, vol. SAC-1, no. 6, 12/83, 1014–1021.

[9] Kasahara, M., H. Shimizu, M. Imaizumi. "Basic Concept of the Digital Network for the Information Network System," *Proceedings of Globecom 83*, 11/83, 969–973.

[10] Kermani, Parviz and Leonard Kleinrock. "Virtual Cut-Through: A New Computer Communication Switching Technique," *Computer Networks*, vol. 3, 1979, 267–286.

[11] Montgomery, Warren A. "Techniques for Packet Voice Synchronization," *IEEE Journal on Selected Areas in Communications*, vol. SAC-1, no. 6, 12/83, 1022–1028.

[12] Rettberg, R., C. Wyman, D. Hunt, M. Hoffman, P. Carvey, B. Hyde, W. Clark and M. Kraley. "Development of a Voice Funnel System: Design Report," Bolt Beranek and Newman, Report No. 4098, 8/79.

[13] Turner, Jonathan S. and Leonard F. Wyatt. "A Packet Network Architecture for Integrated Services," *Proceedings of Globecom 83*, 11/83, 45–50.

[14] Turner, Jonathan S. "Packet Load Monitoring By Trunk Controllers," United States Patent #4,484,326, 11/20/84.

[15] Turner, Jonathan S. "Packet Switching Loop-Around Network and Facilities Testing," United States Patent #4,486,877, 12/4/84.

[16] Turner, Jonathan S. "End-to-End Information Memory Arrangement in a Line Controller," United States Patent #4,488,288, 12/11/84.

[17] Turner, Jonathan S. "Interface Facility for a Packet Switching System," United States Patent #4,488,289, 12/11/84.

[18] Turner, Jonathan S. "Packet Error Rate Measurements by Distributed Controllers," United States Patent #4,490,817, 12/25/84.

[19] Turner, Jonathan S. "Fast Packet Switch," United States Patent #4,491,945, 1/1/85.

[20] Turner, Jonathan S. "Fast Packet Switching System," United States Patent #4,494,230, 1/15/85.

[21] Turner, Jonathan S. "Duplicated Network Arrays," U.S. patent application filed.

[22] Turner, Jonathan S. "Integrated Self-Checking Packet Switch Node," U.S. patent application filed.

[23] Turner, Jonathan S. "Distributed Monitoring of Packet Transmission Delay," U.S. patent application filed.

[24] Turner, Jonathan S. and Leonard F. Wyatt. "Alternate Paths in a Self-Routing Packet Switching Network," U.S. patent application filed.

[25] Weinstein, Clifford J. and James W. Forgie. "Experience with Speech Communication in Packet Networks," *IEEE Journal on Selected Areas in Communications*, vol. SAC-1, no. 6, 12/83, 1022–1028.

MIXING TRAFFIC IN A BUFFERED BANYAN NETWORK

L. T. Wu

Bell Communications Research
Morristown, NJ 07960

ABSTRACT

A model of banyan networks is developed to account for their operational as well as topological characteristics. The model incorporates a description of the states of individual switching nodes and explicitly captures the interconnection patterns between stages. Accordingly, the performance of a network can be evaluated via an iterative procedure. Using the model, we analyzed one class of nonuniform traffic pattern referred to as "point-to-point" traffic that has particular significance to mixed voice, data, and video applications. The results obtained indicate that even a single dedicated channel (with point-to-point traffic) can significantly limit the throughput of the background uniform traffic. The paper concludes with several switch design strategies for switching point-to-point traffic using a self-routing switching network.

1. Introduction

Banyan networks have been used in the past few years to provide interconnections for processor and memory modules [1]-[3]. The approach permits the design of cost-effective communication networks for distributed and parallel multiprocessor systems. More recently, with the advances in VLSI technology, banyan networks are also used as switch networks for general purpose packet-switched networks (see [4], for example) to realize ISDN concepts. It is believed that such switching systems should result in substantial improvements over conventional packet switches both in price and in performance for various types of applications.

Banyan networks are typically composed of 2-inlet and 2-outlet switching nodes connected together in stages (see Figure 1, for example). The routing of a packet inside the interconnection network is performed by individual stages until the packet reaches its destination output link. Each stage performs the same routing function based on information contained in the packet header. At each stage, different packets may be treated differently depending on the state of the switching node as well as congestion conditions at the later stage. The self-routing property of a banyan network permits the construction of a large capacity switch without relying on a complicated control processor. However, since there is only one path connecting each input and each output, the internal traffic congestion problem becomes an important issue to be studied.

The performance of banyan networks have been studied previously using analytical and simulation methods ([5] and [6]). These studies assume that the system enjoys a somewhat ideal traffic mix and that the traffic load is evenly distributed within the interconnection network. Hence, the performance estimates obtained tended to be optimistic because nonuniform traffic loadings could produce internal bottlenecks. Nonuniform traffic may arise in several ways: one typical source of nonuniform traffic is a calling pattern resulting from skewed or focused overloads (e.g., calls to a popular location) and another is typified by wideband services (e.g., transfer of large files or video applications).

In this analysis, we generalize previous models for banyan networks to account for nonuniform traffic patterns in the network. The modeling approach taken is similar to previous models ([5] and [6]) in that the model for the network is divided into smaller pieces so that the performance can be evaluated via an iterative procedure. Individual components of this model, however, are more complex than those of the previous analysis. Each of the two input links of a switching node is modeled by a Markov process that explicitly captures the dependences between two companion data links. Moreover, a binary encoding scheme is used to describe the topology of the interconnection patterns between adjacent stages. The need for such representation distinguishes the current model from those previously studied. This is because under nonuniform traffic, switching nodes are no longer statistically equivalent even if they belong to the same stage.

In the following sections, we first describe the operation of a single-buffered banyan network. We then describe in Section 3 building blocks of the model, i.e., the microscopic model for the switching nodes and the binary number representation for the interconnection pattern. We then study in Section 4 the performance implication of some nonuniform traffic patterns. The results obtained represent answers to typical questions that can be asked and, by no means, constitute a complete assessment of the robustness of banyan networks to handle nonuniform traffic. The solution method, nevertheless, provides a quantitative approach for a more detailed analysis of the problem in the future. Finally, the results are summarized and further work is discussed in Section 5.

2. A Single–Buffered Banyan Network

As depicted in Figure 1, a 2^N input and 2^N output banyan network is composed of 2X2 switching nodes arranged into N stages. Depending on the applications, a banyan network can be designed with or without buffers. In the following analysis, we will focus on the evaluation of single- buffered banyan networks that have one buffer associated with each input of an internal node. However, our approach is completely general and it can readily be applied to other types of banyan networks.

The best way to understand the operation of a banyan network is by examining the events that a packet may encounter when moving through the banyan interconnection network (IN). Once a packet has entered the inlet of a switching node, the node first uses one bit in the destination address in the packet header to select one of its outlets to send the packet to the next stage. If the outlet selected is "free" and the destination node in the next stage is "ready" to receive a packet, the packet is then transmitted immediately. The outlet is *free* when there is no packet passing through. The next stage node is *ready* to receive when its input buffer is empty. At the last stage, however, a packet will be able to move immediately as long as the chosen outlet is not occupied by a contending packet.

On the other hand, packets arriving at a switching node can be *blocked* by the node when either 1) a packet (from the other inlet of the node) is passing through the selected outlet, or 2) the data buffer of the next stage node is congested. When a packet is blocked, it is held in the input buffer of the node temporarily until it can progress again. A blocked packet will prevent other packets from entering the stage through the same inlet. Furthermore, the effect of blocking may propagate all the way back to the first stage and, thus, prevent packets from entering the IN.

Clearly, if an outlet is not available because a contending packet is passing through, the outlet will become free as soon as the packet transmission is completed (unless the packet enters the next stage buffer and must wait). The unblocking of a buffered packet at an internal node generally depends on that of all succeeding stages.

Before describing the model in more detail, let us consider the operating environment of the IN. These environmental assumptions are chosen to reflect the performance limitations inherent to the structure of the interconnection network. Accordingly, the performance estimates obtained should be regarded as the maximum performance that the IN can support under such ideal conditions. We assume that

- All packets have the same length.
- Packets arrive independently at each input port and the time between arrivals is geometrically distributed.
- The IN operates in synchronized mode with the synchronization clock unit taken to be the minimum time required to pass an entire packet through a stage.
- Packets are removed from the IN as soon as all bits arrive at the output port.
- The IN is error-free in the sense that all links are noiseless and all internal nodes are immune to failures.

Since, under these assumptions, the flow of packets is not delayed by long packets, retransmission, or by the traffic congestion at the output port, the resulting performance will be somewhat optimistic. Moreover, when a packet is blocked at a stage, the packet will be delayed for one or more packet times. Thus the model constructed is, at best, an approximation of the real situation.

3. An Iterative Model

The solution procedure for this model requires as input a detailed breakdown of the traffic carried by the incoming and outgoing IN links. It is assumed that we are given not only the steady-state throughput on each incoming link but also the amount destined to each of the outgoing IN links. These input data can be expressed conveniently as a *load matrix*

$$L = [l_{ij}] \quad ,$$

where l_{ij} is the normalized traffic load originating from incoming IN link i and destined for the outgoing IN link j. Normalization is done with respect to the bandwidth of the internal high speed link. Clearly, the steady-state throughput of an incoming or outgoing IN link is the sum of all the entries on the corresponding row or column of the matrix, respectively.

Given a load matrix L, the first step of the solution procedure is to decompose the load matrix into 4X4 matrices representing the steady-state load carried by individual switching nodes. The decomposition uses a binary number representation of the inter-stage interconnection patterns to calculate the load on each node.

The next step of the procedure is to calculate, by iterating through all stages, the delays and blocking probabilities encountered by a packet at each internal node. The iteration starts from the last stage and uses the resulting steady-state probabilities for the preceding stages. Note that, at each stage, the state of a particular switching node depends on the state of the successive nodes connected to it via the specific interconnection pattern between stages.

The last step of the procedure is to calculate the mean and the variance of packet delays encountered along each IN path. The path delay is determined by summing over delays associated with each node along the given path. Again, the binary number encoding of the IN permits us to navigate through the interconnection network to access intermediate results stored away during the iteration. The basic components of the model are described in sections that follow.

3.1 The node model. As discussed in Section 2, there are two discernible states (ready and blocking) that can be associated with each data buffer of a switching node. Thus, a natural way to describe the state of the IN appears to be to list the state of all the distinct data buffers. This approach, however, is impractical, because the size of the resulting state space can be enormous even for interconnection networks with only a few stages. Accordingly, in this section, we model each switching node individually by a simple Markov chain model and describe relationships among different switching nodes through probabilistic means. More precisely, let us name the states of a data buffer as 0 and 1 as follows:

State	Interpretation
0	ready
1	blocking

and define, for *the* ℓ^{th} *node* at the kth *stage*, the probabilities for state transitions as

$$q_{k\ell}(x) = \textit{Steady-state probability that a packet is available for entering the } x^{th} \textit{ buffer at the beginning of the clock period,}$$

and

$$r_{k\ell}(x) = \textit{Steady-state probability that a blocked packet can move during the clock period,}$$

where $x = 0$ or 1 and represents the "upper" or "lower" data buffer of the node, respectively. Then the state behavior of the xth buffer of the node can be represented as a 2-state Markov chain as depicted in Figure 2. Note that the probability $q_{k\ell}(x)$ can be regarded as the *offered load* to the xth buffer of the ℓth node of the kth stage. Note also that, although $q_{k\ell}(x)$ will be referred to as the arrival probability of a packet, an arriving packet can not enter the buffer if the stage is not ready to accept it.

The transition probabilities in Figure 2 are obtained by applying the synchronization assumptions discussed in Section 2. Clearly, when the buffer is in state 0 and there is no packet arriving, the buffer will remain in the state at the next clock tick. On the other hand, if there is an arrival, the packet will be buffered by the node for one or more clock periods depending on whether the packet is allowed to leave at the next clock tick. The transition probabilities for these two cases are $1-q_{k\ell}(x)$ and $q_{k\ell}(x)$, respectively. Also, if the buffer is holding a packet (state 1) at the present clock period, it goes to state 0 at the beginning of the next clock period only when there is no arrival and the blocked packet is allowed to move during the clock period. The transition probabilities for the above cases can be found in Figure 2.

Based on the above transition diagram, the "long run" behavior of the Markov chain can now be expressed conveniently by a set of linear equations. In particular, if we let $p^0_{k\ell}(x)$ and $p^1_{k\ell}(x)$ denote the steady-state state probabilities of the Markov chain, the state diagram can be translated into the following balanced equations:

$$q_{k\ell}(x)\cdot p^0_{k\ell}(x) = [1-q_{k\ell}(x)]\cdot r_{k\ell}(x)\cdot p^1_{k\ell}(x) \qquad (1)$$

satisfying

$$p^0_{k\ell}(x) + p^1_{k\ell}(x) = 1 \ . \qquad (2)$$

Accordingly, the steady-state distribution is determined by calculating parameters $q_{k\ell}(x)$ and $r_{k\ell}(x)$, using known quantities, i.e., the steady-state load carried by individual nodes.

3.2 Movability of a packet. Clearly, solving the above steady-state equations requires a detailed knowledge of interactions between companion buffers of a node as well as dependencies between stages. These relationships can be derived by considering differences in throughput rates assumed by different switching nodes.

To illustrate the steps involved in the derivation, let us decompose the steady-state probability $r_{k\ell}(x)$ into two parts by conditioning on the selected output link. Specifically, let us define

$r_{k\ell}(x,y)$ = *conditional probability that a blocked packet can move given that the packet wants to exit output link y.*

Then, we have

$$r_{k\ell}(x) = r_{k\ell}(x,0)\cdot\frac{\lambda_{k\ell}(x,0)}{\lambda_{k\ell}(x,0)+\lambda_{k\ell}(x,1)} + r_{k\ell}(x,1)\cdot\frac{\lambda_{k\ell}(x,1)}{\lambda_{k\ell}(x,0)+\lambda_{k\ell}(x,1)} \qquad (3)$$

Here $\lambda_{k\ell}(x,y)$ is the traffic load from the xth inlet to the yth outlet and x and y can assume values 0 or 1 indicating upper or lower links. Since $\lambda_{k\ell}(x,y)$ can be computed from the given load matrix as discussed in Section 2, we need only to solve for $r_{k\ell}(x,y)$.

To compute $r_{k\ell}(0,0)$, note that a packet buffered at inlet 0 can leave outlet 0 at the clock tick when there is no contention from inlet 1 and the next stage is ready to receive the packet. Furthermore, there is no contention from inlet 1 when

A) Inlet 1 is in state 0.

B) Inlet 1 is in state 1 and the buffered packet is destined for outlet 1.

C) Inlet 1 is in state 1 and the packet wants to exit outlet 0. However, the contention is resolved in favor of the packet buffered at inlet 0.

The next stage is ready to receive the packet when

D) The destination buffer is empty.

E) The destination buffer is in the blocking state, but its content is moving out.

Summing over probabilities of the above events, we have

$$r_{k\ell}(0,0) = \{\, p^0_{k\ell}(1) + p^1_{k\ell}(1)\cdot\frac{\lambda_{k\ell}(1,1)}{\lambda_{k\ell}(1,0)+\lambda_{k\ell}(1,1)} + 1/2\cdot p^1_{k\ell}(1)\cdot\frac{\lambda_{k\ell}(1,0)}{\lambda_{k\ell}(1,0)+\lambda_{k\ell}(1,1)} \,\} \cdot \{\, p^0_{(k+1)m}(i) + p^1_{(k+1)m}(i)\cdot r_{(k+1)m}(i) \,\} \ . \qquad (4)$$

Note that the second part of the above product assumes that the destination buffer for outlet 0 is the ith buffer of the mth node at the $k+1^{th}$ stage. The exact position of the destination buffer, however, is dictated by the topology of the interconnection network. Repeating arguments as above and assuming outlet 1 is connected to the jth buffer of the nth node, we can represent $r_{k\ell}(0,1)$, $r_{k\ell}(1,0)$, and $r_{k\ell}(1,1)$ with similar expressions.

Finally, we note that the throughput rates sustained on the input links satisfy the following conditions:

$$\lambda_{k\ell}(0,0)+\lambda_{k\ell}(0,1) = q_{k\ell}(0)\cdot[p^0_{k\ell}(0)+p^1_{k\ell}(0)\cdot r_{k\ell}(0)] \qquad (5)$$

and

$$\lambda_{k\ell}(1,0)+\lambda_{k\ell}(1,1) = q_{k\ell}(1)\cdot[p^0_{k\ell}(1)+p^1_{k\ell}(1)\cdot r_{k\ell}(1)] \qquad (6)$$

3.3 Performance Measures. The performance measures considered in this paper are the "maximum throughput" and "packet delay." We define the *channel throughput* to be the average rate at which packets flow through an input link and define the *maximum throughput* to be the maximum channel throughput that the IN can sustain. The *packet delay* is defined to be the average end-to-end IN delay encountered given the input and the output links traversed by a packet.

Since the channel throughput increases with respect to the rate that packets are generated, maximum throughput occurs when packets are generated at a rate that saturates some input links to the switch fabric. The maximum throughput is calculated by gradually increasing the values of the load matrix until the probability $q_{1\ell}(x)$ approaches 1, for some ℓ and some x.

Packet delays are calculated using the iterative scheme described in Section 2. More precisely, note that the node model described in Section 3.1 supplies information for every link of each node in the switch. Thus, to compile this information into a description of the *path* of a particular packet (or call), one must be able to find one's way through the IN using precise directions. In other words, one must be able to determine the internal links traversed by a packet given the addresses of the input and output links to the IN.

More specifically, we label each switching node by a binary number encoding scheme and represent those nodes reachable from an incoming IN link as a binary tree. We then use the given IN configuration (i.e., the connecting patterns between adjacent stages) to traverse the tree until the specified outgoing IN link is reached. The approach used here is similar to the one employed in [2] to describe routing techniques for setting up paths through an interconnection networks.

Finally, note that the node delay encountered by a packet given the input link (x) and output link (y) of a node can be expressed as

$$d_{k\ell}(x,y) = \frac{1}{r_{k\ell}(x,y)} \quad , \tag{7}$$

i.e., the time spent in the buffer or the reciprocal of the rate of emptying the content of a buffer. Thus, the average path delay can be obtained simply by summing over all the node delays at different stages of a path.

4. Performance Results

This section presents results of the performance evaluation of a 4-stage single-buffered banyan network. We illustrate the flexibility of this model by studying one class of nonuniform traffic patterns that has particular significance to integrated voice, data, and video applications.

When taken in the strict sense, the term "uniform traffic" denotes a pattern in which the traffic is evenly distributed among the input and output links, i.e., for a switch with 2^N input links and 2^N output links, each input link supports a (normalized) load of, say, X, a proportion $X/2^N$ of which is directed to each output link. By definition then, *nonuniform* traffic must describe any scenario where this balanced configuration does not exist, and therefore represents a tremendous range of possibilities.

Among the many possible classifications of nonuniform traffic, one may consider *point-to-point* traffic. A point-to-point pattern exists when at least one input link sends all of its load to a unique output link. Thus, the uniformity is destroyed as packets form a pipeline running through the switch fabric. As a special case, we can then define a *strict point-to-point* traffic as every inlet sends traffic to one outlet, and conversely, every outlet receives traffic from exactly one inlet. In addition to a strict point-to-point traffic, we consider here also patterns that are a mixture of point-to-point and uniform loads: there is exactly one dedicated path in the IN and there is uniform traffic elsewhere in the switch.

Point-to-point traffic deserves special attention because different paths of the IN may share common links, thereby placing an extreme load on one or more internal links. The effect of packet collisions on the common link is most severe, of course, when the paths are highly utilized: that is, they are "dedicated" to a traffic source and destination. Such a set-up might exist if, for example, wideband traffic (e.g., file transfer or video) is supported through the IN.

The throughput and delay performance of a 4-stage switch with strict point-to-point traffic patterns are described in Figure 3. Two scenarios are analyzed: nonoverlapping channels and maximum overlapping channels (see Figures 4 and 5, respectively). The nonoverlapping scenario occurs when none of the dedicated paths share internal links. The maximum overlapping scenario corresponds to the situation when four different dedicated paths collide on the middle link (between the second and the third stages).

As depicted in Figure 3, nonoverlapping channels behave somewhat like circuit switched connections in the sense that packets encounter no queueing delays and each can be moved at the maximum clock rate of the switch fabric. This is because packets in a dedicated path arrive back-to-back. Thus, no packet collisions can occur. However, when dedicated paths do share common links as illustrated in Figure 5, packets from different sources may start to collide. For the traffic pattern shown, the maximum throughput dropped quickly from 100% to 25% and the delay almost doubled at the maximum throughput. We also notice that the performance degradation intensifies as the number of the stages increases.

The results summarized in Figure 6 capture the effect of mixed point-to-point/uniform traffic. Here, we assume that there is exactly one dedicated channel and the rest of input/output links form a community of uniform traffic (see Figure 7). If we envision the traffic pattern as representing one single video channel in the presence of packetized voice calls, one interesting question , then, is whether the video channel could have adverse effects on the large number of background narrowband services.

Figure 6 contains three curves representing the throughput and delay relationship among dedicated and uniform channels. We gradually increase the intensity of the point-to-point traffic and calculate packet delays as well as the maximum throughput (of the *worst* uniform channel) achievable on the community of uniform sources. As shown by Curve I, the point-to-point traffic has very little effect on the uniform traffic until its throughput reaches beyond 45%. However, as its load increases further, the maximum throughput of the background traffic decreases almost linearly and becomes zero when the dedicated channel operates at its full speed.

Examining the situation closely, we find that maximum throughput does not always fall on the same source of the uniform community. For low traffic loading on the dedicated path, there is comparatively less contention in the neighborhood of the path. Thus, the point-to-point traffic creates a "pressure release" for adjacent paths and the maximum throughput occurs on paths having no common links with the dedicated path. On the other hand, for high loadings, the point-to-point traffic becomes a roadblock for packets traversing paths overlapping with the dedicated path. Hence, maximum throughput represents what's achievable on paths sharing all internal links with the point-to-point traffic. Furthermore, in this case, the heavy point-to-point traffic effectively reduces the total bandwidth of the IN. Accordingly, the packet delays for both types of traffic drop because of lighter overall traffic load (see Curves II and III).

5. Summary and Conclusions

The objective of this study has been to develop a model that permits the analysis of banyan networks under nonuniform traffic. The objective is accomplished by extending a previous model to account for the topology of an interconnection network. Using the model, we show that point-to-point traffic has significant impacts on background uniform traffic in terms of maximum throughput achievable.

If point-to-point traffic is desired, one must consider the strategy of making sure that dedicated paths are nonoverlapping and lightly loaded (possibly by devising a novel routing scheme and by raising the internal clock rate). For applications that generate a large number of dedicated channels with heavy point-to-point traffic, one possible alternative is to select a nonblocking interconnection network so that packet contentions can be avoided completely. Examples of nonblocking self-routing interconnection network includes self-routing crossbar [4] and banyan networks that synchronize and sort its inputs to create nonoverlapping dedicated paths on slot-by-slot basis.

Beyond the traffic patterns treated in this analysis, the model should prove useful in future investigations of other applications of multistage interconnection networks. The flexibility of the approach may be used to consider transient overload effects because the internal node loads may be arbitrarily specified. With minor changes, the model can also be adapted to study the integration of voice and data traffic, issues of alternate routing, etc. Finally, we would like to point out that the model has only been validated against existing analytical and simulation models for uniform traffic. Experimental results which permit us to establish the validity of the model for nonuniform traffic patterns would be extremely valuable.

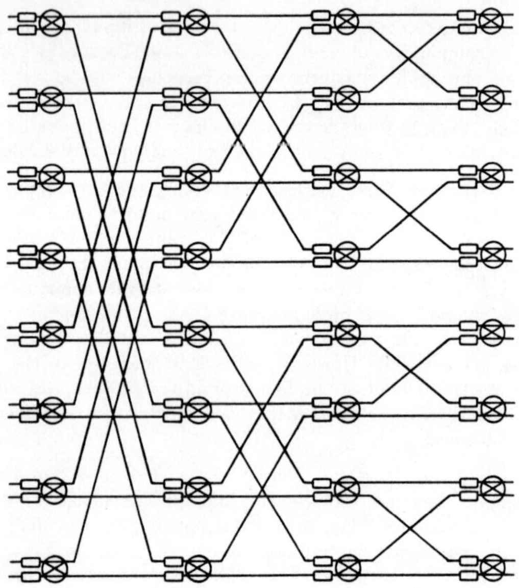

FIGURE 1 A 4-STAGE SINGLE-BUFFERED BANYAN NETWORK

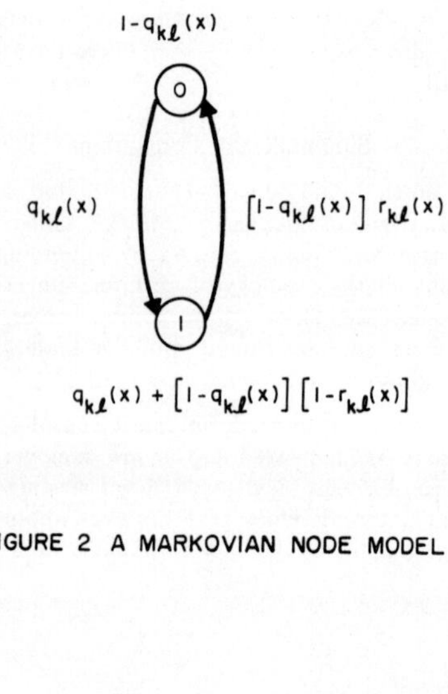

FIGURE 2 A MARKOVIAN NODE MODEL

REFERENCES

[1] D. M. Dias and J. R. Jump, "Packet interconnection networks for modular systems," *Computer 14*, December 1981, pp. 43-53.

[2] C.-L. Wu and T.-Y. Feng, "On a class of multistage interconnection networks," *IEEE Trans. on Computers*, vol. C-29, no. 8, August 1980, pp. 694-702.

[3] P.-Y. Chen, D. H. Lawrie, and P.-C. Yew, "Interconnection networks using shuffles," *Computer 14*, December 1981, pp. 55-64.

[4] J. J. Kulzer and W. A. Montgomery, "Statistical switching architectures for future services," *ISS 1984*.

[5] Y. C. Jenq, "Performance analysis of a packet switch based on single-buffered banyan network," *IEEE Journal on Selected Area in Communications* (on packet switched voice and data communications), December 1983.

[6] D. M. Dias and J. R. Jump, "Analysis and simulation of buffered banyan networks," *IEEE Trans. on Computers*, vol. C-30, no. 4, April 1981, pp. 273-282.

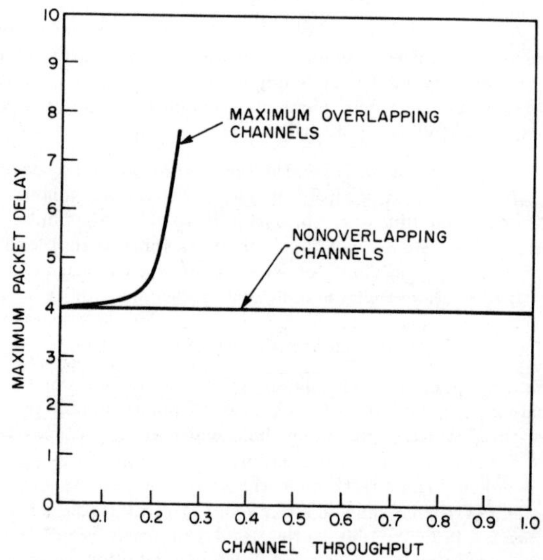

FIGURE 3 DELAY VS. THROUGHPUT STRICT POINT-TO-POINT TRAFFIC

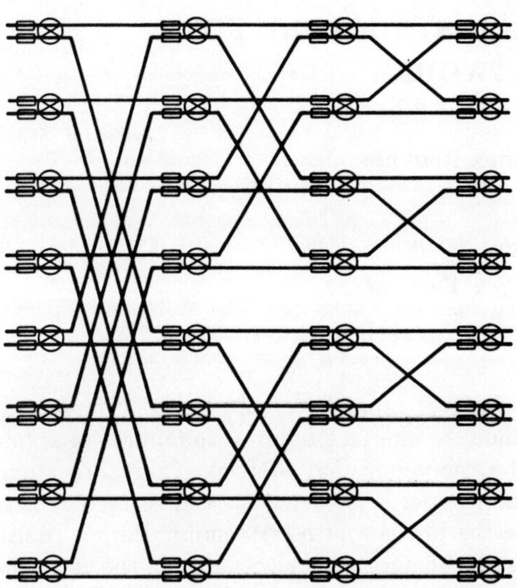

FIGURE 4 A NONOVERLAPPING POINT-TO-POINT TRAFFIC PATTERN

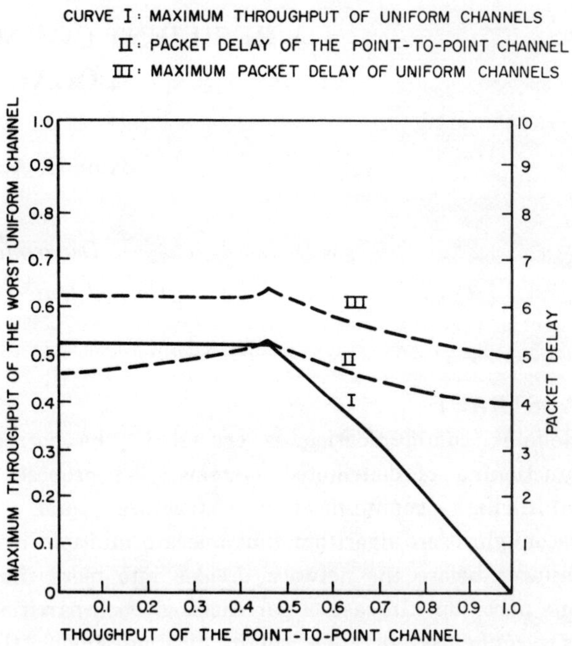

FIGURE 6 DELAY VS. THROUGHPUT MIXED (SINGLE) POINT-TO-POINT WITH BACKGROUND UNIFORM TRAFFIC

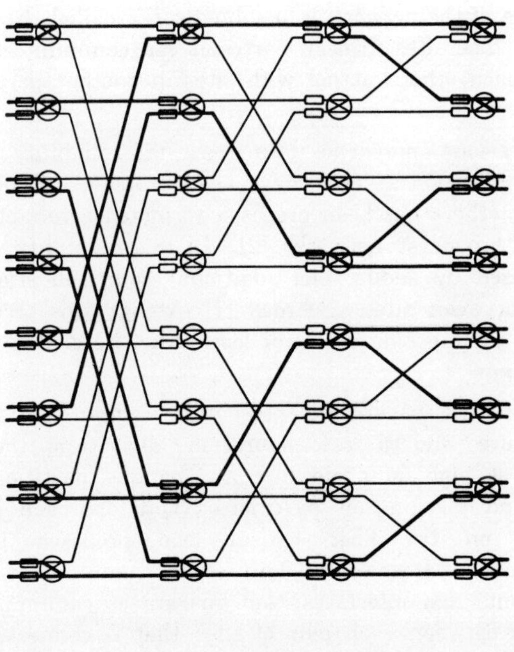

FIGURE 5 A MAXIMUM OVERLAPPING POINT-TO-POINT TRAFFIC PATTERN

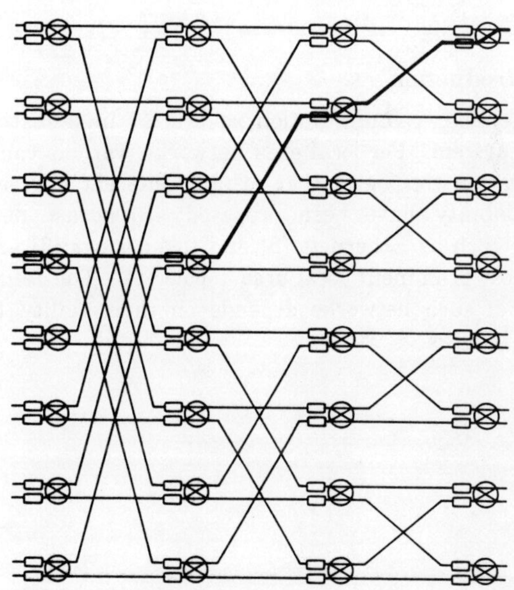

FIGURE 7 ONE DEDICATED CHANNEL WITH BACKGROUND UNIFORM TRAFFIC

A RESILIENT COMMUNICATION STRUCTURE FOR LOCAL AREA NETWORKS

Amr El Abbadi*, Thomas Räuchle

Department of Computer Science
Cornell University, Ithaca, New York, 14853

ABSTRACT

Reliable communication is crucial to the correct functioning of distributed systems. We propose a multi-ring communication structure and a reconfiguration algorithm that tolerate multiple link failures before the network divides into more than one partition. In case of partitioning, each partition is reconfigured to allow communication among the sites within the partition. The algorithm handles recovery of links and merges partitions once links become operational again. The algorithm itself is fault-tolerant, and it is fully distributed and does not require global knowledge about the status of the network at any one site.

1. Introduction

Reliable communication is crucial in a distributed system. For local area networks, various communication topologies that exhibit different degrees of reliability have been proposed. Broadcast networks such as Ethernets [5] and token rings [3] are the most prominent local area topologies. The reliability of such networks depends on their ability to mask failures.

Failures range from noise in the communication medium, to interface failures, to failures of segments of the communication system. There are standard techniques for error detection and correction to overcome the problem of noisy communication channels. These techniques are independent of the communication topology. In Ethernet like networks, failures of the interface only affect the host connected to that interface, while the failure of a segment of the cable causes the entire communication network to stop functioning. In ring networks, the failure of a segment of the medium restricts communication to one-way communication between a subset of the sites in the ring; halted interfaces also lead to such ring discontinuity. Multiple failures result in partitioning of the network into subnetworks called *partitions*. The sites in each partition can communicate with each other but not with sites in another partition.

Various approaches to increase the reliability of communication systems have been proposed. Replicating ethers has been proposed to increase reliability of broadcast networks [2]. In [8], reliability is increased by adding one additional communication ring between sites. Schröck [7] extends this technique by imposing a logical loop structure on top of two rings.

In this paper, we present a communication structure and a reconfiguration algorithm that achieves high fault-tolerance and handles partitioning and merging of partitions. Our approach is based on the fiber optical loop proposed by Schröck[7]. It tolerates failures of communication segments and interfaces, and guarantees communication between each pair of sites that is connected through a functional communication path.

The sites in the system are ordered as a ring. Each neighboring pair of sites (s_i, s_j) is connected by k links. The reconfiguration algorithm tolerates

$2k - 1$ link failures in the worst case before the network partitions into subnets. In case of partitioning, each partition is reconfigured and again tolerates up to $2k - 1$ failures before further partitioning occurs. The algorithm also handles the recovery of links and ensures that partitions are merged correctly.

The algorithm does not require global knowledge about the network and is itself tolerant to failures during its execution. Furthermore, extensions of the network do not require changes to the algorithm.

The paper is organized as follows. In Section 2 we describe the basic communication structure and an algorithm that tolerates one link failure before partitioning occurs. The algorithm is then extended to handle link recovery and partition merging. Section 3 describes an extension to the communication structure and an algorithm to tolerate multiple link failures before partitioning. Section 4 concludes the paper with a summary of the results.

2. The Basic Double-Ring Approach

In this section we present the physical and logical topology of the communication network and then give a reconfiguration algorithm that tolerates one link failure before partitioning occurs.

2.1. The Communication System

Sites access the communication network by connecting to a *switch*. The switches are ordered in a circle and each pair of neighboring switches is connected through a bidirectional *link*. A link consists of two unidirectional *edges*, directed opposite to each other. Thus, the physical communication system forms two unidirectional rings, arranged as depicted in Figure 1a. Messages on the inner ring, called the *receiving ring*, flow in the direction opposite to messages on the outer ring, called the *sending ring*. Message transmission time on a nonfaulty edge is bounded by a constant T; hence the time to send a message around a ring is bounded by nT, where n is the number of segments in the ring.

Logically, the network is configured into a *loop* as depicted in Figure 1b. One of the switches is designated as *head*, the switch immediately preceding the head on the sending ring is designated the *tail*. The head generates a continuous stream of empty message frames on the sending ring. To send a message, a site acquires an empty frame on the sending ring, inserts the data and destination address into the frame, and passes the frame on to the next switch on the sending ring. The access protocol is similar to the slotted ring protocol in [Pie72]. Messages on the sending ring are relayed until they reach the tail. The tail receives messages on the sending ring and sends them back on the receiving ring. A message is received by its destination on the receiving ring. Note that the link between the head and the tail is not used for communication. This redundancy allows for link failures to be tolerated as described below.

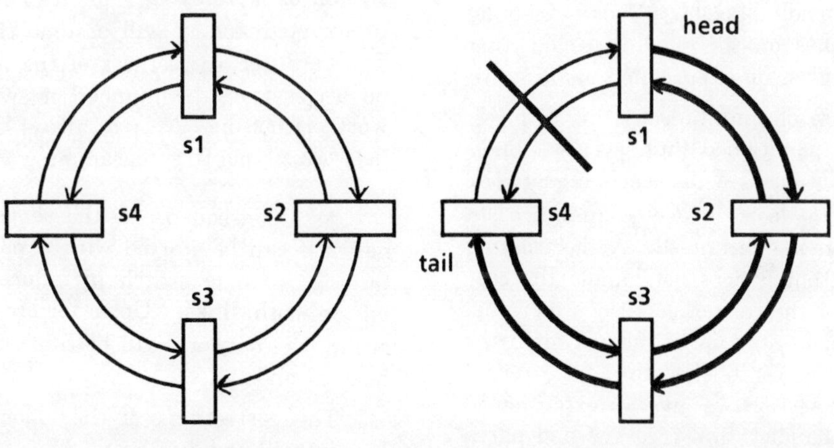

(a) (b)

Figure 1: Physical and Logical Configuration

2.2. The Algorithm

Edges fail by not delivering any messages, losing some messages, or delivering messages too slowly. These failures cause the message stream to be interrupted; thus edge failures can be detected. If the message stream on an edge is interrupted for more than T units of time, the switch at the receiving end of the edge detects the failure. The switch then changes to *failure mode* and executes the reconfiguration algorithm. The algorithm is based on an algorithm given in [7].

In failure mode, a switch s_i first propagates the fact that a failure has been detected. It does so by sending a stream of failure messages, denoted m_f, on both its links.[1] A timer is started when the first failure message is sent. A failure message takes at most T time units to reach a neighboring switch. If this neighbor is not already in failure mode, it will change to failure mode and start sending its first failure message, which also takes at most T time units to reach back to s_i. Thus, if s_i does not receive m_f messages on a link by $2T$ time units after it sent on the link, it declares the link faulty and stops sending over it. Note that a link is faulty if at least one of its edges is faulty. If a link fails during the execution of the protocol, the stream of m_f messages is interrupted. The switch at the incoming end declares that link failed. After faulty links have been detected, switches assume the role of head or tail as follows. If the link that contains the incoming sending edge is faulty, the switch assumes the role of head, if the link with the incoming receiving edge is faulty, it assumes the role of a tail. Each head then starts transmitting empty message frames on the outgoing sending edge. Upon receiving empty message frames on the sending edge, other switches enter normal mode of operation again.

Note that in case of simultaneous link failures, the network will be partitioned into several subnetworks. The algorithm ensures that each subnetwork is configured into a loop; hence communication within each partition is guaranteed. A shortcoming of the algorithm is that it does not handle the case of links recovering or the case of partitions merging. This implies that upon startup of the network, the loop structure has to be established through external means. In the next section we give an extension of the algorithm that handles link recovery and partition merge.

[1] Sending on a link means sending on the outgoing edge of the link.

2.3. Link Recovery and Merging of Partitions

To handle the recovery of links, it is necessary to detect when links become functional again. This is accomplished through the following protocol, executed periodically by the tail switches in each partition. The tail switch sends out a probe message, denoted m_p, on the outgoing edge of the faulty link.[2] Whenever a head switch receives a message m_p on a "faulty link", it sends the message back on the other edge of the same link. If the originating station receives m_p back, the link has recovered and the merge protocol can be initiated.

To merge partitions, the same reconfiguration protocol as for failures is executed, i.e., failure and recovery of links are treated uniformly as "change of configuration." Once a switch detects recovery of a link, it starts sending m_f messages and thus forces its neighbors into failure mode and the reconfiguration algorithm is executed as described above.

This algorithm however fails to establish a head and a tail when all links in the network are operational. In this case, all switches in the network will remain in a state in which they continuously send and receive failure messages on both links. None of the switches has an adjacent faulty link, thus no switch will assume the role of head or tail, and the system will livelock.

To detect this state, a predetermined switch, say s_1, keeps track of the time after it sent the first m_f message. After receiving m_f messages for $2NT$ time units, s_1 will stop sending failure messages and assume the function of head. Consequently, the stream of m_f messages for its neighbor s_n becomes interrupted and s_n will assume the role of the tail. To guarantee correctness of the algorithm, N must be at least n, the number of switches in the network. Choosing $N > n$ allows for extending the network without reprogramming s_1.

As a consequence of the above modification, the network can be started with n partitions, each containing a single switch in failure mode that sends m_f on both links. Upon termination of the algorithm, the network will be configured in a loop with head s_1 and tail s_n.

The correctness of the algorithm follows from Lemmata 2 and 3 in Section 3.

[2] Note that a link is declared faulty once one of its edges is faulty, so sending on a faulty link might be possible.

3. A Multi-Ring Structure that Tolerates Multiple Failures

In this section we present a reliable communication system, the *multi-ring*. In the multi-ring, each pair of neighboring switches in the circle is connected through k links. The communication system consists of k pairs of rings, where the rings in each pair are directed opposite to each other. Figure 2 shows a communication system with 2 pairs of rings. The introduction of additional pairs of communication rings to connect the switches increases redundancy and allows for more fault-tolerance. We extend the previous algorithm to reconfigure a network with k pairs of rings into a single loop. In the worst case, the algorithm tolerates $2k - 1$ link failures before the system is divided into two partitions.

In normal mode of operation, there is a *current loop* that is used for message transmission. In the example in Figure 2b the current loop is depicted by bold edges. As in the simple case, the head generates a continuous stream of message frames on the sending ring of the current loop. The tail receives messages on the sending ring and sends them back on the receiving ring.

Each switch s_i maintains the two sets in_i and out_i. For each i, in_i contains all links l_j, $1 \leq j \leq k$ adjacent to s_i, such that l_j is non-faulty and its sending edge is directed towards s_i. Out_i contains all links l_j, $1 \leq j \leq k$, such that l_j is non-faulty and its sending edge is directed away from s_i. After s_i detects a change in configuration (failure or recovery of a link), it changes to failure mode and starts sending failure messages m_f on all links in $in_i \bigcup out_i$. This causes its neighbors to change to failure mode and send failure messages. A link is deleted from in_i or out_i whenever s_i detects that the link failed. The switch then stops sending on this link. A link is added to the appropriate set when s_i detects that it is operational again. If links fail during the execution of the algorithm, the appropriate *in* and *out* sets reflect this change. If no failures occur for a sufficiently long time, in_i contains the same links as $out_{(i-1) mod\ n}$ for all i.

Failure messages are being sent until one of the two following conditions occurs. Either, there exists a switch s_j with $in_j = \phi$, or, the *in* sets of all switches are non-empty. In the first case, s_j assumes the role of a head. It follows from the observation above, that $out_{(j-1) mod\ n} = \phi$, and hence $s_{(j-1) mod\ n}$ will assume the role of a tail. In the second case, a predetermined switch, say s_1, assumes the role of the head after receiving failure messages for $2NT$ time units. It stops sending on its *in* set, and hence s_n will assume the role of the tail. Note that if the system is partitioned, each partition will contain a head and a tail.

Once a head has been determined, it starts sending empty message frames on the innermost link of its *out* set. A site receives a message over a link from its *in* set, and forwards it on the innermost link of its *out* set. The tail bounces messages back on the innermost link of its *in* set.

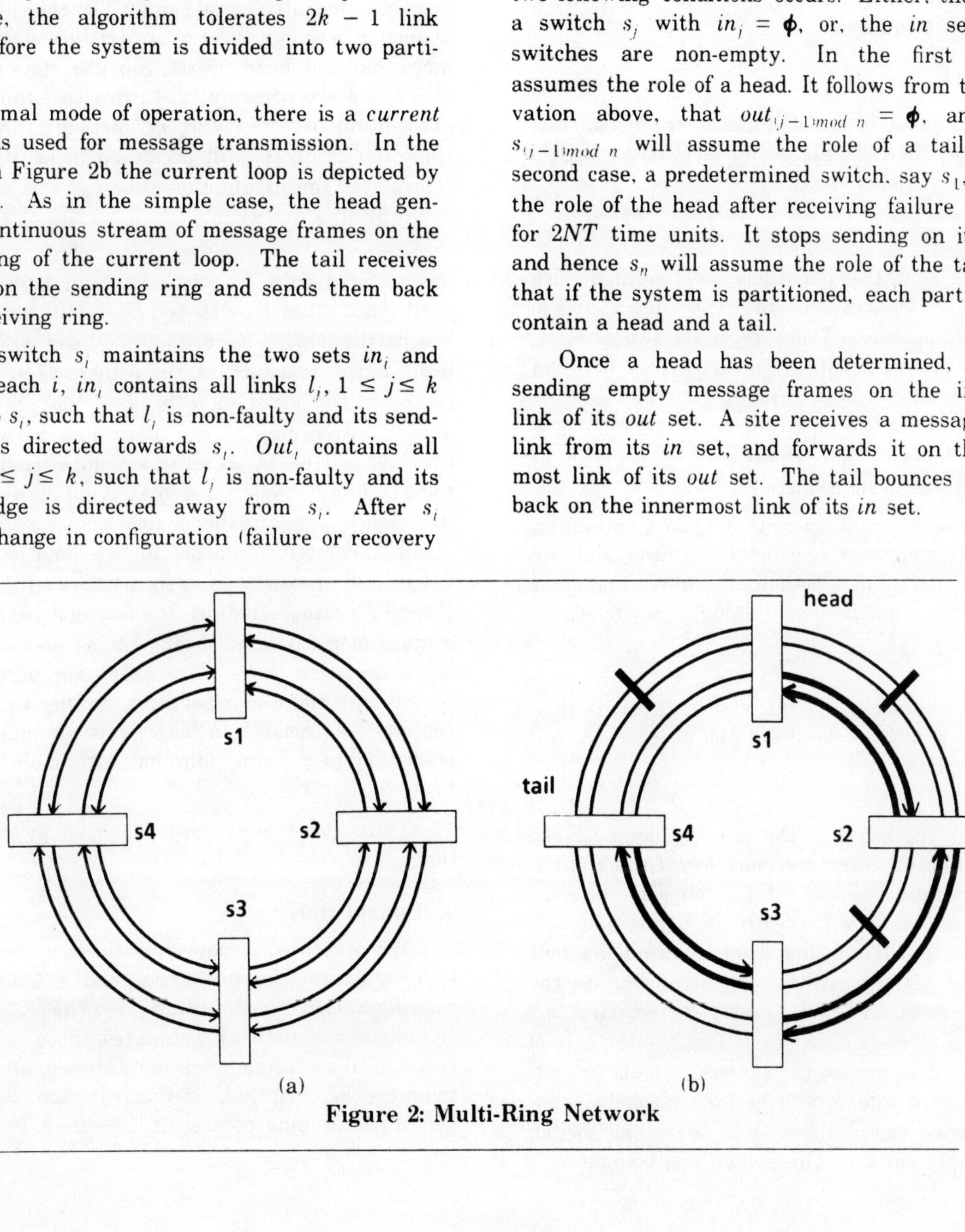

Figure 2: Multi-Ring Network

The algorithm tolerates $2k - 1$ link failures in the worst case before partitioning occurs. Failure of any switch corresponds to the failure of k links. In particular, the correctness of the algorithm does not depend on s_1 being operational.

3.1. Proof of Correctness

In this section we first prove that the network will in the worst case tolerate $2k - 1$ link failures before any partitioning occurs. We then prove that each partition contains exactly one head and one tail.

The following lemma shows that the network tolerates $2k-1$ failures.

Lemma 1:

If the number of link failures in the network is less than $2k$, then the algorithm terminates with a loop that connects all switches.

Proof:

Assume that two partitions have formed with fewer than $2k$ failures such that two sites s_i and s_j cannot communicate. There must be two sites s_p and s_q, $p \neq q$, such that any message sent between s_i and s_j must pass through s_p or s_q and $in_p = in_q = \phi$. A set in_i is empty, if and only if, k links have failed. Thus, there must have been at least $2k$ failures; a contradiction. □

Lemmata 2 and 3 prove that upon termination of the algorithm, there is exactly one head and one tail in each partition. Note that a fully functional network forms a partition containing all switches.

Lemma 2:

Upon termination of the reconfiguration algorithm there is at least one head and at least one tail in each partition.

Proof:

We give the proof for the tail, the proof for the head is analogous. Assume there exists some partition that contains no tail. First consider the case where there is no switch with an empty *out* set. This implies that failure messages m_f are being sent continuously to all switches. But according to the algorithm, after $2NT$ time units of receiving m_f messages on a receiving edge, s_1 assumes the role of a head and does not send messages on its *in* set. Hence $out_n = \phi$ and s_n will be tail. Secondly, consider the case where there is at least one switch with an empty *out* set. This switch will become a tail $2T$ time units after its *out* set became empty. In both cases we obtain a contradiction to the assumption that there is no tail in a partition. Therefore each partition contains at least one tail. □

Lemma 3:

Upon termination of the reconfiguration algorithm, there is at most one head and at most one tail in each partition.

Proof:

We prove that there is at most one head, the proof for the tail is analogous. For the sake of contradiction, assume there is a partition that contains more than one head. First, consider the case where all *in* sets are non-empty. In this case, all switches are continuously receiving m_f messages. After $2NT$ time units, only s_1 will decide on being head. This contradicts the assumption that there is more than one head in a partition.

Secondly, consider the case where the system has suffered from link failures, and hence may be partitioned. Let $P = (s_i, s_{i+1}, \cdots, s_l)$ be the smallest partition with more than one head where s_i is a head and s_l is a tail. Assume there is at least one switch s_j, $i < j \leq l$ such that $in_j = \phi$, i.e., s_j is also a head. According to the algorithm, a switch that does not receive m_f on its *in* set stops sending m_f on these links and becomes a head. Since $in_j = out_{(j-1) \bmod n}$, the switch $s_{(j-1) \bmod n}$ becomes a tail. Therefore, s_j is not in the same partition as s_i, i.e., P consists of subpartitions P_1 and P_2. P_1 and P_2 either contain at most one head or they contain more than one head. In the first case, there is no partition with more than one head, in the second case there is a partition smaller than P that contains two heads. In both cases we obtain a contradiction to P being minimal with more than one head.

Thus, there is at most one head in each partition. □

4. Conclusions

We presented a new fault-tolerant communication structure, the multi-ring and a fault-tolerant reconfiguration algorithm. The algorithm reconfigures a network connected through k double rings to allow communication between all sites. It tolerates $2k - 1$ link failures in the worst case before partitioning occurs. If more than $2k - 1$

failures cause the network to partition, the algorithm ensures communication within each partition. In terms of network connectivity, this compares favorably with the resilient broadcast network proposed in [2], which tolerates $k - 1$ link breakdowns in the worst case before partitioning. The kinds of failures our system tolerates are interface failures, link breakdowns, lossy links and slow links.

After adding the ability to detect link recovery, we achieved a uniform treatment of link failures and recovery. Unlike the approach presented in [8], our algorithm treats all failures uniformly; furthermore, it merges partitions once links become operational again. Consequently, the network can now be started with each switch in a singleton partition. After termination of the algorithm, communication will be possible between all sites in the network.

The logic required in the switches is very simple and can thus be made very reliable without high cost. The algorithm is fully distributed and depends only on the ability to communicate between neighboring switches in the network. Thus, there is no single central switch on whose reliability the reliability of the network depends.

5. Acknowledgements

We would like to thank Thomas Joseph and Özalp Babaoğlu for their comments on an earlier version of this paper.

6. References

[1] S. B. Akers, B. Krishnamurthy, "Group Graphs as Interconnection Networks", *GE Technical Information Series 84crd156*, July 1984

[2] O. Babaoğlu, R. Drummond, "Communication Architectures for Fast Reliable Broadcasts", *Proceedings 4th Symposium on Fault-Tolerant Distributed Computing*, Silver Spring, MD, 1984

[3] D. J. Farber, K.C. Larson, "The System Architecture of the Distributed Computer System - the Communications System", *Symp. on Computer Networks*, Polytech. Institute of Brooklyn, 1972

[4] L. Lamport, "Using Time Instead of Timeout for Fault-Tolerant Distributed Systems", *ACM Trans. on Prog. Languages and Systems, 6, 2*, April 1984

[5] R. M. Metcalfe, D. R. Boggs, "ETHERNET Distributed Packet Switching for Local Computer Networks", *CACM, Vol 19*, 1976

[6] J. R. Pierce, "Network for Block System of Data", *Bell System Technical Journal, Vol. 51, No. 6*, July - August 1972

[7] W. Schröck, "A Fault-Tolerant Loop-Structured Local Area Network on the Base of Fibre-Optical Transmission", *Proceedings 2.GI/NTG/GMD Conference on Fault-Tolerant Computing Systems*, Bonn, September 1984

[8] J. J. Wolf, et al, "Design of a Distributed Fault-Tolerant Loop Network", *Proceedings 9th IEEE Fault-Tolerant Computing Symposium*, Madison, WI, 1979

ALPHA TRANSPORT

Bahram Enshayan *

Amdahl Communications Systems Division
2500 Walnut Avenue, Marina del Rey, CA 90291

ABSTRACT

The control of the resources of a packet switching network is a very difficult problem to solve.[1] Past attempts to solve this problem have been handicapped primarily by lack of an adequate means to quantify the user demands and then assign distributed resources to the demands in a globally coherent way.

This paper presents a new approach to control the resources of packet switching networks. The key to this control is a new network transport scheme called alpha transport. Alpha transport combines the TDM circuit switching concept and packet switching technology. It is easy to implement and particularly cost effective to carry the subnet traffic of meshed packet switching networks. Alpha transport is robust, enables high utilization of the subnet bandwidth, reduces the subnet delay, and provides fairness.

1. INTRODUCTION

Most of the control in a circuit-switched network concerns the establishment of a circuit between a pair of users. Once the resources to carry the traffic on a circuit are dedicated, the control becomes simple. Indeed, this dedication of resources makes possible a perfect phasing (i.e., synchronization) between resources carrying the traffic. By contrast, in a packet switching environment where the resources should only carry the actual traffic characterized by random interarrival time and random packet size, a perfect phasing of the same resources becomes impossible. This paper presents a way to achieve a high degree of phasing in a packet-switched network.

Consider the travel of a packet from source to destination in store and forward fashion. Perfect phasing occurs if, as soon as the packet is completely received over an incoming link, it is processed and immediately transmitted over the outgoing link. Consequently, the phasing consists of coordinating memory, processing, and capacity resources along the path.

Among the three resources, the strategy of control is determined by the relative cost of processing versus capacity -- memory resources are only used in compensating the lack of phasing between processing and capacity resources. With the present economics, processing resources are cheaper than capacity resources. With such economics, the processing resources are overallocated and the strategy of control becomes the optimization of the utilization of the capacity resources. This paper assumes such strategy; however, it will be shown later that the mechanisms of control will work equally well when the network is CPU bounded.

The lack of phasing is manifested by idle trunk capacity and by collisions. The capacity idling (i.e., underutilization) happens when, despite the user's demands, the capacity cannot be used because of at least one of the following reasons:

- Unavailability of the upstream resources to carry the existing traffic to the idle trunk
- Restraint applied by the control mechanisms to support more traffic because of the risk of congestion (e.g., in the conventional packet switching, trunk utilization of more than, say, 50 percent, is considered excessive)
- Utilization of the trunk capacity was neutralized because of lack of memory resources in the downstream node

A collision occurs when two or more packets are competing for the same resources -- in this case memory and trunk capacity. Collisions cause an increase in the size of outgoing queues. When the size goes beyond a certain level, congestion occurs. To prevent capacity underutilization and congestion, a high degree of phasing is required.

Given the stochastic nature of the demands on a packet-switched network, the major obstacle to achieving a high degree of phasing is the distributed environment of the network.[1] In the distributed environment of the packet-switched network, assigning backbone capacity to rapidly changing demands through the conventional control approach (that is, the exchange of control information between nodes) is ineffective. Indeed, the control information,

* Bahram Enshayan is currently employed at Citicorp TTI, Santa Monica, CA.

besides having a high overhead,[1] must be carried within the packet-switched network and is subject to large stochastic delays which make the information inappropriate for a high degree of phasing. This paper proposes to overcome this "natural" obstacle (i.e., distributed environment) by achieving phasing through minimal exchange of control information.

The remaining parts of this paper portray the general principles of alpha transport. Section 2 explains the principles of control that alpha transport has used. Section 3 describes the way to achieve phasing. Finally, Section 4 quantifies the phasing and compares its results with ideal results.

2. CONTROL: BASIC CONCEPT

This section presents the principles of control that have been applied to achieve a high degree of phasing in a distributed environment.

2.1 Scope of Control

This subsection identifies the part of the network that requires control in a distributed environment.

To begin with, consider a wideband meshed packet-switched network composed of n nodes. Each node supports a set of DTEs and is directly connected to some other nodes. We call trunk one or several communication lines connecting two adjacent nodes. This paper views a node of such network as two separate parts: the access part and the backbone part. The access part directly supports the DTEs and uses the backbone part to exchange traffic with remote nodes. It is assumed that the access part views the backbone part as n-1 virtual channels where n is the number of network nodes. From the control point of view, the access node does not distinguish between local and remote traffic and exercises flow control by regulating the incoming traffic to not exceed the outgoing traffic. Since the control mechanisms in the access part can be achieved purely locally, in the remaining part we consider only the backbone parts.

Proceeding further, this paper divides the backbone into a relatively limited number of separate entities. These entities are circuits, which we call alpha circuits. A pair of nodes only communicate through one or several alpha circuits. An alpha circuit is a virtual circuit. It can be a simplex or full duplex circuit; this paper assumes that it is a simplex. An alpha circuit can be established, terminated, and modified at run-time. Once established between a pair of nodes, an alpha circuit carries the traffic by multiplexing the user packets at the source node and demultiplexing them at the destination node.

Such circuit-oriented backbone networks can be viewed as two distinct levels: routing and transport. The routing level can be considered as a set of control points, called hereafter routing control points or RCPs, where an RCP controls the distribution of the load over all alpha circuits routed to a given destination node. Note that RCPs are not unique to alpha transport: they exist in any circuit-oriented backbone network to distribute the load over backbone circuits. Note also that if perfect phasing had been achieved between RCPs, they would share the transport resources without any conflict. The transport level, called alpha transport, consists of all control mechanisms necessary to carry the traffic on alpha circuits. Alpha transport views the RCPs as a set of users competing for the backbone capacity resources. As it will be shown later, alpha transport eliminates the need for the RCPs to be coordinated: thus, the latter, though in a distributed environment, do not need to communicate. The control functions of an RCP become local: it has to decide locally how to distribute traffic over its alpha circuits. How RCPs achieve optimal distribution of flow is beyond the scope of this paper. In the following, we only examine how alpha transport achieves a high degree of phasing of its capacity resources to support the flow submitted by RCPs.

2.2 Goal and Principles of Control

The goal of control of the backbone resources is to optimize the utilization of the capacity resouces while at the same time preserving the robustness of the backbone network against fluctuations of flow and resources -- due to failures.

In a circuit-oriented backbone network, the robustness can be achieved by assigning -- except for very short durations -- the same capacity along the path of an alpha circuit. This paper defines the optimization of the backbone resource as achieving fairness among all the alpha circuits and then maximizing the global network throughput. At this point, one should define the criteria of fairness.

This paper associates (but does not physically assign) certain capacity resources, that we call nominal capacity, to each alpha circuit. Actually, the nominal capacity is what distinguishes an alpha circuit from a network virtual circuit. The nominal capacity is defined during the topology design and corresponds to the capacity necessary to support the average estimated flow of the traffic. For instance, in the network depicted by Figure 1, it is assumed that the topology design has defined three circuits with nominal capacity 10Kbps, 50Kbps, and 5Kbps to support the traffic between a pair of nodes, i and j. We call an alpha circuit underloaded or overloaded if its flow needs less (or more) capacity than the nominal capacity. For the sake of clarity and without loss of generality, this paper assumes that the transmission lines are perfect and that the capacity of a trunk is the sum of the nominal capacity of all the circuits supported by this trunk. Note that the nominal capacity associated with an alpha circuit is backed up by actual physical resources.

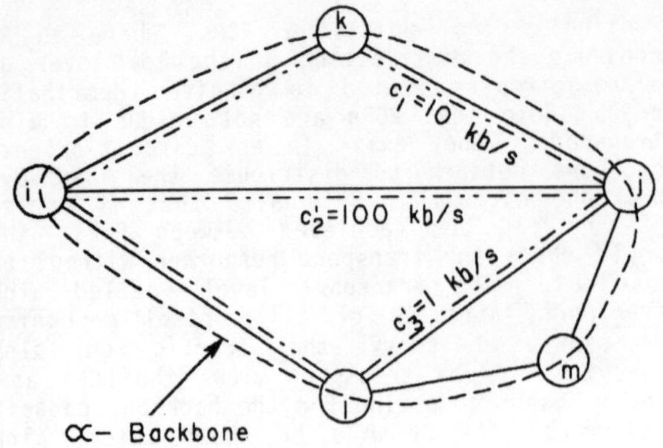

Figure 1: Structure of Alpha - Subnet.

This paper defines fairness as follows: if an alpha circuit is underloaded, then alpha transport must fully satisfy its request. The residual capacity (i.e., unused capacity) will be assigned among the overloaded circuits (if any) in proportion to their nominal capacity.

For example, in Figure 2-a, the three circuits are overloaded and fully utilize their nominal capacity, while in Figure 2-b, circuit 1 is underloaded and its capacity is distributed among alpha circuits 2 and 3 in proportion to their nominal capacity.

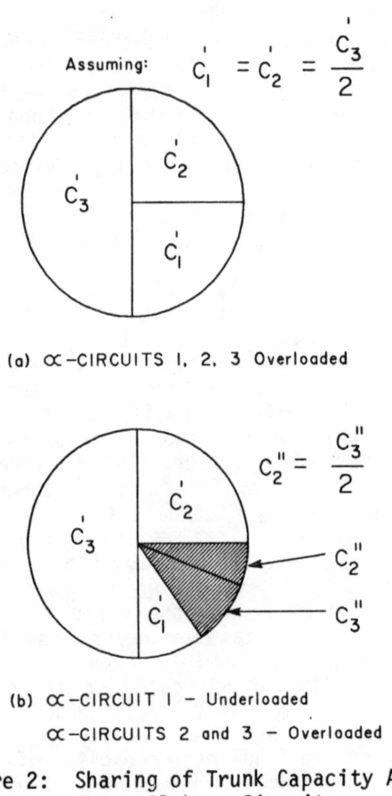

Figure 2: Sharing of Trunk Capacity Among Three Alpha - Circuits.

To explain how optimization of the capacity resources and robustness can be achieved, this paper starts examining how it can be achieved in a hypothetical -- ideal -- environment where exchanging control information is free and instantaneous. The hypothetical environment assists in identifying the intrinsic requirements of phasing. Once these requirements are identified, the paper will determine how they can be achieved in a real environment. Assume that at each end of a trunk there is a process that is called capacity assigner or CA, which assigns the capacity of the trunk according to the above mentioned fairness rules, among all the alpha circuits supported by this trunk. Assuming that any CA has instantaneous knowledge of:

- Flow submitted at the entry of the alpha circuits, and
- Capacity computed by all other CAs supporting the same alpha circuits

Then for a given circuit, a CA determines the assigned capacity as:

$c^\wedge(t) = \min(c(o,t), c^\sim(i,t))$ where:

$c(o,t)$ is the capacity necessary to support the flow submitted at the entry of the alpha circuit,

$c^\sim(i,t)$ is the capacity offered -- but not assigned -- by the CA of the nth trunk supporting the alpha circuit,

$c^\wedge(t)$ is the capacity assigned -- by all the concerned CAs -- to the alpha circuit.

In the remaining part of this paper, the flow supported by $c^\wedge(t)$ is called theoretical -- because of hypothetical environment -- flow. It can be shown that for a given submitted flow to the alpha transport (i.e., to the entry of the alpha circuits), the CAs distribute them in a unique way.

Note that unlike conventional packet switching in which shorter circuits tend to capture a larger share of the network resources,[2] alpha transport guarantees that each circuit will have at least its nominal capacity.

Until now, it was assumed that the CAs were operating in a hypothetical environment and had instantaneous knowledge of the state of the system. In the real environment, the CAs can use flow of traffic as a means to exchange information about the submitted flow and the distribution of the trunk capacity by other CAs. Indeed, certain regulation of flow that this paper calls capacity alignment, will communicate -- not instantaneously but in a relatively short period, T -- minimum $(c(o,T), c^\sim(i,T))$ to all concerned CAs.

3. PHASING

Alpha transport achieves a high degree of phasing through two mechanisms: capacity alignment and capacity assignment. This section presents how the alpha transport implements each of them.

3.1 Capacity Alignment

The capacity alignment can be achieved through a rather conventional flow regulation scheme. Alpha circuit propagates the capacity of a bottleneck trunk to all upstream CAs through backpressure, and attracts maximum flow through rapid delivery of acknowledgment packets.

Alpha transport implements its backpressure through a classical credit scheme.[3] The backpressure is applied by the receiver node through lack of renewal of the credits -- called hereafter storage credits -- for the sender node. To speed up the backpressure, the amount of storage credits is kept small. Assuming that a node stores a large packet into several buffers and an upstream node knows how many buffers are required to store a packet in the next downstream node, using buffer credits instead of window credits is suggested. The major advantage of buffer credits over window credits is that they perform well regardless of the size of the packets. One way to implement the credit scheme is for the sender nodes to keep track of the number of buffers necessary to store a packet in the downstream node. The receiver node then acknowledges the packet as if there is a window credit.

Transport mechanisms of an alpha circuit can attract maximum flow by providing: a) rapid renewal of credits; b) larger number of credits. Rapid renewal of credit can be achieved through: a) assigning higher priority to the transmission of acknowledgment packets than to data packets; b) hop by hop acknowledgment. To simplify the alignment of the capacity along a path, the acknowledgment packets are not generated before the link level has accepted the packet. To speed up the delivery of acknowledgment packets, it is assumed that an acknowledgment is generated as soon as the packet is given to the link level for transmission. The capacity alignment has been achieved through transport of data from source to destination and the transport of acknowledgment in the opposite direction (see Figure 3). Note that except for acknowledgment packets, no other control information has been exchanged to achieve capacity alignment.

To support an alpha circuit, a node needs KxL storage resources, where K is a storage coefficient -- simulation shows that for a K=1.5 the throughput is maximized -- and L is the number of buffers to support a maximum packet size. This paper assumes that alpha transport adopts buffer preallocation as storage strategy. The reason is that the maximum number of buffers to support worst case traffic is known and is relatively small. Note that the source node of an alpha circuit does not need to preallocate buffers. Also, in VC networks, an alpha subnet eliminates the need of supporting large numbers of data structure of VCs. It can be shown that the average size of memory that a node needs to support the data traffic of alpha transport is:

$k \times L \times p \times (q-1) \times (n-1)$ Kbytes; where:

- p is the average number of alpha circuits between a pair of nodes
- q is the average number of nodes supporting an alpha circuit
- n is the number of nodes in the network

By reducing its memory requirements, alpha subnet enables the access part of a node to have more memory (it is assumed that the access part and backbone part are sharing the same buffers) and thus accept a greater flow of traffic. The acceptance of the flow at the source will boost the potential of the alpha subnet to have a high utilization of the trunk capacity.

The fact that alpha transport can reasonably adopt preallocation buffer strategy makes alpha subnet a congestion free subnet and consequently robust against fluctuation of traffic and resources -- due to failures.

The small size of storage resources significantly reduces the delay and its fluctuation within the alpha subnet (see Section 4, Simulation Results).

3.2 Capacity Assignment

At a CA, the program that applies the capacity assignment rules is called CA algorithm. A way to implement the CA algorithm follows.

To each circuit i, a variable B ^ i, called current used bandwidth, is associated. It is assumed that when the link level is ready to transmit another packet, it signals the CA. The latter then selects the next alpha circuit that has a packet to transmit. Each time an alpha circuit i transmits a data packet p bytes long, the variable B ^ i will be incremented by p*Ki

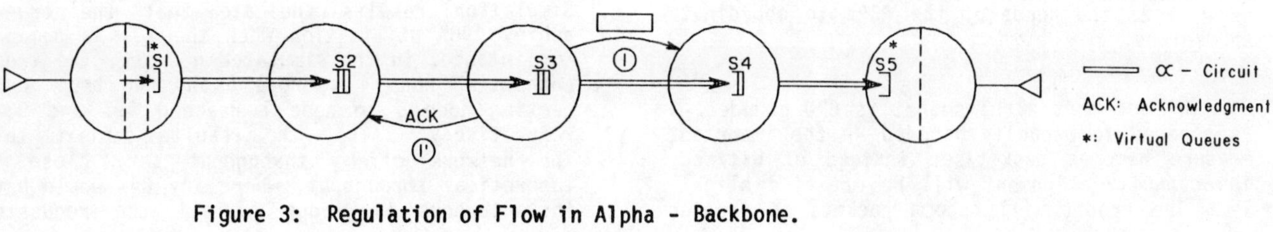

Figure 3: Regulation of Flow in Alpha - Backbone.

where Ki is the credit coefficient of the alpha circuit i. The credit coefficients are determined such that when B ^ i = B ^ j, the alpha circuits i and j have used the trunk bandwidth in proportion to their nominal capacity. In that context, the alpha circuit with the smallest used credit B ^ i is the one that has used, relative to other alpha circuits, the least of its nominal capacity. The credit coefficient is determined at the establishment of the circuit and whenever the nominal capacity changes. Each node computes its coefficient at run-time (see Appendix).

To achieve fairness, the CA algorithm tries (upon reception of the link level request for the next packet) to equalize all B ^ i (i=1, 2, ..., n) by selecting the alpha circuit i that has the smallest B ^ i and at least one packet waiting in the transmission queue. The CA algorithm then gives the first packet at the head of the transmission queue to the link level.

While it is desirable that an alpha circuit preserves its right to transmit for a certain period of time -- just enough to compensate the lack of very tight phasing between any two tandem trunks -- the algorithm as it is defined until now has a major drawback. Indeed, the problem with such an algorithm is that an idle alpha circuit preserves indefinitely its right to transmit. So, an alpha circuit which has been idle for a long time and then has a large number of packets to send may choke off all other alpha circuits supported by these trunks.

To eliminate this deficiency, an upper value B is imposed on the variables B ^ i. When a selected alpha circuit i already has a B ^ i larger than B, then all non-null B ^ i (i = 1, 2, ..., n) will be reduced by the value b where b is the smallest non-zero value of B ^ i (i = 1, 2, ..., n).

Notes

A. Except for acknowledgment packets, no other control information has been exchanged in the alpha transport.

B. Alpha transport assists the RCPs in the adequate distribution of the flow by ensuring the availability of an alpha circuit if its RCP is not momentarily using it. Even though alpha transport may allocate the capacity of an underloaded alpha circuit to other alpha circuits, it allows the RCP of the underloaded alpha circuit to fully utilize the capacity of the alpha circuit as soon as it is needed. So alpha transport eliminates the risk of oscillations[4] as well as the needs of the RCPs to coordinate themselves.

C. In case that alpha subnet is CPU bounded -- as opposed to capacity bounded -- the important measure becomes packet/sec instead of bit/sec. The capacity alignment will be modified slightly. The credit will become packets as opposed to buffers. The acknowledgment will be sent once a packet is processed -- as opposed to transmitted. The CA will be replaced by an equivalent algorithm called Processing Allocator, which will assign the processing resources to the incoming traffic. The algorithm of fairness is almost the same: instead keeping track of the amount of data, it will keep track of the number of packets.

4. SIMULATION RESULTS

This section presents the results of simulation regarding the throughput and delay of alpha transport.

4.1 Throughput

For studying the throughput of alpha transport, a 7-node network has been simulated. Figure 4 depicts the topology of the network, and Figure 5 presents the characteristics of the load and the results. Overall, we simulated the traffic of 28 alpha circuits, and for sake of clarity, we considered only 3 one-hop alpha circuits. The simulated traffic is stationary and randomly chosen as underloaded and overloaded. The packet length distribution is exponential and truncated at four times the mean value. The interarrival distribution is exponential. To present the impact of the storage credit on the throughput, we simulated the network with L and 1.5 L, where L is the maximum packet size.

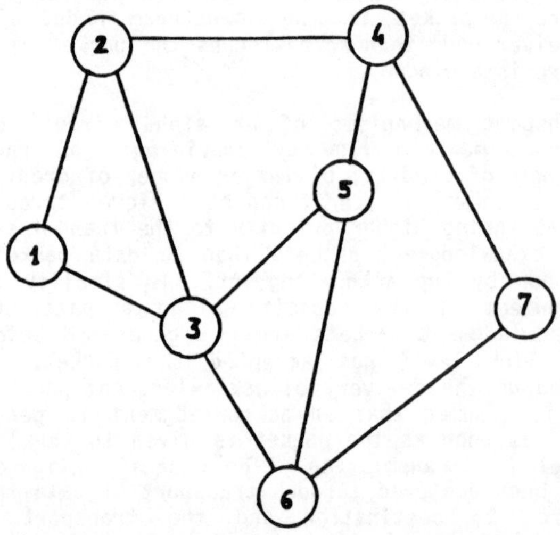

Figure 4: Topology of Simulated Network.

Simulation results indicate that the trunks achieve 98% utilization when there is a demand. For intance, in the simulated network, the trunk connecting node 4 to node 5 and the trunk connecting node 3 to node 1 reached 99% and 98%, respectively. Also, the results confirm that the network offers throughput very close to theoretical throughput, where any CAs would have instantaneous knowledge of all the requested

Alpha Circuit Index	Path	Nominal Capacity	Requested Traffic	Theoretical Throughput Value	Simulation 1 Throughput/Delay (kb/s) /(ms)	Simulation 2 Throughput/Delay (kb/s) /(ms)
1	1-2	300	202	202	201/0	202/0
2	2-1	300	400	302	302/0	302/0
3	3-5	100	200	200	200/0	200/0
4	5-3	100	50	50	50/0	50/0
5	1-2-4	100	97	97	97/2.	97/2
6	4-2-1	100	200	100	100/26	100/48
7	1-3-5	400	500	500	424/2	474/2 *
8	5-3-1	400	500	404	391/5	403/9
9	2-3-5	300	50	50	49/1	49/1
10	5-3-2	300	400	350	338/1	341/2
11	4-2-3	200	50	52	52/1	52/1
12	3-2-4	200	196	195	196/3	196/3
13	5-6-7	100	100	100	100/6	100/5
14	7-6-5	100	50	60	60/4	60/4
15	4-5-6	200	500	200	196/4	196/4
16	6-5-4	200	194	194	191/5	194/6
17	2-3-5	300	400	305	302/9	305/16
18	6-3-2	300	101	101	101/1	101/1
19	1-2-4-7	200	47	47	47/3	47/3
20	7-4-2-1	200	200	200	194/5	196/18
21	2-3-6-7	200	400	200	200/27	200/43
22	7-3-6-2	200	51	51	50/2	51/2
23	1-3-5-7	100	100	100	100/25	100/45
24	7-6-3-1	100	100	100	92/16	91/32
25	3-6-7-4	400	400	400	388/10	394/16
26	4-7-6-3	400	500	500	419/4	494/5 *
27	1-3-5-4	300	200	200	200/5	200/4
28	4-5-3-1	300	400	300	293/6	299/4

Simulation 1: All storage capacities are L = 41ℓ(minimal value allowed)

Simulation 2: All storage capacities are 1.5L = 61ℓ
Notice circuits 7 and 25 (*): the difference with theoretical throughput disappears with a larger storage.

Trunk capacities:

trunk 1-2 : 600	trunk 1-3 : 800	trunk 2-3 : 1000
trunk 2-4 : 500	trunk 3-5 : 1100	trunk 3-6 : 1000
trunk 5-6 : 300	trunk 6-7 : 800	trunk 4-7 : 600
trunk 4-5 : 500		

Figure 5: Simulation Results

traffic and the results of other CAs. Column 5 displays the computed theoretical throughput, and columns 6 and 7 display the results of simulation with storage capacity at L and 1.5 L. The simulation results show that in most of the cases, the storage capacity equal to L gives results very close to the theoretical throughput -- the only exceptions are circuits 7 and 26 (see Figure 5). These two circuits achieve their theoretical throughput with storage capacity respectively equal at 2.5xL and 2xL. The storage value 1.5xL has minor effect in further increasing the throughput (compare results of column 6 and 7). It has been observed that the interarrival and packet length distributions have minor effects over alpha transport.

The simulation has shown two situations when an alpha circuit cannot reach its theoretical throughput.

The first case is when the alpha circuit wants to use a large part of the capacity of a trunk and the reverse trunk is highly utilized. Because the transmission of the acknowledgment packets is delayed by the high utilization of the reverse trunk, and the transmission window is too small to compensate this delay, the loss of throughput occurs. Figure 6-a depicts a network simulating this situation, and Figure 6-b shows that by increasing storage credit to 2L, the alpha circuit can reach its theoretical throughput.

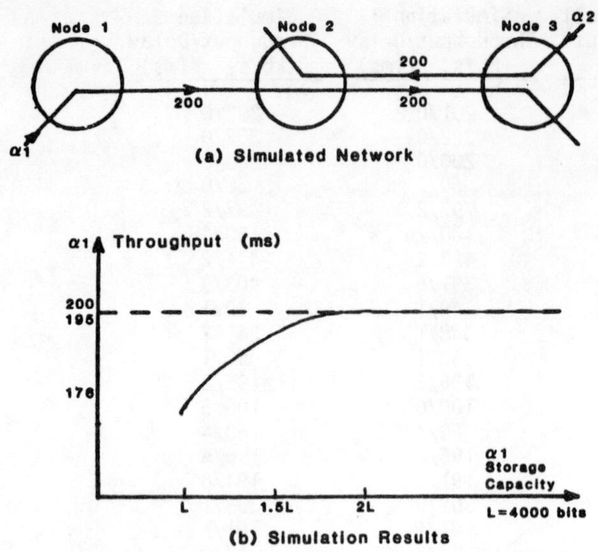

Figure 6: Effect of Line Utilization of the Reverse Trunk on the Delay of a Balanced Alpha - Circuit.

The second case is when an alpha circuit may lose its throughput in favor of another one and only happens in the following exceptional situations. A trunk supports only two alpha circuits, 1 and 2, and alpha circuit 1 has a relatively -- to alpha circuit 2 -- small nominal capacity. Alpha circuit 1 is overloaded and alpha circuit 2 is balanced (see Figure 7-a). In this scenario, alpha circuit 2 loses throughput in favor of alpha circuit 1. This is caused by multiple consecutive resets where, in some resets, alpha circuit 2 loses forever some of its right to transmit. Note that a trunk

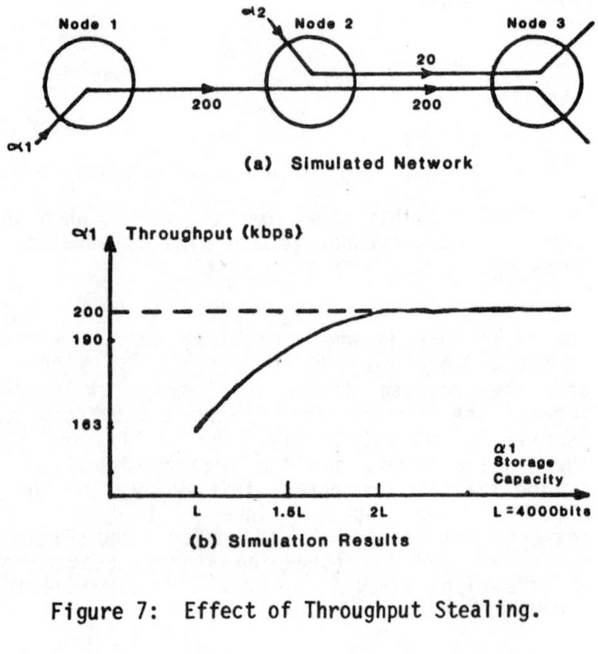

Figure 7: Effect of Throughput Stealing.

supporting only two alpha circuits is rather an exceptional case. This drop of throughput can be eliminated by increasing storage credits to a value which most of the time is 2L (see Figure 7-b).

Except for the above -- very rare -- described situation, the capacity assignment achieves the fairness requirements.

4.2 Delay

Delay is defined by the mean waiting time in an intermediate node. For a given alpha circuit, this delay has an upper limit which is defined simply by the ratio of the storage credits to the actual capacity of the alpha circuit. The delay characteristics of an alpha circuit depend on whether it is underloaded or overloaded.

Underloaded Alpha Circuit. To study the delay of an underloaded alpha circuit, the simple network depicted by Figure 8-a has been simulated. This simple network supports three alpha circuits. Alpha circuit 1 is the underloaded circuit being measured for delay. Alpha circuits 2 and 3 modify the utilization of the direct (respectively reverse) trunk.

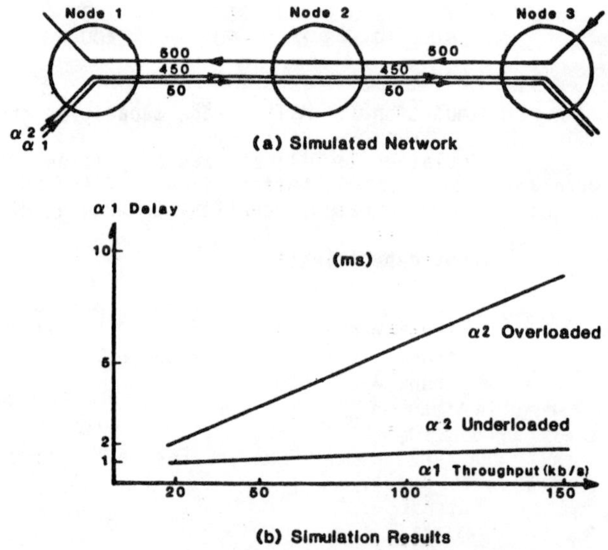

Figure 8: Effect of Line Utilization on the Delay of an Underloaded Circuit.

For an underloaded alpha circuit, the delay increases with the utilization of the direct trunk. Figure 8-b shows the effect of utilization of the direct trunk on the delay of alpha circuit 1. In this figure, the underloaded (overloaded) direct trunk is characterized by a 125Kbps and 500Kbps. Simulation results have also shown that the utilization of the reverse trunk does not influence the delay of an underloaded alpha circuit. Note that for an underloaded alpha circuit, increase of the storage credits does not influence the delay.

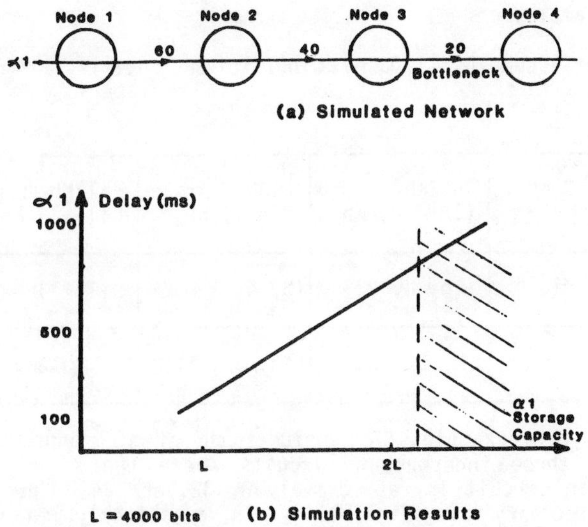

Figure 9: Effect of Bottleneck on Delay.

Overloaded Alpha Circuit. To study the delay of an overloaded alpha circuit, the simple network depicted by Figure 9-a has been simulated. This simple network supports only one alpha circuit which was submitted to a load larger than 60Kbps. The capacities of the trunks are arranged to create a bottleneck on the trunk connecting node 3 to node 4. Simulation results show that the delay behind the bottlenecked trunk is the same for every node and increases with the storage credits (see Figure 9-b). The dashed line separates the area where the increase of storage credits does not increase the throughput but rather the delay.

5. ACKNOWLEDGMENTS

I wish to thank Dr. P. Ziv and J. Rinde for their encouragement and support to pursue this work, and D. Horton for encouraging me to present this paper. I am indebted to M. Heard for many intense and productive discussions and to J. Rivals for his outstanding contributions in simulating alpha transport and his valuable comments. I would like to thank Dr. A. Brandwajn, Dr. Yung-Li Jow, and J. M. Chauvet for their analytical and simulation help on alpha transport. I wish also to thank Professor M. Gerla for his valuable suggestions in organizing this paper. Finally, I am grateful for the assistance provided by Amdahl CSD, which made this paper possible.

REFERENCES

(1) L. Kleinrock, "Principles and lessons in packet communication," Proc. of the IEEE, Vol. 66, No. 11, pp. 1320-1329, November 1978.

(2) W. Chou and M. Gerla, "A unified flow and congestion control model for packet networks," Proc. Third Int. Conf. on Computer Communication, ICCC, Toronto, Canada, August 1976.

(3) V. Cerf and R. Kahn, "A protocol for packet network intercommunication," IEEE Trans. on Communication, COM 22, May 1974.

(4) M. Gerla, "Routing and flow control in computer networks," IEEE Magazine, November 1984.

APPENDIX

This appendix first provides a way to determine the coefficient Ki and then a way to compute them by each node at the run-time.

A. Determination of Ki

Let:

- Ci be the nominal capacity of the ith alpha circuit,
- D be the largest common divisor of (C1, C2, ..., Cn)
- M be the smallest common multiplier of (d1, d2, ..., dn) where di = Ci/D

Ki is defined as M/di

Example:

i	1	2	3	4	5
Ci	30Kbps	20Kb	60Kb	40Kb	12Kb
D	2Kbps				
di	15	10	30	20	6
M	60				
Ki	4	5	2	3	5

This formulation of Ki has two disadvantages. First, the value of Ki may become very large. Second, the run-time addition of an alpha circuit or run-time modification of the nominal capacity of an alpha circuit may require the recomputation of all Ki of the CAs of all trunks supporting the alpha circuit. In other words, the run-time introduction or modification of the nominal capacity becomes a very complex task and seriously compromises the flexibility of alpha transport.

Consequently, during the establishment of a new alpha circuit or during the modification of the nominal capacity of an existing alpha circuit, it is highly desirable that all concerned nodes can set up the new alpha circuit or modify (the nominal capacity of) an existing alpha circuit rapidly and easily. The second formulation makes it possible that a node do this easily.

B. Computation of Ki

In this formulation, each nominal capacity is decomposed into a binary expansion:

$$Ci = \sum_{j} Aij * 2 \quad \text{where Aij are either 0 or 1.}$$

CA considers each non-null Cij as a separate circuit; in addition, it takes D and M as network constants.

Example:

Let assume D = 1000 bits and M = 2^7 = 128

i	1	2	3
C i	22Kb (16+4+2) Kb	13Kb (8+4+1) Kb	17Kb (16+1) Kb
di	(16, 4, 2)	(8, 4, 1)	(16+1)
Ki	(8, 32, 64)	(16, 32, 128)	(8, 128)

In this example, CA considers the first circuit as three independent circuits where the Ki of each circuit is respectively 8, 32, and 64. The price of this flexibility is a slightly increased computational load for the CA algorithm.

Panel 5

Remote Procedure Calls Versus Problem-Oriented Protocols

Chairperson

L.J. Miller
Xerox

Panelists

P. Leach, *Apollo Computer*
A. Freier, *Xerox*
R. Lyon, *SUN Microsystems*
F. Parr, *IBM Yorktown Heights*
AT&T Bell Laboratories

Session 6

Naming, Addressing, Routing

Chairperson
D. Oran
Digital Equipment Corporation

ISSUES IN USING DARPA DOMAIN NAMES FOR COMPUTER MAIL

Douglas E. Comer
Purdue University

Larry L. Peterson
University of Arizona

ABSTRACT

Although the principle goal of the *DARPA* Domain Name project is to provide a distributed name resolution mechanism for host names in the *DARPA* Internet, it is envisioned that the domain naming scheme will also support names for mailboxes in a *global* mail system that extends to mail networks beyond the *DARPA* Internet. In addition to overviewing the fundamental ideas behind the domain name scheme and outlining a proposal for using domain names as an umbrella naming system for computer mail, we discuss the administrative and technical problems involved in using domain names in the global mail system.

1. OVERVIEW OF DOMAIN NAMES

From its inception, hosts in the *DARPA* Internet have been identified by alphanumeric names chosen from a *flat* name space. The database of bindings of host names to Internet addresses (called the *host table*) is maintained by a central authority that oversees changes to the table, and distributes copies of the table to every host in the Internet. Because the table must be distributed to all hosts in the Internet whenever changes are made, the centralized host table approach assumes that changes happen infrequently, and that the number of hosts remains relatively small.

As the number of hosts in the *DARPA* Internet has grown larger in recent years (resulting in more frequent changes to the host table), the flat name space has become impractical. The domain name project offers a solution by replacing the flat name space with a hierarchical one [4, 5]. Structuring the name space in a hierarchy allows the central authority to delegate authority over the collection of name to address bindings; the single host table is partitioned and distributed over multiple databases, thus making the name space more manageable.

Domain names for hosts are composed of a sequence of simple names separated by a delimiter ("."). For example, a host named *state-red* in the flat naming scheme might have the domain name *red.CS.State.EDU* (*EDU* is the proposed abbreviation for "Education"). Domain names are resolved into Internet addresses by a hierarchy of naming authorities, called *domains*, each of which defines an interpretation of a set of simple names. That is, each domain partially resolves a full domain name.

In [3] we abstractly model naming hierarchies with a directed acyclic graph, with leaf vertices denoting addresses, and interior vertices representing contexts. Our model applies to domain names when leaf nodes denote Internet addresses and interior nodes correspond to domains. The edges of the graph represent the bindings of simple names to either another domain, or to an Internet address. (Our description of the naming graph differs from that presented by Mockapetris.) Conceptually, the resolution mechanism resolves the example domain name *red.CS.State.EDU* by starting at the root domain and moving down the hierarchy, processing the name from right to left. Figure 1 shows an example naming tree, where the resolution mechanism interprets the right-most simple name at each domain by moving to the next domain based on a match between the simple name and an edge label. The resolution mechanism concludes when it reaches a leaf containing an Internet address. (Note that the subtree beginning at d_{EDU} contains educational organizations, while the subtree beginning at d_{COM} contains commercial organizations, as prescribed by [9].)

When implemented, the single conceptual naming graph is distributed by partitioning the set of domains into *zones*, such that the bindings contained in the domains of a given zone are managed by a *name server*.† For example, the subtree with a root corresponding to domain

† For efficiency, name servers cache information from other zones, and for reliability the bindings in a given zone may be contained in multiple physical name servers.

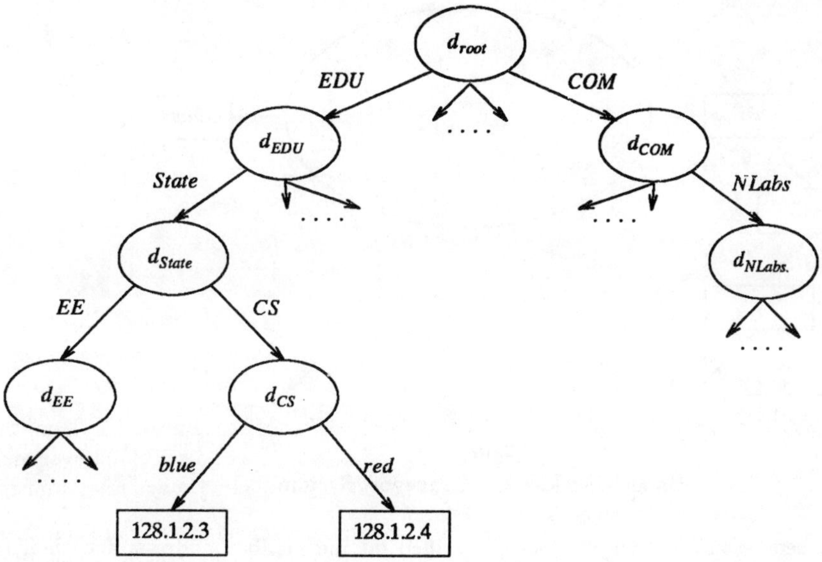

Figure 1
Conceptual Domain Name Graph

d_{state} might constitute a zone that is implemented by a name server (called the State University name server), where an administrator at State University is responsible for the contents of the name server (rather than the central Internet authority). Other name servers also exist for other organizations (e.g. National Labs), as well as for the upper levels of the hierarchy (i.e. a root name server might implement a zone consisting of d_{root}, d_{EDU}, and d_{COM}). Thus, to resolve our example domain name, a *resolver* program running on the behalf of a user first contacts the root name server, which in turn contacts the State University name server, which returns the appropriate Internet address.

We note two final observations regarding name servers. First, because organizations are required to be of a certain size before they are permitted to operate their own name server, a *guardian* authority (e.g. the root domain), implements a name server for a collection small organizations. Second, because domains are abstract entities, and in practice name servers define how names are resolved, we note that name servers are the ultimate authority for names in the domain name space.

Based on the understanding of domain names just presented, we identify two fundamental principles of the domain name scheme. First, the domain name solution to naming is based on the premise that authority over naming should be delegated. That is, control over names defined by the name space should be partitioned along administrative boundaries. Second, domain names provide a *logical* way to name hosts, premised on the belief that resources should be named independent of the underlying communication network.

The rest of this paper addresses the design problems involved in applying domain names to computer mail. In other words, we view domain names as providing a solution to naming, and we think of the way in which mailboxes are identified in computer mail as an application of such a naming scheme. The purpose of this paper is to discuss how well the solution matches the application. Specifically, the next section briefly describes the underlying structure of computer mail, while Section 3 explains how domain names can be used in computer mail. Section 4 then investigates three issues concerning the desirability and workability of using domain names for computer mail.

2. COMPUTER MAIL

The underlying mail system consists of a confederation of interconnected delivery systems. Users on a single computer share a *local* mail delivery system. If computers communicate through shared protocols, they may cooperate to form a *network* mail system. Finally, when computers are connected to more than one network, network-wide systems overlap to form a *global* mail system [1]. Hosts that connect networks and pass mail from one to another are called *relays*. Although the global mail system extends over several networks, we simplify our discussion by assuming a global mail system that consists of the *DARPA* Internet [8], and the *UUCP* network [6], as illustrated in Figure 2.

On the *DARPA* Internet, users identify recipients with a mailbox address that specifies a user on a certain host. For example, a sender might identify John Doe, a computer scientist at the National Laboratory, as *jxd@nlabs*. On *UUCP*, however, the mailbox address must specify all the intermediate computers through which the

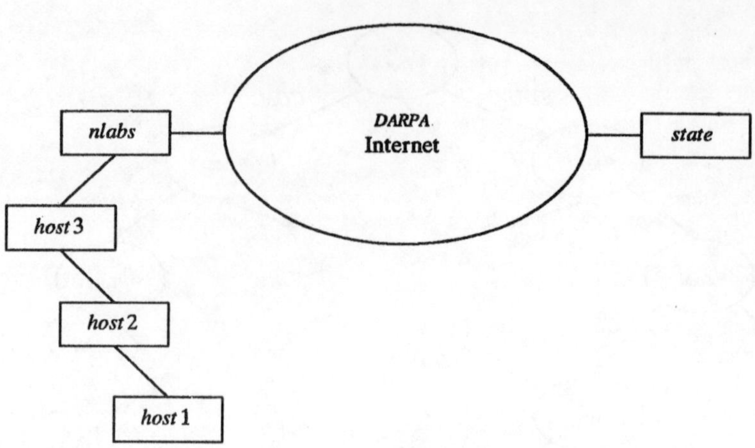

Figure 2
Underlying Message Transport System

message must travel. Thus, a sender on *host1* might identify John Doe's mailbox as *host2!host3!nlabs!jxd*. *UUCP*-style addresses are called *source-route* addresses [11] because the route through the network is specified at the source computer. Finally, when the recipient's mailbox exists on a remote network, the address must identify the relay through which the message must pass. For example, the sender in the previous example might address John Smith at State University as *host2!host3!nlabs!jxs@state* (i.e. *nlabs* serves as a relay).

3. DOMAIN NAMES APPLIED TO MAIL

We now describe the use of the domain name system to identify mailboxes. Initially, we note that changing the style of host names has an obvious effect on mailbox addresses in the *DARPA* Internet: a mailbox address of the form *user@host* becomes *user@host.domain*. In this section we consider three additional ways in which domain names can be used to replace the mailbox addresses described in Section 2.

First, in the same way as universal name servers like the Clearinghouse [7] support uniform names for all objects in a distributed system, the domain name system provides a general mechanism for naming any resource in the *DARPA* Internet. That is, in addition to naming hosts, the scheme directly supports names for mailbox objects (called *mailbox binding* [4]), thereby allowing a user's mailbox to be identified with a domain name like *user.red.CS.State.EDU*, rather than by the conventional address (e.g. *user@red.CS.State.EDU*).

Resolving a domain name for a mailbox produces the Internet address of the host on which the user's mailbox is **currently** implemented. Because the domain name for a mailbox can be rebound to a new Internet address if the mailbox is moved (i.e. the owner of the mailbox moves to a new host), using domain names for mailboxes avoids the situation in which a message is sent to the host specified int the mailbox address, and then *forwarded* to the host at which the mailbox actually resides. The ultimate destination is determined at resolution time rather than at delivery time.

Second, domain names provide an *umbrella* naming mechanism for the entire global mail system. In other words, rather than forcing users to remember a different syntax for identifying mailboxes in different networks, and requiring that users specify a path through relays to remote networks, domain names can be extended to provide a single naming convention for both hosts and mailboxes throughout the global mail system.

One approach to extending domain names is to represent each network mailing system in the global system with a top-level domain (an exception is made for the *DARPA* Internet). For example, *UUCP* could be a top-level domain, such that a sender in the *DARPA* Internet could identify a user on host *host1* in the *UUCP* network with the domain name *user@host1.UUCP*. In this situation, resolving the domain name produces the Internet address of a relay between the *DARPA* Internet and the *UUCP* network. The message is then sent to that relay, which is in turn responsible for forwarding the message to *host1*. In practice, the relay might execute the *USENET pathalias* [2] program to generate a route to *host1*.

A more appealing approach is to use logical names like *EDU* rather than network names like *UUCP*, thereby hiding the physical network to which the recipient's host is connected. In other words, the recipient in the previous example might be named as *user@host1.EDU*. Such a name is still bound to the Internet address of the relay to the *UUCP* network.

Finally, domain names allow users to identifying mailboxes by specifying a user at an *organization* instead of forcing the sender to specify the host on which the mailbox resides. Thus, a mailbox named *user@red.CS.State.EDU* would be nameable as *user@State.EDU*. The domain name servers resolve names of the form *user@organization* into the Internet address of a host at the organization that is willing to accept messages for the user and take the responsibility of forwarding those messages to the host within the organization where the user currently reads his mail. This use of domain names is called *agent binding* [4].

4. DESIGN ISSUES

This section addresses the design issues involved in identifying mailboxes with names drawn from the domain name space described in Section 1. First, we argue against a single naming mechanism that resolves names for mailbox objects as well as hosts. Second, we consider the implementation difficulties associated with expanding the domain name system to the non-*DARPA* Internet portions of the global mail system. Finally, we identify a major strength of domain names: they permit users to identify users with respect to organizations rather than hosts.

4.1 Administrative Authority

Consider the difference between identifying a mailbox with a domain name of the form *user.host.domain*, and a conventional mailbox name of the form *user@host.domain*. In the first case, the same mechanisms that implement domain names for hosts (i.e. domain name servers and resolver programs), manage the *user* part of the name. In the second case, each individual host decides how to interpret the *user* portion of the mailbox address outside the realm of the domain name system. The "@" is said to provide a *firewall* between the network's name for a host, and the host's name its users' mailboxes. We argue against the extension of domain names to the *user* portion of mailbox names for administrative reasons. Extending domain names to mailbox objects contradicts the premise of domain names that authority should be delegated by forcing the domain name syntax and conventions on the way hosts name users. We foresee two consequences.

First, whereas retaining the firewall delegates authority over the *user* part of a mailbox address to the administrator at each host (and in some case to the individual users on each host), removing the firewall may result in a given host's mailbox names being managed at a remote location. Consider the following. If a given organization is not allowed to operate its own name server (i.e. a guardian authority implements a name server for a small organization), then mailbox names for the organization are managed by a name server at a remote location. Furthermore, the guardian name server is faced with managing the names of a very large number of mailboxes, and therefore encounters frequent update problems similar to those currently faced by the central host table.

Second, we understand that the *user* portion of a mailbox name usually identifies a person. For example, John Doe's mailbox might be named as *jxd.nlabs.COM*, *doe.nlabs.COM*, or even *john.doe.nlabs.COM*. In all three cases, the collection of names to the left of *nlabs* identifies the user John Doe. Extending domain names to the *user* part of a mailbox address forces the domain name's hierarchical approach to naming onto human names, and is therefore potentially restrictive. Retaining the firewall, however, implies that the user portion of a name is treated as uninterpetted data by the network's resolution mechanism for host names. For example, the *user* part of the name could contain a tuple [10] that is treated as a database query. Database tuples are more powerful then hierarchical names if the sender does not know the full hierarchical name of the recipient, but does know enough to uniquely identify the recipient. To illustrate the point, suppose that a sender does not know a recipient's last name (therefore a name like *john.doe.nlabs.COM* will fail), but he does know the recipient's first name, where the first name uniquely identifies the recipient in the organization. Hierarchical names fail in this case, while non-hierarchical names provide a more flexible solution.

A final note. While we realize that domain names for mailboxes merely provide an alternative to conventional mailbox addresses, and do not directly replace them, we understand that there exists an implicit effort to achieve uniformity in naming that may lead to the wide-spread adoption of domain names for mailboxes. We conclude that such use of domain names infringes upon the administrative autonomy of each organization, and therefore contradicts the domain name premise that authority should be delegated on administrative boundaries. Because we support the premise, we believe that domain names should not be extended to mailbox objects.

4.2 Implementing the Umbrella

As mentioned in Section 3, the naming umbrella allows users to identify message recipients with uniform domain names rather than network-dependent names. We now consider the routing difficulties involved in extending domain names to provide an umbrella naming mechanism for the global mail system.

Recall that when a sender in the *DARPA* Internet identifies a recipient in another mail network with a domain name, a name server in the *DARPA* Internet returns the Internet address of a relay to the adjacent network, and the message is forwarded to the relay for further processing. The approach assumes that the resolver program running on the sender's host is able to contact the name server; an assumption that holds only within the *DARPA* Internet.

When the sender is **outside** of the Internet, the local resolver program does not have direct access to a name server. For example, if a name server resides on the host *nlabs* in the system depicted in Figure 1, then the resolver program on *host2* cannot directly contact the name server. Thus, when a sender on *host2* identifies a recipient as *smith.State.EDU*, there is no way to determine the relay to which the message should be forwarded. The problem is compounded when there are several alternatives (i.e. several multiply connected networks.)

One solution is to have each non-Internet host maintain a list of *smarter* hosts. Messages are then forwarded to one of these hosts, which is assumed to be better able to interpret the recipient's address. Unfortunately, when the smart host is several hops away, the solution may be inefficient; forwarding a message to a smart host may move the message farther away from the destination. For example, if *nlabs* serves as a smart host for the *UUCP* network, mail addressed to *user@host1.EDU* originating at *host2* is first forwarded to *nlabs*, and then sent back to *host1*.

The only way to avoid the problem is to have the mail program (i.e. *SENDMAIL* [1]), on each *UUCP* host circumvent the domain name resolution scheme by recognizing messages destined for other *UUCP* hosts. The host would then generate its own route instead of depending on the relay. If *UUCP* mailbox names are of the form *user@host.UUCP* then having the mail program recognize *UUCP* names is trivial. If logical names of the form *user@host.EDU* are used, however, then each host must maintain a list of hosts for which it is willing to generate the route.

Similar routing problems occur with mail travelling from the Internet to a *UUCP* host. Consider the following example. Although the *DARPA* Internet and the *UUCP* networks are connected at many points, the domain name servers only identify a small number of those hosts as official relays. Without loss of generality, suppose the domain name server reports that a *DARPA/UUCP* relay is located on the west coast. If a user on an east coast *DARPA* host sends a message to an east coast *UUCP* host, the message will be sent to the west coast relay and then forwarded back to the east coast destination. There exists a problem of selecting the relay that results in the best route to a host on a remote network. That is, in addition to translating a domain name into an address, name servers must determine the best such address to return for the situation.

In general, the problem is one of selecting a route to the recipient's host (i.e. a path through the underlying transport system), based on two pieces of information: the name of the recipient (i.e. the mailbox address), and each host's knowledge of the network's connectivity. We can characterize a given network's resolution mechanism along a spectrum. At one extreme, names only contain information identifying the destination, and each host has complete knowledge of the network topology. At the other extreme, routing information is encoded into the name, and each host knows only about its diretly connected neighbors. In the middle of the spectrum, hosts know enough about the network topology to make an intelligent decision about the next step in a route.

In the conventional mail system for example, each host only knows about the hosts to which it is directly connected, and users explicitly specify the route that a message is to travel in the mailbox address. In the *UUCP* network, users provide a source-route to the destination, while in the global mail system they identify the relays through which a message must pass. Because users control the route, they employ heuristics when chosing the mailbox address. For example, to send a message from a *DARPA* Internet host on the east coast to a *UUCP* host on the west coast, the sender might route the message over the *DARPA* Internet to a west coast relay (assuming the Internet is faster and less expensive than *UUCP*), which then forwards the message to the destination.

Unlike conventional mailbox addresses that contain information identifying a route through the underlying physical network, domain names contain information that identifies a hierarchy of administrative authorities. Using logical domain names does not cause any problems within the *DARPA* Internet because all hosts are logically connected by the lower-levels of the network architecture; selecting a route in the Internet is trivial because for the purpose of mail all routes are logically one hop in length (i.e. routing is handled at lower levels, independent of application programs like mail).

Using domain names as an umbrella naming scheme for the global mail system, however, means that routing through the transport system is the responsibility of the name servers and relays. They must select the best route across networks for a given domain name. In other words, while domain names are founded on a logical hierarchy, and they do not provide a direct mechanism for determining routes, a route must be selected when transporting a message from the sender to the recipient over a global mail system. Thus, a fundamental problem faced by the umbrella naming system (i.e. the name servers and relays), is to translate a domain name into a reasonable route.

As stated above, the routing decision is based on both the information contained in the mailbox address, and the name server's knowledge of the connectivity of the underlying network (i.e. the global mail system). One solution is to encode into the name of non-*DARPA* hosts information that will aid in selecting the correct relay. For example, the *UUCP* network could be partitioned into geographical sectors, where a *DARPA/UUCP* relay is associated with each sector. The domain name for a *UUCP* host would then identify the geographical sector in which the host resides. For example, if *nlabs* is the relay for the "North Eastern United States", and *host3* is geographically located in the northeast, then a recipient at *host3* might be identified as *user@host3.NE.UUCP*. Unfortunately, encoding routing information into domain names violates the premise that domain names should logically identify hosts, and that they should not contain infor-

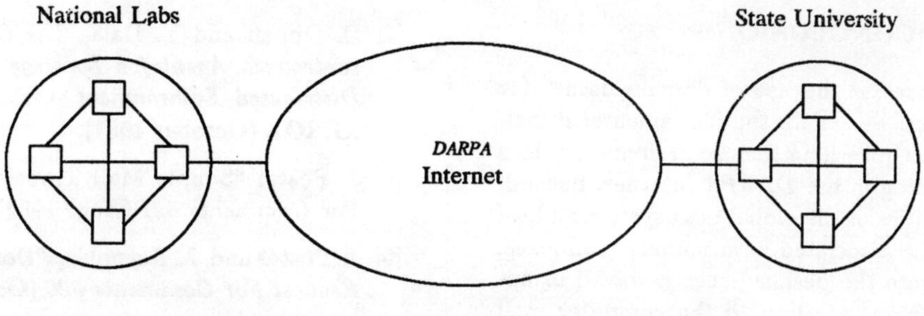

Figure 3
Hierarchical View of Message Transport System

mation linking them to physical networks. We therefore reject this approach.

At the other extreme, the name servers could examine the source address as well as the destination address, and then determine the best route from the source to the destination based the connectivity of the transport system. Because name servers would have to maintain information about the entire underlying network, however, the solution contradicts the assumption that the Internet and global mail system has grown too large (and changes too frequently), for schemes that store knowledge of the entire system centrally.

If we assume, however, that the path from the relay to the destination host in a remote network is the critical portion of the route that a message must travel (i.e. we consider the cost of sending a message from a host in the Internet to any relay in the Internet as being equal), then a compromise solution is possible. The name server selects the best relay based only on the length of the route from each official relay to the destination host. For example, if the root name server knows of *n* relays to the *UUCP* network, then the name server computes the route from each of the *n* relays to the destination (using *pathalias*), and returns the address of the relay with the shortest route. The name server needs to know about the connectivity of the *UUCP* network — information that is already maintained at each relay. In fact, if *UUCP* is a top-level domain, then the name servers and relays could be implemented on the same hosts.

4.3 User at Organization

Although in general domain names do not support routing through the global mail system, we now consider a special case in which the domain name system does support routing. When a sender identifies a recipient with a name of the form *user@organization*, the domain name servers return the Internet address of a host at the organization that is willing to accept the recipient's messages and forward them to the host within the organization that currently implement's the recipient's mailbox.

Initially, we note that sending mail to a user at an organization supports the premise that authority should be delegated; not only does it give hosts control over naming users, but it gives the organization control over identifying which host within the organization servers as the user's *home* machine. In addition, we believe that such use of domain names is desirable based upon the following two claims. First, we argue that it is natural for users to associate organizations with people. In other words, people already use the "person at organization" paradigm for identifying each other (possibly because of communication mechanisms outside the realm of computer mail; e.g. paper mail). Second, we claim that a hierarchical routing scheme can be imposed on the interconnection of hosts and networks that make up the global mail system.

Consider for example, the transport system depicted in Figure 3. Typically, the hosts at a given organization are closely linked together by a local area network, and the organization as a whole is connected to one or more long haul networks. (We simplify our discussion by assuming that an organization is confined to a single *site*. In general, it may be necessary to identify recipients with names of the form *user@site.organization.*) Thus, a natural route from a sender at National Labs to a recipient at State University is first from the sender's host at National Labs to a *mail agent* host at State University, and then from the mail agent host to the recipient's home host. In practice, even though the sender's and the recipient's hosts may be directly connected by the long haul network, little additional cost is incurred by sending a message to a single mail agent host and then forwarding the message over the local area network to the recipient's home machine. (In some cases the sender's host and the recipient's host are not directly connected and the message must be routed through the intermediate mail agent host anyway.)

Thus, although we understand that domain names provide a logical hierarchy for identifying resources, we conclude that the organization defines a point at which the logical hierarchy and the underlying transport system are "aligned" (assuming that organizations are limited to single sites). In other words, the hierarchical routing scheme for mail coincides with the logical hierarchy defined by the domain name tree. Because *user@organization* is a natural way for users to identify each other, and because the hierarchical routing scheme does not lead to a significant increase in cost, we favor its use.

5. CONCLUSIONS

This paper addresses the use of domain names for computer mail. First, we isolate the fundamental principles and assumptions governing the use of domain names as a solution to naming in the *DARPA* Internet. Second, we describe the realities of the global mail system and we identify the problems associated with naming mailboxes. Finally, we investigate the design issues involved in applying the domain name solution to the computer mail problem.

Specifically, we argue against using domain names to identify mailbox objects (i.e. mailbox bindings), because it compromises the principle of domain names that authority should be delegated across administrative boundaries. Second, we describe the routing difficulties involved in providing an umbrella naming scheme for the global mail system, and we discuss possible solutions. We point out that the routing problems arise because of an inherent conflict between the premise that domain names logically identify hosts, and the nature of routing messages through the underlying message transport system. Finally, we discuss names of the form *user@organization.* Using such names takes advantage of a match between the logical naming hierarchy provided by domain names and a hierarchical routing strategy through the underlying transport system.

ACKNOWLEDGEMENTS

In addition to [4], the extensions to domain names described in the third section have been discussed by the *namedroppers* mailing list.

REFERENCES

[1] E. Allman, "SENDMAIL — An Internetwork Mail Router," *UNIX Programmer's Manual,* 4.2 Berkeley Software Distribution, Vol. 2 (Aug. 1983).

[2] S. Bellovin and P. Honeyman, "Pathalias," *UNIX Programmer's Manual,* 4.2 Berkeley Software Distribution, Vol. 1 (Aug. 1983).

[3] D. E. Comer and L. L. Peterson, "Naming in Distributed Systems," *ACM Transactions on Computer Systems* (1984), submitted for publication.

[4] P. Mockapetris, "Domain Names — Concepts and Facilities," *Request For Comments 882* (Nov. 1983).

[5] P. Mockapetris, "Domain Names — Implementation and Specification," *Request For Comments 883* (Nov. 1983).

[6] D. A. Nowitz and M. E. Lesk, "Implementation of a Dial-up Network of UNIX Systems," *UNIX Programmer's Manual,* 4.1 Berkeley Software Distribution, Vol. 2 (1979).

[7] D. Oppen and Y. Dala, *The Clearinghouse: A Decentralized Agent for Locating Named Objects in a Distributed Environment,* Office Products Division, XEROX (October 1981).

[8] J. Postel "Simple Mail Transfer Protocol," *Request For Comments 821* (Aug. 1982).

[9] J. Postel and J. Reynolds, "Domain Requirements," *Request For Comments 920* (Oct. 1984).

[10] M. Sirbu and J. Sutherland, "Naming and Directory Issues in Message Transfer Systems," *Computer-Based Message Services,* Edited by H.T. Smith, North-Holland, New York (1984), 15-37.

[11] Carl Sunshine, "Source Routing in Computer Networks," *ACM SIGCOMM Computer Communication Rev. 7,* 1 (Jan. 1977), 29-33.

A PATH-ORIENTED ROUTING STRATEGY FOR PACKET SWITCHING NETWORKS WITH END-TO-END PROTOCOLS

R. AUBIN - P. NG

BNR, P.O. Box 3511, Station C
Ottawa, Ontario, K1Y 4H7 Canada

ABSTRACT

A path-oriented routing strategy is proposed for packet switching networks with end-to-end internal protocols. It allows switch pairs to communicate over multiple paths (for better network throughput), while maintaining knowledge of user connections at the network's endpoints only. The most significant aspect of this strategy lies in its flow assignment method. A distributed loop-free shortest path algorithm assigns a number to a path at the time it is created and this number remains valid through shortest path changes. Consequently, existing traffic can be maintained on existing paths, while new traffic is assigned to the current shortest paths. Stable multiple path routing is thus achieved without packet disordering. Abnormal conditions such as trunk failure and recovery and trunk congestion are dealt with by tagging routing updates with update causes. Simulation of this routing strategy shows that maximum network throughput (under a certain congestion constraint) can be increased substantially compared to a single path routing strategy.

INTRODUCTION

Today's packet switching networks are characterized by an increasing diversity and utilization of switching and transmission equipment. Making optimal use of such diverse and burdened resources is a challenge at a time when the service matures and network operators become more cost conscious.

Routing strategies have a major role to play in optimizing the use of network resources. Yet, current routing methods have been criticized for failing to sometimes even approach optimality[3]. In particular, switch pairs often communicate over single paths, when optimal routing would typically require multiple paths in order to even out traffic over the network. This effect was measured by Seligman[10]. Two experiments were particularly indicative of the improvement possible, one with mixed trunk speeds and the other with sparse traffic matrices and high network connectivity. Optimal multiple path routing achieved respectively 122 percent and 186 percent better network throughput than single path routing (under a certain delay constraint). Mixed trunk speeds, sparse traffic matrices, and high connectivity are precisely typical of the trends seen in packet switching networks.

This paper proposes a path-oriented routing method for packet switching networks with end-to-end internal protocols. This method permits the establishment of multiple paths between any two switches in a network in a fashion that allows the packets of the same connection to follow the same path. Packet disordering is thus minimized and tracing of user connections is possible. Some earlier strategies such as isolated load-splitting[4] were not path-oriented. They had the disadvantage of reacting slowly to distant events and increasing packet disordering. More recently, a path-oriented method was developed for the datagram-based SITA network[6]. Nevertheless, the static approach to routing taken by this strategy applies only to traffic that is easily predictable.

The fundamental property of end-to-end architectures can be preserved even if routing is path-oriented. Namely, state information about user connections can still be kept only at the network's entry and exit entities that perform the internal embedding X.25-like protocol. User connections can be switched from one path to another without explicit reconnect procedures; the decision to do so resides strictly with the connections' endpoints and is based on significant changing network conditions. Furthermore, transactions involve no more than the network's entry and exit points.

Path-oriented routing is presented in the context of a comprehensive experimental routing system called THESEUS. The paper starts by giving the general model underlying THESEUS. The methods used for metric evaluation and shortest path computation are then outlined. This is followed by a more elaborate explanation of multiple path flow assignment. Finally, the results of a simulation are presented.

ROUTING MODEL

A routing scheme is normally characterized by its performance criterion, its path computation method, and its packet forwarding strategy. More precisely, given a performance criterion, one can identify a routing metric, that is, a function that measures the lengths of network trunks with respect to performance optimization. This metric can then be used to compute optimal paths. Finally, the knowledge of optimal paths allows one to assign traffic flow. These three steps are illustrated in Figure 1.

A dimension not visible in Figure 1 is system responsiveness. It is customary to identify three increasing levels of routing responsiveness: static, quasi-static, and dynamic. Most systems react dynamically to the failure of a network trunk. However, reacting to changes in traffic is more controversial. With static routing, the feedback loop is performed off-line, while quasi-static routing controls traffic on-line. Both try to track statistically significant trends, but at a quite different pace. Dynamic routing is a limit case suitable to traffic with no predictable patterns; it tries to adapt to the instantaneous network state.

The method of computing routing tables is yet another dimension of the routing model. Computation can be centralized or distributed. This question can realistically be posed for quasi-static routing only; indeed, static routing is normally centralized and dynamic routing is normally distributed.

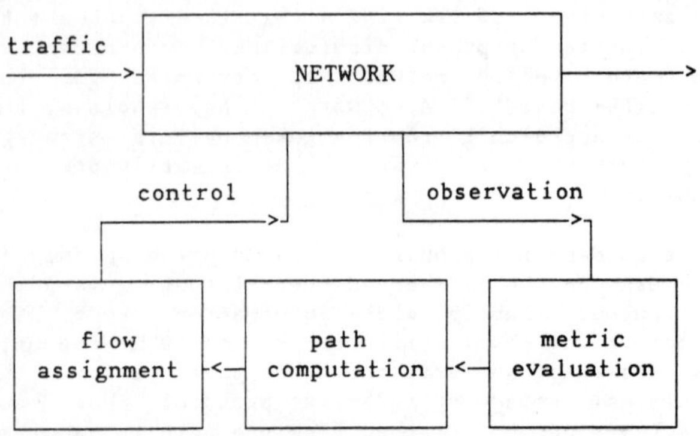

Figure 1. Basic Routing Model

THESEUS follows a *quasi-static distributed* approach to routing table computation. Better performance is thus expected. In this context, the following sections expand on the three basic routing components for THESEUS: metric evaluation, shortest path computation, and flow assignment.

METRIC EVALUATION

THESEUS adopts a tractable and approximate metric formula that takes into account processing, transmission, and propagation delay, as well as any necessary waiting delay. It is proportional to *marginal delay*, which is the first derivative of average network delay with respect to trunk throughput; this is a necessary condition for optimality. If propagation delay is ignored, then network throughput is maximized; otherwise, network delay is minimized[2]. Choosing a metric is not sufficient, however; one also has to decide how frequently to evaluate it and generate routing updates.

THESEUS proposes that every switch of a network evaluate the parameters of the metric (i.e., processor and line utilizaton, packet size, and packet throughput) periodically for each trunk. A new metric value results from taking the exponential average of the current value and the instantaneous measured value. However, routing updates are not necessarily generated every time the metric is evaluated. THESEUS favors *aperiodic* generation of routing updates, at the risk of converging more slowly and less accurately. In practice, control traffic due to frequent periodic updating has been observed to increase network congestion[7]; this phenomenon is not factored into periodic routing optimization methods. THESEUS sends routing updates only when certain thresholds are crossed. These thresholds depend non linearly on trunk congestion as captured by trunk utilization. Typical thresholds may be 60, 75, and 82 percent, for an increase in utilization, and 43, 69, and 79 percent for a decrease. Hysteresis is applied for stability.

With this approach, THESEUS is increasingly adaptive as load goes up. At very high load, however, when trunk utilization exceeds a certain maximum threshold, congestion control takes precedence over routing as will be explained later.

SHORTEST PATH COMPUTATION

The next step after deciding on a metric is the selection of a method for computing optimal paths. Classical distributed shortest path algorithms of the Tajibnapis family[11] are prone to transient routing loops during convergence, after a trunk length increase. As will be seen in the next section, THESEUS proposes to remember paths as soon as they are created. Consequently, in order to avoid the establishment of infinite paths, THESEUS requires a shortest path algorithm that provides *loop-free paths at all times*.

THESEUS has adapted an algorithm proposed by Jaffe and Moss[5] that not only maintains loop-free paths at all times, but also gives the best failure recovery performance. Furthermore, it is particularly well suited to aperiodic updating. Another loop-free algorithm by Merlin and Segall[9] is better suited to periodic updating. The enhanced version of the Jaffe-Moss algorithm used by THESEUS can be summarized as follows. The basic procedure is similar to the Tajibnapis algorithm. For decreasing trunk lengths and for trunk recovery, only this basic procedure is used. However, for increasing trunk lengths and for trunk failure, the Tajibnapis algorithm is preceded by a more coordinated phase that prevents loops.

Let the shortest paths from all sources to a given destination be represented by a spanning tree rooted at that destination (an arc in this tree represents one direction of a full-duplex trunk). When a length increase occurs on a trunk, special routing updates propagate to the switches uptree of this trunk as well as the neighbors of these uptree switches. The recipients of the special updates change only their distance to the destination without modifying the structure of the spanning tree. Upon completing this first phase, no uptree neighbor of a switch can be immediately chosen by this switch to be part of a new shortest path because none presents the shortest distance to the destination. No loops can then be formed and the second, Tajibnapis phase can safely be started by the switch that initiated the first phase. The Jaffe-Moss algorithm recovers from trunk failure in O(h) steps, where h is the height of the spanning subtree rooted at the point of failure.

FLOW ASSIGNMENT

The shortest path algorithm always delivers a single path for every destination. One simple way to attain multiple path routing is to assign new user connections only to this new path and keep existing connections on existing paths. Such a method, although not optimal, converges to a neighborhood of the optimum and then oscillates within that neighborhood[3].

Packet Forwarding Principle

THESEUS identifies each path at the time it is created. When a user connection is established, each endpoint obtains the identifier of the current shortest path to the other endpoint. Every packet sent over the connection is tagged with the appropriate path identifier instead of the destination switch identifier. Because these identifiers remain valid even after the creation of new shortest paths, the same paths can be used by a connection throughout its lifetime (barring any failure).

Routing tables also are indexed by path identifiers instead of destination identifiers. In addition to a trunk identifier, each entry contains the new path identifier to be assigned to the packet. Indeed, unlike a destination identifier, a packet's path identifier must typically be changed at every relay switch, akin to a logical channel number, since path identifiers are independently selected by every switch. At destination, the packet's path identifier is mapped to a special end-of-path indicator and further routing information in the packet is used to locate the connection's endpoint. If the path originally set for a packet terminates prematurely, e.g., because of a failure, the packet is discarded and the end-to-end network protocol ensures recovery.

Basic Extension to the Shortest Path Algorithm

THESEUS extends the distributed shortest path algorithm given in the previous section to incorporate the establishment of paths. The basic idea is to let routing updates carry a unique path identifier in addition to a destination and a distance (for the Tajibnapis phase only). Sending and receiving these new updates gives rise to the following new events:

- Suppose switch S receives over trunk t a routing update for destination D which contains path identifier p. This identifier is recorded by S along with the usual distance from S to D via t.
- Suppose switch S establishes a new shortest path to D via t. Then, S selects a path identifier q (see below for how q is chosen) and updates its routing table so that any packet arriving with path identifier q will be sent over t with path identifier p. Finally, switch S sends over its trunks routing updates containing the new shortest distance to destination D as well as path identifier q.

In order to trigger the algorithm when a trunk comes up, a switch sends over that trunk routing updates containing its current shortest distance and path identifier to every destination. A switch always has at least a valid path identifier to itself.

This general path assignment mechanism is best seen with an example. Figure 2 depicts a certain path assignment throughout the minimal spanning tree for a destination E just before an update takes place. The '*' indicates that a path is undefined or leads to a destination other than E; the '#' is an end-of-path indicator.

Suppose that as a result of a routing update, switch A has detached itself from its successor switch B in the minimal spanning tree and attached itself to a new successor switch D (see Figure 3).

Switch A still receives with path identifier 1 the packets that should go down the old path through B. However, it needs a new path identifier, say, 2, for the packets that are to go down the new shortest path through D. Meanwhile, both switches B and D still expect to receive packets with path

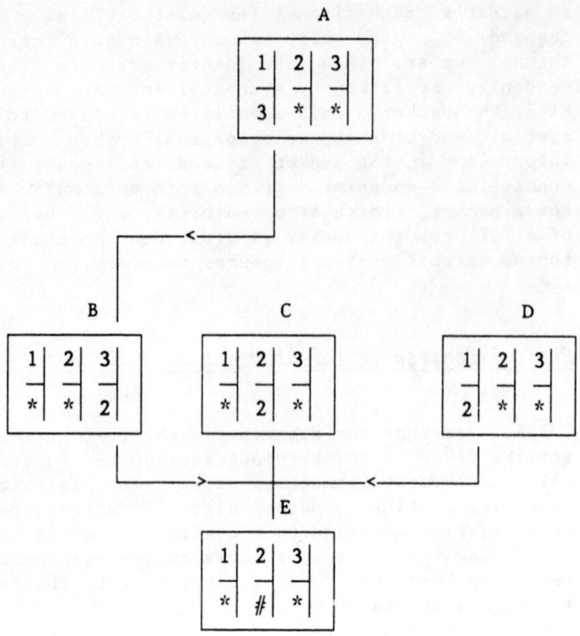

Figure 2. Initial path assignment

identifiers 3 and 1 respectively, because they were unaffected by the routing update. Consequently, switch A must change a packet's path identifier 1 to 3 before forwarding it to B and

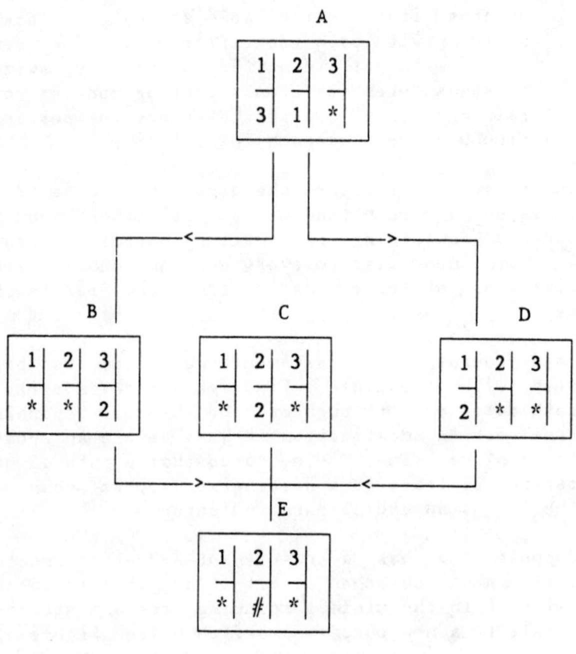

Figure 3. Path assignment after first updating

must also change a packet's path identifier 2 to 1 before forwarding it to D.

Figure 4 shows the state of the paths as a result of another perturbation, after switch D has attached itself to switch C and switch A has reattached itself to switch D. At this point, three path identifiers are in use in switch A.

The exact method for selecting new path identifiers at a switch for use in routing updates was left open. In reality, it divides into four cases. Suppose that switch S has chosen trunk t for a new shortest path to destination D, which happens to carry path identifier p:

- If the distance from switch S to destination D is infinite or if p is the undefined path identifier, then S selects the undefined path identifier. (No traffic can be carried on a path labeled with the undefined identifier.)
- If there is already a path identifier q mapping to p in the routing table of switch S, then S reuses path identifier q.
- Otherwise, if there is a path identifier q that is unassigned in the routing table of switch S, then S chooses this path identifier.
- Otherwise, path identifiers are exhausted at switch S; in this case, S uses the undefined path identifier.

A network should be configured with enough path identifiers to avoid the latter kind of congestion. It must be noted to the credit of

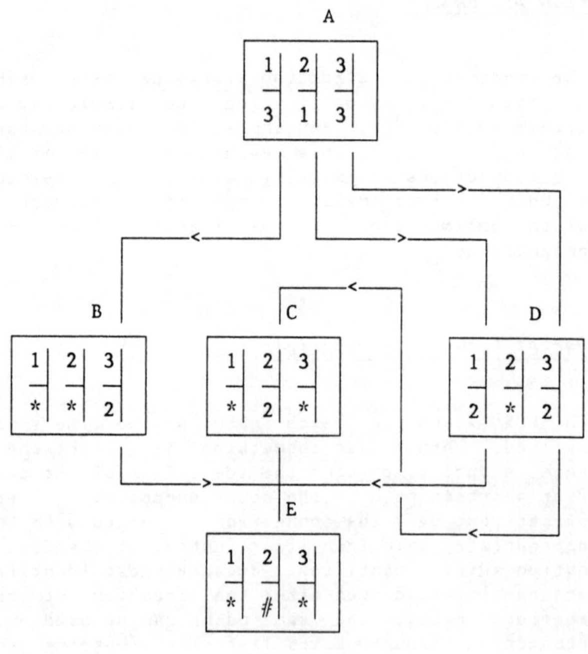

Figure 4. Path assignment after second updating

THESEUS that since it is based on spanning trees, it requires only O(n) path identifiers, where n is the number of switches in a network. Source-destination path assignment, such as routes in the Systems Network Architecture (SNA)[1], demands $O(n^2)$ identifiers.

Further Extensions to the Shortest Path Algorithm

The strategy of assigning new traffic to new paths and keeping old traffic on old paths must be adjusted to handle abnormal conditions such as trunk failure and recovery and trunk congestion. The general approach taken by THESEUS is to tag routing updates with an update cause. Such a cause is determined at the point of perturbation by the switch responsible for generating the first routing updates. The cause is then simply inherited from one update to the next. A switch can receive updates with one cause interleaved with updates bearing another cause without confusion. Some typical update causes are considered in the remainder of this section. In general, update causes are not limited to those.

Trunk Failure: When a trunk fails, messages are sent to erase all the paths using this trunk and free up their identifiers. The user connections generating traffic on these paths are forced to stop for a while. In parallel, routing updates are disseminated to establish a new shortest path. However, these updates are tagged with the 'trunk failure' cause. This is an indication that the user connections affected by the trunk failure must be switched to the new path. Furthermore, 'trunk failure' updates can be given higher priority than other kinds of updates.

Trunk Recovery: Another possible update cause is 'trunk recovery'. When a switch creates a new shortest path on the basis of a routing update carrying this cause, THESEUS proposes to move permanent virtual circuits to this path, but not established virtual calls. In general, permanent virtual circuits should be moved to a new path created as a result of a trunk length decrease; this ensures that they converge back to their intended paths, following a trunk recovery.

Trunk Congestion: When trunk utilization exceeds a certain maximum threshold, e.g., around 80 percent or more, THESEUS sends routing updates with the 'trunk congestion' cause. No new user connections can be assigned to a shortest path built from such updates. A drop in trunk utilization can later trigger new routing updates and recreate an open shortest path. This simple on/off congestion control mechanism allows for the detection and control of bottlenecks in a network. It is reminiscent of choke packets[8]; however, THESEUS completely integrates this warning to the shortest path algorithm.

SIMULATION RESULTS

THESEUS was validated by extensive simulation. The sample network studied contained 32 switches interconnected with 57 trunks and represented a real operating network. The simulator generated traffic at the user connection rather than packet granularity. The establishment of user connections was Poisson. For every source, the choice of a destination switch was random and biased according to the actual traffic matrix of the sample network, which was fairly sparse and unbalanced. All simulation runs were carried with the same offered traffic pattern.

The simulator reacted immediately to the mean connection throughput. For example, if a 120 b/s connection was established from source A to destination D through switches B and C, then the traffic over the A-B, B-C, and C-D trunks was increased immediately by 120 b/s. Furthermore, if a utilization threshold was crossed as a result of this increase, routing updates were generated immediately. Routing updates were scheduled as events in the simulator and interleaved with all the other events.

Some of the simulation parameters are tabulated as below:

packet size: exponential with mean = 400 bits
connection packet throughput: exponential with mean = 0.3 packets/s of mean size

source-destination selection: random, according to the traffic matrix

connection establishment rate: Poisson with rate = 8 connections/s

trunk processors capacity: 300 packets/s

trunk utilizaton thresholds: 60%, 75%, and 82% for an increase; 43%, 69%, and 79% for a decrease.

The main aspect studied by the simulation was maximum network throughput (under a congestion constraint). The results for a network with homogeneous 56 kb/s trunks show that THESEUS delivers 46 percent better throughput than a single path routing strategy whose metric is inversely proportional to trunk capacity (for stability). For both strategies, no connections were released and the simulation was terminated as soon as a trunk in the network was congested, i.e., over 82 percent utilized.

At the time the simulation stopped, an average of 3.75 paths (and a maximum of 12) had been created per switch pair, out of which an average of 1.72 path (and a maximum of 9) was carrying traffic. Average network delay was lower even when these extra paths involved more hops because long waiting times were avoided. Figure 5 plots the network average number of hops against the applied load, where the average number of hops is the ratio of internal to external network throughput.

A version of the network with mixed trunk speeds was also studied. Switches located in the same city were connected with 112 kb/s trunks while the speed of inter city trunks was maintained at 56 kb/s. With single path routing, the higher speed trunks tend to attract too much traffic at the expense of the lower speed trunks. Again, THESEUS avoids this by diverting traffic over alternate paths. In such a network, THESEUS achieved 99 percent better throughput than single path routing, when the simulation was stopped as soon as a trunk was congested.

More stress was placed on THESEUS by pushing the simulation until 70 percent of all trunks were congested in the homogeneous network. A well-known phenomemon was then observed: network throughput became worse for multiple path routing. This is because the traffic committed at lower applied loads over paths with extra hops prevents additional traffic from entering the network when higher loads are offered. But such levels of congestion are highly improbable and if a system should be able to cope with improbable situations, it should not necessarily be optimized for them. Note that when 35 percent of the trunks were congested, multiple path routing was still superior.

Finer aspects of THESEUS were also examined. For example, the number of thresholds certainly determines how active the routing system is; it presumably also affects performance. The simulation indicated that by increasing the number of thresholds (e.g., from 3 to 7), only a slight improvement in network throughput was observed (e.g., less than 4 percent).

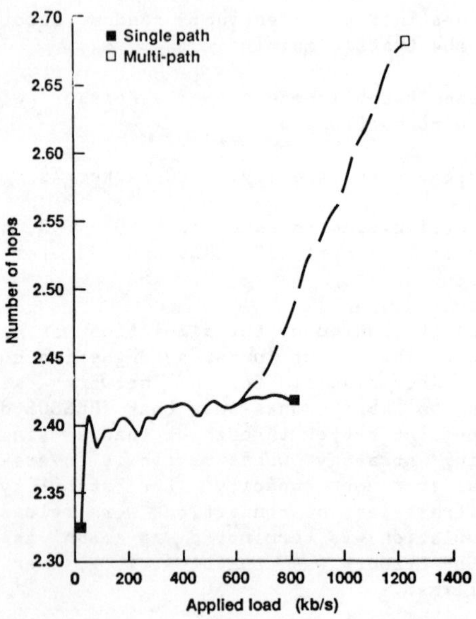

Figure 5. Traffic Spread

Besides maximum throughput, the other major aspect studied by the simulation was steady state throughput. This was achieved by loading the simulated network with connections of exponential duration with various means. The results are shown in Figure 6 for mean connection durations of 5, 10, and 15 minutes. In steady state, network throughput should oscillate around:

connection establishment rate
* connection mean throughput
* connection mean duration

Finally, regarding control information, the excess traffic due to routing updates was found to be negligible (less than 0.1 percent) and did not affect the simulation results. Furthermore, the average number of path identifiers used by a switch of the homogeneous network at the end of the simulation was 120, with a maximum of 159. Out of this number, an average of 55 identifiers only were assigned to paths carrying traffic, with a maximum of 72. This indicates that the demand on path identifiers is not excessive, even though there is a case for reusing the identifiers of idle or quasi-idle paths.

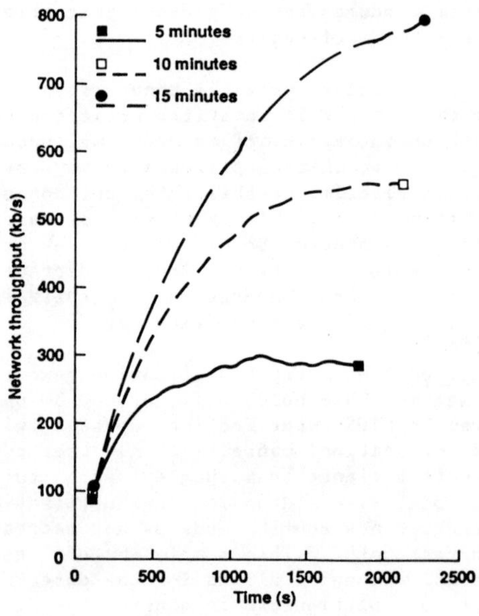

Figure 6. Steady State Throughput.

CONCLUSION

The work described in this paper began with the objective of increasing the maximum throughput of packet switching networks with mixed trunk speeds, sparse traffic matrices, and high connectivity. Theoretically, a necessary condition was known: multiple path routing had to be used in order to even out traffic over the network. The problem was to find a practical approach that would converge reasonably close to the optimum for quasi-static traffic conditions while preserving an end-to-end architecture and without introducing adverse effects such as packet disordering and excessive control information.

The cornerstone of the solution developed in the foregoing sections is a management scheme for multiple paths that is completely integrated to a loop-free distributed shortest path algorithm. In the context of this scheme, two design decisions were made to cut down on the amount of control information required, at the risk of converging less accurately. The first one said that routing update generation would be aperiodic and based on trunk utilization thresholds. The second one concerned flow assignment. The simplest scheme was adopted for which new user connections would be assigned to current shortest paths while existing user connections would be maintained on existing paths.

Simulation has shown that these practical choices were not overly simplifying. The increase in throughput obtained for a particular network under a congestion constraint was impressive compared to a single path routing strategy, i.e., up to 100 percent. However, it is not known how far from the optimum the better network throughput lies. It was further observed that just a few thresholds were needed to achieve this level of improvement; adding more thresholds yielded only a small marginal gain.

Inasmuch as the results of the simulation can be transposed to an operating network, this path-oriented routing strategy is stable, increases network throughput substantially, and does not require an excessive amount of control information to be stored and exchanged.

ACKNOWLEDGEMENTS

We would like to thank W.S. Lai for his technical advice and F. Mellor for his constant support throughout this project.

REFERENCES

1. Ahuja, V. "Routing and Flow Control in Systems Network Architecture." *IBM Syst. J. 18*, 2 (1979): 298-314.

2. Aubin, R. "Including Processing and Propagation Delay to Routing in Packet Switching Networks." *Proc. Int. Conf. Communications*, Chicago, 1985, pp. 107-111.

3. Bertsekas, D.P. "Optimal Routing and Flow Control Methods for Communication Networks." In *Analysis and Optimization of Systems*, pp. 615-643. Edited by F. Bensoussan and J.-L. Lions. Berlin: Springer-Verlag, 1983.

4. Davies, D.W.; Barber, D.L.A.; Price, W.L.; and Solomonides, C.M. *Computer Networks and Their Protocols*. Chichester: John Wiley, 1979.

5. Jaffe, J.M. and Moss, F.H. "A Responsive Distributed Routing Algorithm for Computer Networks." *IEEE Trans. Communic. COM-30*, 7 (July 1982): 1758-1762.

6. Kaliszewski, J.M. "The New Routing Algorithm on the SITA Data Transport Network." In *Performance of Computer-Communication Systems*, pp. 415-431. Edited by H. Rudin and W. Bux. Amsterdam: North-Holland, 1984.

7. Kleinrock, L. "Principles and Lessons in Packet Communications." *Proc. IEEE 66*, 11 (November 1978): 1320-1329.

8. Majithia, J.C.; Irland, M.; Grange, J.-L.; Cohen, M.; and O'Donnell, C.O. "Experiments in Congestion Control Techniques." In *Flow Control in Computer Networks*, pp. 211-234. Edited by J.-L. Grange and M. Gien. Amsterdam: North-Holland, 1979.

9. Merlin, D.M. and Segall, A. "A Failsafe Distributed Routing Protocol." *IEEE Trans. Communic. COM-27*, 9 (September 1979): 1280-1287.

10. Seligman, D.R. "Traffic Routing in a Computer Network." *Performance 7*, 2 (April 1984): 59-64.

11. Tajibnapis, W.D. "A Correctness Proof of a Topology Information Maintenance Protocol for Distributed Computer Networks." *C. ACM 20*, 7 (July 1977): 477-485.

Host Groups: A Multicast Extension for Datagram Internetworks

David R. Cheriton
Stephen E. Deering

Computer Systems Laboratory
Stanford University

Abstract

The extensive use of local networks is beginning to drive requirements for internetwork facilities that connect these local networks. In particular, the availability of multicast addressing in many local networks and its use by sophisticated distributed applications motivates providing multicast across internetworks.

In this paper, we propose a model of service for multicast in an internetwork, describe how this service can be used, and describe aspects of its implementation, including how it would fit into one existing internetwork architecture, namely the US DoD Internet Architecture.[1 2]

1. Introduction

Multicast is the transmission of a datagram packet to a set of zero or more destination hosts in a network or internetwork, with a single address specifying the set of destination hosts. For example, hosts A, B, C and D may be associated with multicast address X. On transmission, a packet with destination address X is delivered with datagram reliability to hosts A, B, C and D.

Multicast has two primary uses, namely distributed binding and multi-destination delivery. It is useful for binding when one or more of a set of hosts contain the desired object but particular host addresses are not known, only a multicast address. For example, in a distributed file system, all the file servers may be associated with one multicast address. To bind a file name to a particular server, a client sends a query packet containing the file name to the file server multicast address, which is delivered to all the file servers. The server that recognizes the file name then responds to the client, allowing subsequent interaction directly with that server host. This also illustrates the use of multicast for *logical addressing*. The multicast address for a group of hosts can denote *function* rather than location. One can similarly associate the group of time servers, name servers, computation servers and so on each with their own multicast address.

Multi-destination delivery is useful to several applications, including:

- distributed, replicated databases[1, 2].
- conferencing[3].
- distributed parallel computation, including distributed gaming[4].

Ideally, multicast transmission to a set of hosts is not more complicated or expensive for the sender than transmission to a single host. Similarly, multicast transmission should not be more expensive for the network than traversing the shortest path tree that connects the sending host to the hosts identified by the multicast address.

Multicast, transmission to a set of hosts, is properly distinguished from from *broadcast*, transmission to *all* hosts on a network or internetwork. Broadcast is not a generally useful facility since there are few reasons for communicating with all hosts. In fact, it is best viewed as an "accident of the technology" for broadcast networks in the same way that self-modifying programs are an accident of the technology for stored program machines; just because the technology provides it does not mean it is efficient or safe to use. A proper multicast facility allows efficient transmission to multiple hosts while avoiding unnecessary loading of the network and receiving hosts that arises with broadcast.

Multicast is now available in standard local networks[5]. For example, the Ethernet[6] provides 2^{47} multicast addresses. Sending a packet to an Ethernet multicast address delivers it (with datagram reliability) to the set of hosts listening to that multicast address. A variety of local network applications and systems make use of this facility. For instance, the V distributed system[7] uses network-level multicast for implementing efficient operations on groups of processes spanning multiple machines. Similar use is being made for replicated databases[1] and other distributed applications[8]. Providing multicast in the internetwork environment would allow porting such local network distributed applications to the internetwork, as well as making some existing internetwork applications more robust and portable (by, for example, removing wired-in lists of addresses, such as gateway addresses).

In current internetwork environments, an application logically requiring multicast must send individually addressed packets to each recipient. There are two problems with this approach. Firstly, requiring the sending host to know the specific addresses of all the recipients defeats its use as a binding mechanism. For example, a diskless workstation needs on boot to determine the network address of a disk server and it is undesirable to "wire in" specific network addresses. With a multicast facility, the multicast address of the disk servers (or name servers that holds the address of the disk server) can be *well-known*, allowing the workstation to transmit its initial queries to this address.

Secondly, transmitting multiple copies of the same packet makes inefficient use of network bandwidth, gateway resources and sender resources. For instance, the same packet may repeatedly traverse the same network links and pass through the same gateways. Furthermore, the network level cannot recognize multi-destination delivery to take advantage of multicast facilities that the underlying network technologies may provide. For example, local-area bus, ring, or radio networks and even satellite-based wide-area networks can provide efficient multicast delivery directly. Besides using excessive communication resources, the use of multiple transmissions to effect multicast severely limits the amount of parallelism in transmission and

---

[1]This work was sponsored in part by the Defense Advanced Research Projects Agency under contract N00039-83-K-0431 and National Science Foundation Grant DCR-83-52048.

processing that can be achieved compared to an integrated multicast facility.

In this paper, we describe a model of multicast service we call *host groups* and discuss aspects of implementing this service in a datagram internetwork environment. We argue that it is feasible to implement this facility in an internetwork as an extension of the existing "unicast" internetwork datagram model and mechanism.

We restrict ourselves to the communication environment of a datagram-based internetwork, like the IP[9] or XNS[10] internetwork architectures. In these architectures, all hosts employ a common internetwork datagram format and a common internetwork addressing convention to identify the sources and destinations of datagrams. On transmission, an internetwork datagram is delivered to its destination address with "best efforts" reliability, via the transmission services of the underlying networks and the relaying services of the gateways. This service best corresponds to OSI layer 3 or the network level in providing host-to-host delivery. Reliable delivery, including error handling and flow control, is handled by higher-level protocols that operate in terms of internetwork datagrams.

Figure 1 illustrates a heterogeneous collection of independent networks interconnected by hosts that serve as store-and-forward *gateways* typical of datagram internetworks.

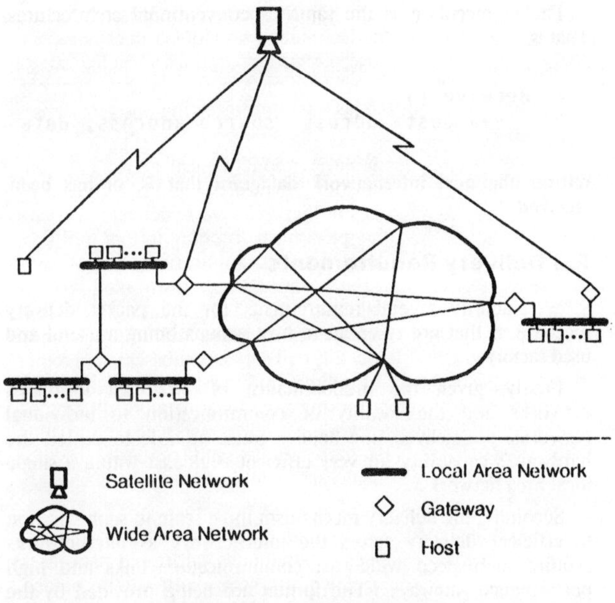

Figure 1 A Typical Internetwork

In Figure 1, a satellite network and a wide area, store-and-forward network connect several local area networks as well as individual hosts. The combination of broadcast and point-to-point technology plus the usual complications of different speeds, delay and maximum transmission unit make an efficient implementation of multicast a challenge.

The next section describes the host group model of multicast service. Section 3 describes the implementation strategy we propose. Section 4 describes how this extension fits into the current US DoD Internet architecture and briefly touches on other internetwork architectures. Section 5 illustrates how this facility can be used by a variety of applications. Section 6 relates this model to other proposals. Finally, we conclude with remarks on the status of our experimental prototype implementation of host groups and our future directions for investigation.

2. The Host Group Model

In an internetwork designed in the host group model, each internetwork address identifies a host group. A *host group* is a set of zero or more hosts in one internetwork.[3] · When an internetwork packet is sent, it is delivered with "best efforts" datagram reliability to all members of the host group identified by the internetwork address in the packet destination field.

The sender need not be a member of the destination group. We refer to such a group as *open*, in contrast to a *closed* group where only members are allowed to send to the group. We chose to provide open groups because they are more flexible and more consistent as an extension of conventional unicasts models (even though they are harder to implement).

Dynamic management of group membership provides flexible binding of internetwork addresses to hosts. Hosts may join and leave groups over time. A host may also belong to more than one group at a time. Finally, a host may belong to no groups at times, during which that host is unreachable within the internetwork architecture. In fact, an internetwork host need not have an individual internetwork address at all. Some hosts may only be associated with multi-host group addresses. For instance, there may be no reason to contact an individual time server in the internetwork, so time servers would not require individual addresses. Similarly, a bank of shared processors may be identical from the standpoint of clients and only acquire individual internetwork addresses while they are serving individual clients.

Internetwork addresses are dynamically allocated for *transient* groups, groups that often last only as long as the execution of a single distributed program. A range of host group identifiers is reserved for identifying *permanent* groups. One use of permanent host groups identifiers is for host groups with standard logical meanings such as "name server group", "boot server group", "internetwork monitor group", etc. Permanently assigned addresses are also used for conventional single-host addresses.

The host group model of internetwork generalizes the binding of internetwork addresses to internetwork hosts by allowing one address to bind to multiple hosts on multiple networks, more than one address to be bound (in part) to one host, and the binding of an address to host to be *dynamic*, i.e. possible to modify under application control. For performance reasons, the conventional case of single-member groups is handled specially as an optimization. A range of internetwork addresses are reserved for designating groups of at most one internetwork host, allowing the delivery mechanism to make appropriate optimizations. Moreover, if the internetwork address is statically bound to a host permanently attached through one network, a network identifier can be embedded as a subfield of its internetwork address in order to simplify gateway routing. As should be apparent, this special case corresponds to the unicast facility provided by several current datagram-based internetwork architectures, including IP and XNS. Thus, the host group model is a compatible extension of these architectures.

The following subsections provide further details of the model.

2.1 Host Group Management

Dynamic binding of internetwork addresses to hosts is managed by the following three operations available to higher-level protocols or applications:[4]

[3] In reality, the internetwork address is bound to network interfaces or host access ports, not the host machine per se.

[4] In this procedure call notation, the arguments for an operation are listed in parentheses after the operation name, and the returned values, if any, are listed after a --> symbol.

```
CreateGroup ( type )
  --> outcome, group-address, access-key
```

requests the creation of a new transient host group with the invoking host as its only member. The **type** argument specifies either a general group or a one-member-only group plus whether the group is restricted or unrestricted. A restricted group restricts membership based on the access-key. Only hosts presenting a valid host access-key are allowed to join. All unrestricted host groups have a null access-key. **outcome** indicates whether the request is approved or denied. If it is approved, a new transient group address is returned in **group-address**. **access-key** is the protection key (or password) associated with the new group. This should fail only if there are no free transient group addresses.

```
JoinGroup ( group-address, access-key )
  --> outcome
```

requests that the invoking host become a member of the identified host group (permanent or transient). **outcome** indicates whether the request is approved or denied. A request may be denied if the access key is invalid.

```
LeaveGroup ( group-address )
  --> outcome
```

requests that the invoking host be dropped from membership in the identified group (permanent or transient). **outcome** indicates whether the request is approved or denied.

There is no operation to destroy a transient host group because a transient host group is deemed to no longer exist when its membership goes to zero.

Note that in conventional internetworks allocation and binding of internetwork addresses is typically performed statically by internetwork administrators

2.2 Packet Transmission

Transmission of a packet in the host group model is controlled by two parameters of scope, one being the destination internetwork address and the other being the "distance" to the members in the group. In particular,

```
Send ( dest-address, source-address,
                       data, distance )
```

transmits the specified data in an internetwork datagram to the hosts in the host group specified by **dest-address** that are within the specified **distance**. The destination address is thus similar to conventional networks except that delivery may be to multiple hosts; the distance parameter requires further discussion.

Distance may be measured in several ways, including number of network hops, time to deliver and what might be called administrative distance. Administrative distance refers to the distance between the administrations of two different networks. For example, in a company the networks of the research group and advanced development group might be considered quite close to each other, networks of the corporate management more distant, and networks of other companies much more distant. One may wish to restrict a query to members within one's own administrative domain because servers outside that domain may not be trusted. Similarly, error reporting outside of an administrative domain may not be productive and may in fact be confusing.

Besides limiting the scope of transmission, the distance parameter can be used to control the scope of multicast as a binding mechanism and to implement an expanding scope of search for a desired service. For instance, to locate a name server familiar with a given name, one might check with nearby name servers and expand the distance (by incrementing the distance on retransmission) to include more distant name servers until the name is found.

To reach all members of a group, a sender specifies the maximum value for the distance parameter. This maximum must exceed the "diameter" of the internetwork.

The distance parameter can be viewed as an extension of the time-to-live or hop count parameters that are used in several internetwork architectures to prevent infinite routing cycles. In those cases, the distance parameter basically ensures that the delivery mechanism only expends a finite amount of work in delivery and therefore discards a packet caught in a routing loop. The distance parameter in the host group model refines this finite bound into further gradations.

Rather than define specific semantics of the distance parameter in the model, we see it having a refinement of the semantics of the time-to-live or hop count parameters specific to each internetwork architecture. However, in all cases, there is a need for well-known boundaries values that coincide with administrative domains. For instance, there is a need for a distance value that corresponds to "not outside this local network".

Packet reception is the same as conventional architectures. That is,

```
Receive ()
  --> dest-address, source-address, data
```

returns the next internetwork datagram that is, or has been, received.

2.3 Delivery Requirements

We identify several requirements for the packet delivery mechanism that are essential to host groups being a useful and used facility.

Firstly, given the predominance of broadcast local-area networks and the locality of communication to individual networks, the delivery mechanism must be able to exploit the hardware's capability for very efficient multicast within a single local-area network.

Secondly, the delivery mechanism must scale in sophistication to efficient delivery across the internetwork as internetworks acquire high-speed wide-area communication links and high performance gateways. The former are being provided by the introduction of high-speed satellite channels and long-haul fiber optic links. The latter are made feasible by the falling cost of memory and processing power plus the increasing importance in controlling access to relatively unprotected local network environments. A host group delivery mechanism must be able to take advantage of these trends as they materialize.

Finally, the delivery mechanism must avoid "systematic errors" in delivery to members of the host group. That is, a small number of repeated transmissions must result in delivery to all group members within the specified distance, unless a member is disconnected or has failed. We refer to this property as *coverage*. In general, most reliable protocols make this basic assumption for unicast delivery. It is important to guarantee this assumption for multicast as well or else applications using multicast may fail in unexpected ways when coverage is not provided. For efficiency, the multicast delivery mechanism should also avoid regularly delivering multiple copies of a packet to individual hosts.

Failure notification is not viewed as an essential requirement given the datagram semantics of delivery. However, a host group extension of internetwork architectures such as IP and XNS should provide "hint"-level failure notification as the natural extension of their failure notification for unicast.

3. Implementation

In this section, we sketch a design for implementing the host group model in a datagram internetwork. This description of the design is given to further support the feasibility of the host group model as well as point out some of the problems yet to be addressed.

Implementation of host groups involves implementing a binding mechanism (binding internetwork addresses to zero or more hosts) and a packet delivery mechanism (delivering a packet to each host to which its destination address binds). This facility fits most naturally into the gateways of the internetwork and the switching nodes of the constituent point-to-point networks (as opposed to separate machines) because multicast binding and delivery is a natural extension of the unicast binding and delivery (i.e. routing plus store-and-forward). That is, a multicast packet is routed and transmitted to multiple destinations, rather than to a single destination.

A gateway in a host group internetwork is thus viewed as a "communication server", providing multicast delivery and host group management. The multicast delivery service is invoked implicitly by sending packets addressed to host groups, with unicast delivery as a special case. The group management service is invoked explicitly using a request-response transaction protocol between the client hosts and the server gateways. In addition to the operations for creating transient host groups and adding and deleting host memberships in groups (Section 2.1), the gateway supports operations for administrative allocation of permanent group addresses, including static, single-host group addresses (i.e. unicast addresses).

In the following description, we start with a basic, simple implementation that provides coverage and then refine this mechanism with various optimizations to improve efficiency of delivery and group management.

3.1 Basic Implementation

A host group defines a *network group*, which is the set of networks containing current members of the host group. When a packet is sent to a host group, a copy is delivered to each network in the corresponding network group. Then, within each network, a copy is delivered to each host belonging to the group.

To support such multicast delivery, every internet gateway maintains the following data structures:

- *routing table*: conventional internetwork routing information, including the distance and direction to the nearest gateway on every network.
- *network membership table*: A set of records, one for every currently existing host group. The *network membership record* for a group lists the network group, i.e. the networks that contain members of the group.
- *local host membership table*: A set of records, one for each host group that has members on directly attached networks. Each *local host membership record* indicates the local hosts that are members of the associated host group. For networks that support multicast or broadcast, the record may contain only the local *network-specific multicast address* used by the group plus a count of local members. Otherwise, local group members may be identified by a list of unicast addresses to be used in the software implementation of multicast within the network.

A host invokes the multicast delivery service by sending an internetwork datagram to an immediate neighbour gateway (i.e. a gateway that is directly attached to the same network as the sending host). Upon receiving a datagram from a directly attached network, a gateway looks up the network membership record corresponding to the destination address of the datagram. For each of the networks listed in the membership record, the gateway consults its routing table. If, according to the routing table, a member network is directly attached, the gateway transmits a copy of the datagram on that network, using the network-specific multicast address allocated for the group on that network. For a member network that is not directly attached and is within the distance constraint specified in the datagram, the gateway creates a copy of the datagram with an additional inter-gateway header identifying the destination network. This inter-gateway datagram is forwarded to the nearest gateway on the destination network, using conventional store-and-forward routing techniques. At the gateway on the destination network, the datagram is stripped of its inter-gateway header and transmitted to the group's multicast address on that network. Member networks that are beyond the datagram's distance constraint are ignored.

The network membership records and the network-specific multicast structures are updated in response to group management requests from hosts. A host sends a request to create, join, or leave a group to an immediate neighbour gateway. If the host requests creation of a group, a new network membership record is created by the serving gateway and distributed to all other gateways. If the host is the first on its network to join a group, or if the host is the last on its network to leave a group, the group's network membership record is updated in all gateways. The updates need not be performed atomically at all gateways, due to the datagram delivery semantics; hosts can tolerate misrouted and lost packets caused by temporary gateway inconsistencies, as long as the inconsistencies are resolved within normal host retransmission periods. In this respect, the network membership data is similar to the network reachability data maintained by conventional routing algorithms, and can be handled by similar mechanisms.

In many cases, a host joins a group that already has members on the same network, or leaves a group that has remaining members on the same network. This is then a local matter between the hosts and gateways on a single network: only the local host membership table needs to be updated to include or exclude the host;

This basic implementation strategy meets the delivery requirements stated at the end of Section 2. However, it is far from optimal, in terms of either delivery efficiency or group management overhead. One simple improvement is to recognize the important special case of static, one-member-only groups. This again corresponds to the conventional unicast provided in (for example) IP and XNS. In this case, the internetwork address for the single-host group encodes within it the network of the one host so there is no need to maintain a separate group membership record for that group. Consequently, the number of group membership records in the gateways is greatly reduced. Also, delivery to these groups degenerates to conventional unicast techniques such as currently used in IP and XNS implementations. Below, we discuss some further refinements to the basic implementation.

3.2 Multicast Routing Between Networks

Multicast routing among the internetwork gateways is similar to store-and-forward routing in a point-to-point network. The main difference is that the links between the nodes (gateways) can be a mixture of broadcast and unicast-type networks with widely different throughput and delay characteristics. In addition, packets are addressed to networks rather than hosts (at the gateway level).

We use the extended reverse path forwarding algorithm of Dalal and Metcalfe[11]. Although originally designed for broadcast, it is a simple and efficient technique that can serve well for multicast delivery if network membership records in each gateway are augmented with information from neighbouring gateways. This algorithm uses the *source* network identifier, rather than a *destination* network identifier to make routing decisions. Since the source address of a datagram is a general group address, it cannot be used to identify the source network of the datagram; the first gateway must add a header specifying the source network. This approach minimizes redundant transmissions when multiple destination networks are reachable across a common intergateway link, a problem with the basic implementation described earlier.

Note that we eliminated from consideration techniques that fail to deliver along the branches of the shortest delay tree rooted at the source, such as Wall's center-based forwarding[12] because this compromises the meaning of the multicast distance parameter and detracts from multicast performance in general. We also rejected the approach of having a multicast packet carry more than one network identifier in its inter-gateway header to indicate multiple destination networks because the resulting variable length headers would cause buffering and fragmentation problems in the gateways.

3.3 Multicasting Within Networks

A simple optimization within a network is to have the sender use the local multicast address of a host group for its initial transmission. This allows the local host group members to receive the transmission immediately along with the gateways (which must now "eavesdrop" on all multicast transmissions). A gateway only forwards the datagram if the destination host group includes members on other networks. This scheme reduces the cost to reach local group members to one packet transmission from two required in the basic implementation[5] so transmission to local members is basically as efficient as the local multicast support provided by the network.

A similar opportunity for reducing packet traffic arises when a datagram must traverse a network to get from one gateway to another, and that network also holds members of the destination group. Again, use of a network-specific multicast address which includes member hosts plus gateways can achieve the desired effect. However, in this case, hosts must be prepared to accept datagrams that include an inter-gateway header or, alternatively, every datagram must include a spare field in its header for use by gateways in lieu of an additional inter-gateway header.

3.4 Distributing Membership Information

A refinement to host group membership maintenance is to store the host group membership record for a group *only* in those gateways that are directly connected to member networks. Information about other groups is cached in the gateway only while it is required to route to those other groups When a gateway receives a datagram to be forwarded to a group for which it has no network membership record (which can only happen if the gateway is not directly connected to a member network), it takes the following action. The gateway assumes temporarily that the destination group has members on *every* network in the internetwork, except those directly attached to the sending gateway, and routes the datagram accordingly. In the inter-gateway header of the outgoing packet, the gateway sets a bit indicating that it wishes to receive a copy of the network membership record for the destination host group. When such a datagram reaches a gateway on a member network, that gateway sends a copy of the membership record back to the requesting gateway and clears the copy request bit in the datagram.

Copies of network membership records sent to gateways outside of a group's member networks are cached for use in subsequent transmissions by those gateways. That raises the danger of a stale cache entry leading to systematic delivery failures. To counter that problem, the inter-gateway header contains a field which is a hash value or checksum on the network membership record used to route the datagram. Gateways on member networks compare the checksum on incoming datagrams with their up-to-date records. If the checksums don't match, an up-to-date copy of the record is returned to the gateway with the bad record.

This caching strategy minimizes intergateway traffic for groups that are only used within one network or within the set of networks on which members reside, the expected common cases. Partial replication with caching also reduces the overhead for network traffic to disseminate updates and keep all copies consistent. Finally, it also reduces the space cost for data in large internetworks with large numbers of multiple host groups.

We have not addressed here the problem of maintaining up-to-date, consistent network membership records within the set of gateways connected to members of a group. This can be viewed as a distributed database problem which has been well studied in other contexts. The loose consistency requirements on network membership records suggest that the techniques used in Grapevine[13] might be useful for this application.

4. Integration into the DoD Internet

To show how the host group model can be supported by straightforward extension of an existing internetwork architecture, we outline how it might fit into the US DoD Internet.

The current Internet provides unicast datagram delivery between hosts on a wide variety of networks, both local-area and wide-area, broadcast and point-to-point. An Internet address is a 32-bit value consisting of two subfields: a network number and a host-within-network number. Every Internet gateway maintains a routing table that specifies the distance and direction to every network in the Internet, relative to the gateway. Thus, given a datagram, a gateway can determine from the network number subfield of its destination address, where to send it next on the path towards its destination. When the datagram reaches a gateway into its destination network, that gateway maps the host-within-network number to a local network address for final delivery.

The existing architecture supports our model of static, one-member-only groups. We extend this architecture to support multiple host groups by reserving a single network number to identify all such groups. Each multiple host group is distinguished by a unique value in the host-within-network subfield of its internet address. The Internet gateways are augmented with the data structures and procedures discussed in Section 3 to support internet multicast.

An IP datagram contains a "time to live" field which is decremented by the gateways once a second and on every network hop. If the time to live goes to zero before the datagram reaches its destination, the datagram is discarded. In the host group implementation, this field is used to limit the delivery distance of multicasts.

Other datagram internetwork architectures yield to similar extensions. For example, the Xerox Network Systems architecture is essentially identical to the DoD Internet with regards to address encoding (network, host-within-network) and contents of routing tables. XNS datagrams contain a hop count field that can be used for multicast scope control.

The proposed ISO internetwork protocol[14] provides the same style of internetwork datagram service as IP or XNS. The draft proposal for ISO internetwork addresses[15] specifies a much more complex structure than the fixed-length, two-level hierarchical addresses of IP and XNS. A more sophisticated, possibly hierarchical, distribution of the network membership records would be appropriate for the enormous potential size of the ISO

[5]One unicast transmission from sender to gateway and one multicast transmission from gateway to local group members

"world network".

5. Use of Multicast

A number of applications that can use multicast have been cited earlier in the paper, including distributed databases, conferencing, distributed computation and locating internetwork services. Rather than describe these applications in greater detail, we focus on some general issues that were identified in previous work[7]. (This work dealt with the use of local network multicast in a distributed operating system to support the concept of interprocess group communication where process groups are distributed across host groups.)

A key issue is providing reliable communication as required by the application. Firstly, some applications, such as real-time conferencing, do not need reliable delivery, assuming the periodic updates are generally received. Secondly, binding applications, such as locating a name server, do not require delivery to all but simply a positive response from at least one host. Retransmission with possibly expanding scope of search until a response is received provides the required semantics.

As an aside, one might argue that the binding use is only really required to locate a name server. While true in theory, it may be simpler for some applications to locate other servers directly using this simple search protocol. Then they do not need to implement the protocol to lookup a name in the name server as well as this simple search protocol to locate the name server in the first place. For example, the PROM network loader for diskless workstations might be simpler if it can locate a boot server using a boot server group address directly rather than going through a name server.

For applications requiring reliable delivery, there are basically two approaches. The most common approach is to place the onus for reliable delivery on the sender. Here, the sender knows the membership of a group and retransmits to the group until it has received acknowledgements from each group member. As an optimization, the sender can use unicast to retransmit to particular group members if the number of missing acknowledgements is relatively small compared to the cardinality of the host group.

The second approach places the onus on the receivers to implement reliable delivery, what we call *publishing*. It is so named because it mimics real world publishing. That is, information to be sent to a group, the *subscribers*, is filtered though the *publisher*, which collates and numbers the information before issuing it to the subscribers. A subscriber noticing a missing issue by a gap in the issue numbers or a new issue not being received in the expected time interval requests the *back issue* from the publisher. Thus, instead of automatic retransmission until the receiver acknowledges the message, the receiver must request retransmission if it is required.

A family of reliable multicast protocols is specified by Chang and Maxemchuk[16] that combines both techniques built on top of an unreliable broadcast or multicast network. They describe a protocol that guarantees not only that all group members receive all messages, but also that they all receive the messages in the same order, regardless of the number of senders. Furthermore, this strong level of reliability is achieved with only one acknowledgement per message in the normal case, no single point of failure, and survival in the face of multiple host failures and recoveries. In another paper[1], Chang describes the use of this protocol to support a distributed, replicated database.

In general, the problem is not implementing reliable delivery for multicast delivery but choosing the right trade-off between cost, performance and reliability as required by the application. We have briefly described some basic techniques. However, further study is required to understand these trade-offs with various applications and internetworking parameters.

6. Related Work

There is relatively little published work on the use or implementation of internetwork multicasting.

Wall's thesis[12] presents several mechanisms for performing efficient broadcast and multicast delivery in point-to-point networks. His results can be applied to providing multicast within point-to-point networks that are constituents of an internetwork, and to the problems of multicast routing to "network groups" of gateways.

Boggs, in his thesis[8], describes a number of distributed applications that are impossible or very awkward to support without the flexible binding nature of broadcast addressing. Although he recognizes that almost all of his applications would be best served by a multicast mechanism, he advocates the use of "directed broadcast" because it is easy to implement within many kinds of networks and can be extended across an internetwork without placing any new burden on internetwork gateways. Unfortunately, broadcasting has the undesirable side effect of delivering packets to more hosts than necessary, thus incurring overhead on uninvolved parties and possibly creating security problems. Furthermore, directed broadcasting supports simple communication with unknown destinations on directly connected networks only; for destinations on more distant networks, the sender must know their network numbers or perform a search using gateway routing tables.

Recent proposals by Mogul[17] and Aguilar[18] have addressed the issue of multi-destination delivery within the DoD Internet. Mogul proposes an implementation of Bogg's directed broadcast facility. Aguilar suggests allowing an IP datagram to carry additional destination addresses, which are used by the gateways to route the datagram to each recipient. Such a facility would alleviate some of the inefficiencies of sending individual datagrams to a group, but it would not be able to take advantage of local network multicast facilities. More seriously, Aguilar's scheme requires the sender to know the individual IP addresses of all members of the destination group and thus lacks the flexible binding nature of true multicast or broadcast.

Blaustein et al[19] discuss a variety of protocols for reliable multicast delivery based on various (inter)network characteristics (e.g. point-to-point or broadcast or both, clusters of fast networks joined by slower networks, degree of multicast support provided by the networks, etc.). As well as making a case for unreliable multicast services at the internetwork level, their work suggests ways of achieving efficient multicast among gateways in a heterogeneous internetwork.

7. Concluding Remarks

We have described a model of multicast communication for datagram-based internetworks. As an extension of existing internetwork architectures, it views unicast communication and time-to-live constraints as special cases of the more general form of communication arising with multicast. We have argued that this model is implementable in current and future internetworks and that it provides a powerful facility for a variety of applications. In some cases, it provides a facility that is required for certain applications to work in the internetwork environment. In other cases, it provides a more efficient, robust and possibly more elegant way of implementing existing internetwork applications.

We are currently implementing a prototype host group facility as an extension of IP. For practical reasons, this prototype implements all group management functions and multicast routing outside of Internet gateways, in special hosts called *multicast agents*. The collection of multicast agents in effect provides a second gateway system on top of the existing Internet, for multicast purposes. The major costs of this separation are redundancy of routing tables between gateways and multicast agents and the increased delay and unreliability of extra hops in the delivery path. Much of the routing information in the multicast agents must be "wired-in" because they do not have access to the gateways' routing tables. However, this rudimentary implementation provides an environment for evaluating the interface to the multicast service and for investigating group management and multicast routing protocols for eventual use in the gateways. It also serves as a testbed for porting multicast-based distributed applications to an internetwork from the V distributed operating system.

For now, we are restricting group membership to local networks that already have a broadcast or multicast capability, such as the Ethernet. We feel that, in the future, any network that is to support hosts other than just gateways must have a multicast addressing mode. Efficient implementation of multicast within point-to-point or virtual circuit networks deserves investigation.

A significant issue raised by the host group model is authentication and access control in internetworks. Gateways must control which hosts can create and join host groups, presumably making their decision based on the identity of the requestor (thus requiring authentication) and permissions (access control lists). This issue does not arise in conventional internetwork architectures because host addresses are administratively assigned with no notion of dynamic assignment and binding as provided by host groups. We believe that access control should be recognized as a proper and necessary function of gateways so as to protect the hosts of local networks from general internetwork activity. Thus, group access control can be subsumed as part of this more general mechanism, although more investigation of the general issue is called for.

On a philosophical point, there has been considerable reluctance to make open use of multicast on local networks because it was network-specific and not provided across internetworks. We were originally of that school. However, we recognized that our "hidden" uses of multicast in the V distributed system were essential unless we resorted to dramatically poorer solutions - wired-in addresses. We also recognized, as described in this paper, that an adequate multicast facility for internetworks was feasible. As a consequence, we now argue that multicast is an important and basic facility to provide in local networks and internetworks. Higher levels of communication, including applications, should feel free to make use of this powerful facility. Networks and internetworks lacking multicast should be regarded as deficient relative to the future (and present) requirements of sophisticated distributed applications and communication systems.

Acknowledgements

The original proposal for host groups was developed as a proposal for extending the US DoD Internet architecture. Work is continuing develop a specific Internet extension in conjunction with the DARPA Internet task force on end-to-end protocols, headed by Bob Braden. This task force has provided valuable input in the development of the host group model.

References

1. J-M. Chang, "Simplifying Distributed Database Design by Using a Broadcast Network," *SIGMOD '84*, ACM, June 1984, .

2. F. Cristian et al, "Atomic Broadcast: from simple message diffusion to Byzantine agreement," *15th International Conference on Fault Tolerant Computing*, , Ann Arbor, Michigan, June 1985, .

3. H. Forsdick, "MMCF: A Multi-Media Conferencing Facility", personal communication

4. E. J. Berglund and D. R. Cheriton, "Amaze: A distributed multi-player game program using the distributed V kernel," *Proceedings of the Fourth International Conference on Distributed Systems*, IEEE, May 1984, .

5. IEEE, "Standards for Local Area Networks: Logical Link Control," The Institute of Electrical and Electronic Engineers, Inc., New York, New York, 1985,.

6. Digital Equipment Corporation, Intel Corporation, and Xerox Corporation, "The Ethernet, A Local Area Network: Data Link Layer and Physical Layer Specifications," Version 1.0,September, 1980

7. D. R. Cheriton and W. Zwaenepoel, "Distributed Process Groups in the V Kernel," *ACM Transactions on Computer Systems*, Vol. 3, No. 3, May 1985, .

8. D. R. Boggs, *Internet Broadcasting*, PhD dissertation, Stanford University, January 1982.

9. J. Postel, "Internet Protocol," Tech. report RFC 791, SRI Network Information Center, September 1981.

10. Xerox Corporation, *Internet Transport Protocols*, XSIS 028112, Xerox System Integration Standard, Stamford, Connecticut, December, 1981.

11. Y. K. Dalal and R. M. Metcalfe, "Reverse Path Forwarding of Broadcast Packets," *Communications of the ACM*, Vol. 21, No. 2, December 1978, pp. 1040-1047.

12. D. W. Wall, "Mechanisms for Broadcast and Selective Broadcast," Tech. report 190, Computer Systems Laboratory, Stanford University, June 1980.

13. A. D. Birrell et al, "Grapevine: an exercise in distributed computing," *Communications of the ACM*, Vol. 25, No. 4, April 1982, pp. 260-274.

14. ISO TC97 SC6, "Information Processing Systems - Data Communications - Protocol for Providing the Connectionless Network Service," Draft Proposal N2971,.

15. ISO TC97 SC6, "Addendum to the Network Service Definition Covering Network Layer Addressing," Draft Proposal N3444,.

16. J-M. Chang and N. F. Maxemchuk, "Reliable Broadcast Protocols," *ACM Transactions on Computer Systems*, Vol. 2, No. 3, August 1984, .

17. J. Mogul, "Broadcasting Internet Datagrams," Tech. report RFC 919, SRI Network Information Center, October 1984.

18. L. Aguilar, "Datagram Routing for Internet Multicasting," *ACM SIGCOMM '84 Communications Architectures and Protocols*, ACM, June 1984, pp. 58-63.

19. B. Blaustein et al, "Notes on the Reliable Broadcast Protocol (The Distributor)", Computer Corporation of America. working paper.

Panel 6

Integrated Services Digital Network (ISDN)

Chairperson
D. Clark
MIT

Panelists
S. Newman, *Bell Communications Research*
D. Eigen, *Siemens*
A. Clark, *Advanced Micro Devices*
J. Mackie, *Mitel*
J. Kaufeld, *AT&T Bell Laboratories*

Session 7

Application Level Issues

Chairperson

J. Limb

Bell Communications Research

AN APPROACH FOR INTERCONNECTING SNA AND XNS NETWORKS

Kenneth O. Zoline

Amoco Corporation
200 East Randolph Street
Chicago, Illinois 60601

William P. Lidinsky

AT&T Bell Laboratories
Naperville-Wheaton Road
Naperville, Illinois 60566

ABSTRACT

Interest in computer internetworking has resulted from the proliferation of wide area and local area networks. The CCITT, DARPA/DoD, and ISO/ECMA internetworking models, which have become widely accepted for doing this, do not address the pragmatic problem of interconnecting computer networks that are based upon closed-system, vendor-proprietary network architectures. This paper presents an approach for interconnecting private data networks that are based upon IBM's System Network Architecture (SNA) and Xerox's Network System (XNS). Paramount to the solution of the SNA/XNS network interconnection problem is the identification and definition of a number of gateway-provided data transport and application protocol support services, which can be incorporated into a multi-function, multi-level, protocol translation gateway.

INTRODUCTION

Interest in computer internetworking has resulted from the rapid proliferation of computer resource sharing networks that provide various kinds of service and coverage. The principle reasons for interconnecting such networks have been based upon the following requirements: (1) the need to exchange information between geographically dispersed, autonomous computer systems; (2) the need to share computer resources without regard for the physical location of the resource or its user; (3) the need to access application services that are available on other networks; and (4) the need to exchange information between spatially separated people.

Three approaches for interconnecting computer resource sharing networks have been discussed extensively in the literature. These approaches are distinguished by their use of virtual circuits or datagrams to achieve network interconnection, and the architectural layering upon which each approach is based (see Figure 1):

1. The CCITT X.75 approach [21] proposes the formation of an end-to-end internetwork connection through the concatenation of a series of intranetwork and internetwork virtual circuits. This approach requires all interconnecting networks to utilize the following CCITT defined recommendations: the X.75 interface for packet network interconnection, the X.25 packet network interface, and the X.121 packet network addressing scheme.

2. The DARPA/Dod Internet approach [5,17] proposes that a singular datagram-based "internet" protocol be used to facilitate the interconnection of heterogeneous packet transport networks. Unlike the virtual circuit-based CCITT approach, the DARPA/DoD model presumes that a variety of packet transmission networks will be implemented within the internetworking environment, each with its own unique packet network interface.

3. The ISO/ECMA Internetworking approach [4] like the DARPA/DoD Internet model, proposes the use of datagrams to provide a minimal data transfer service capability between all of the underlying networks that comprise the global network. The principle difference between the two approaches is however an architectural one. The ISO/ECMA approach also differs from the CCITT approach: although both models are based upon the same architectural layering (i.e., the 7-layer OSI Reference Model), they differ with respect to their internetwork interface -- datagrams versus virtual circuits.

Unfortunately, the three aforementioned internetworking approaches do not address the pragmatic problem of interconnecting computer networks that are based upon vendor-proprietary, closed-system, network architectures. This problem exists for two reasons. First, vendor-proprietary network architectures may differ significantly from those prescribed by the CCITT, DARPA/DoD, and

ISO/ECMA models. Thus, an internetworking approach that is based upon any one of these three models would not be applicable to interconnecting mixed-vendor network product offerings, due to profound differences in their architectural layering and operation. Second, the three internetworking approaches also presume that a homogeneous set of end-to-end protocols will be used to facilitate interprocess cooperation. In a mixed-vendor networking environment however, the opposite should be assumed; that is, a heterogeneous set of end-to-end protocols would likely exist. Therefore, end-to-end protocol translation would be required to facilitate interprocess cooperation in a mixed-vendor networking environment.

This paper will present an approach for interconnecting private data networks that are based upon IBM Corporation's System Network Architecture (SNA) and Xerox Corporation's Network System (XNS). The SNA/XNS network interconnection problem has been chosen for the following two reasons:

1. SNA and XNS are representative of heterogeneous network architectures that are based upon vendor-proprietary, closed-system designs.

2. SNA and XNS (i.e., Ethernet in particular) are also representative of wide area and local area networking, respectively. The architectures and operating characteristics of local (LAN) and wide area networks (WAN) tend to differ greatly causing the LAN-WAN interconnection problem to present more issues than either LAN-LAN or WAN-WAN interconnection.

The SNA/XNS network interconnection problem will be functionally decomposed into a data transport problem and an application cooperation problem. An approach for interconnecting SNA and

OSI Layer	ISO (OSI) 7-Layer Reference Model	CCITT Adopted Standards	DARPA/DoD Internet Model	ISO/ECMA Internetwork Model	OSI Layer
(7)	Application	/ / / / / / / /	Application	Application	(7)
(6)	Presentation	X.3 X.28 X.29	Utility	Presentation	(6)
(5)	Session	/ / / / / / / /		Session	(5)
(4)	Transport	/ / / / / / / /	Transport Protocols (e.g., TCP)	Transport	(4)
(3)	Network	X.75 Internet (VC)	Internet Protocol (Datagram)	Internetwork (Datagram)	(3c)
				Convergence	(3b)
		X.25 Packet Level	Any of a Wide Variety of Data Transmission Protocols	Subnetwork Access	(3a)
(2)	Data Link	HDLC		Data Link	(2)
(1)	Physical	X.21		Physical	(1)

Figure 1. Comparison of the ISO (OSI), CCITT, DARPA/DoD, and ISO/ECMA Architectural Layering

XNS networks, which is based upon a multi-level, multi-function, protocol translation gateway, will be presented.

THE SNA/XNS INTERCONNECTION PROBLEM

Systems Network Architecture (SNA) is IBM Corporation's specification for distributed data processing networks [12,13,14]. Since its introduction in 1974, SNA has evolved from an architecture which supported only single host networks to one which supports multi-host networking as well as the interconnection of multiple SNA networks [2].

Like IBM's SNA, the Xerox Network System (XNS) is Xerox Corporation's specification for distributed communication systems [22,23]. XNS was introduced in 1981, shortly after the commercial introduction of Xerox's Ethernet local area networking system. The architecture of XNS is based upon the experimental PUP internetwork system, which was developed at the Xerox Palo Alto Research Center [3].

The most notable architectural difference between SNA and XNS is their protocol layering. SNA defines a 7-layer protocol hierarchy, which resembles the OSI Reference Model. In contrast, XNS defines a 4-layer protocol hierarchy that is fashioned in part after the DARPA/DoD Internet model. These differences are illustrated in Figure 2.

As would be expected, SNA and XNS also differ with respect to protocol message units. These differences are shown in Figure 3. Other differences exist between SNA and XNS. These differences are summarized in Table 1.

Together, all of the aforementioned differences and incompatibilities are representative of the general kinds of problems that must be overcome when two networks are interconnected. Additional problem areas that must

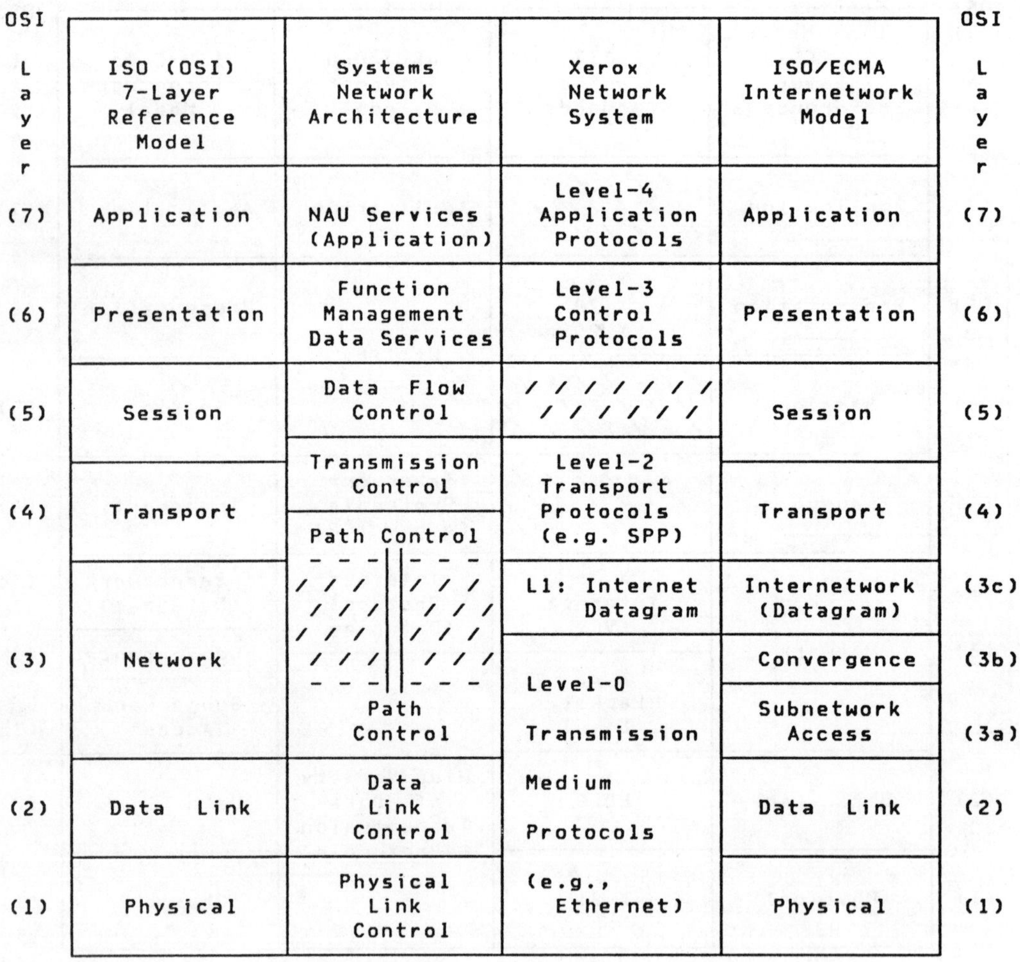

Figure 2. Comparison of the ISO (OSI), SNA, XNS, and ISO/ECMA Architectural Layering

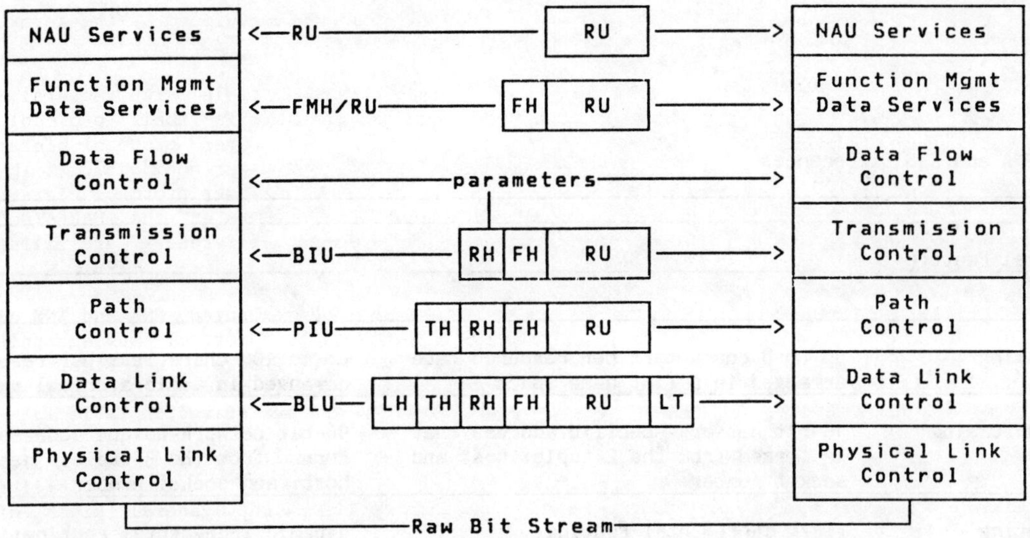

Legend:

BIU - Basic Information Unit
BLU - Basic Link Unit
FH - Function Management Header (FMH)
LH - Link Header
LT - Link Trailer
PIU - Path Information Unit
RH - Request/Response Header
RU - Request/Response Unit

Figure 3a. SNA Protocol Message Units

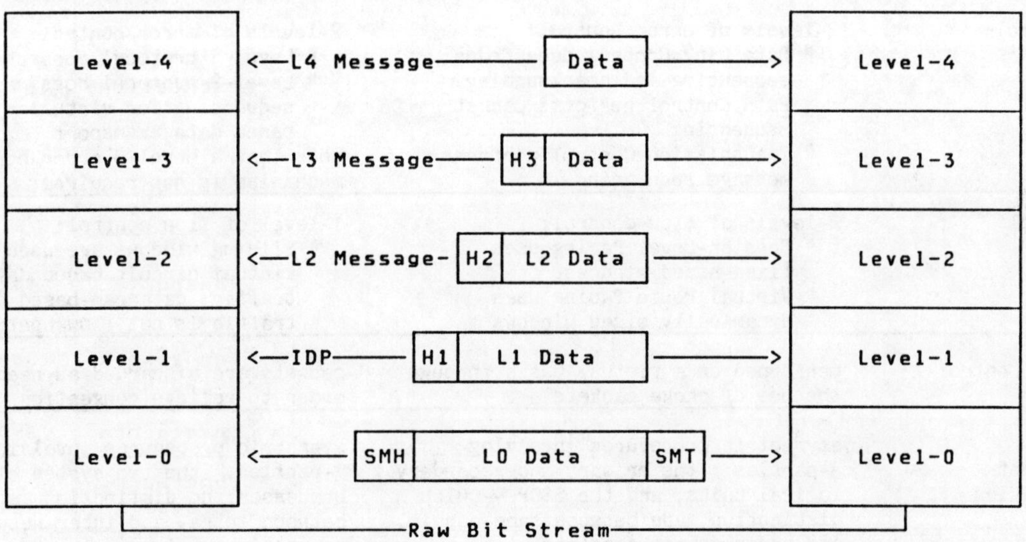

Legend:

IDP - Internet Data Packet (Level-1 Message Unit)
H1 - Level-1 Protocol Message Header
H2 - Level-2 Protocol Message Header
H3 - Level-3 Protocol Message Header
SMH - Subnetwork-dependent Message Header
SMT - Subnetwork-dependent Message Trailer

Figure 3b. XNS Protocol Message Units

Table 1. SNA and XNS Differences

Architectural Feature	SNA	XNS
Resource Naming	up to 8 characters per resource name arranged in a flat name space	up to 100 characters per resource name arranged in a hierarchical name space
Resource Addressing	16-bit network-specific address that is formed from the 2-tuple: host and socket number	96-bit network-unique address that is formed from the 3-tuple: network, host, and socket number
Network Routing	fixed incremental routing	dynamic incremental routing; routing algorithm is distance sensitive
Communications Subnetwork Interface	virtual circuit based data transport service interface	virtual circuit or datagram based data transport service interface
Communications Subnetwork Structure	subnetwork internally uses virtual circuits	subnetwork internally uses datagrams
Typical Inter-Node Transmission Speeds	9.6 - 56 thousand bits per second	1.5 - 3.0 million bits per second
Packet Fragmentation	not performed	optionally performed by Level-0 protocols
Error Control	3-levels of error control: * Data Link Control does frame sequencing and checksumming * Path Control performs packet sequencing * Transmission Control performs message sequencing	2-levels of error control: * Level-1 protocol does checksumming * Level-2 protocol does packet sequencing for virtual circuit based data transport Note that a Level-0 error control mechanism is not required.
Flow Control	2-levels of flow control: * Session-Level Pacing uses fixed-sized windows * Virtual Route Pacing uses dynamically sized windows	1-level of flow control: * Sliding windows are used for virtual circuit based data traffic; datagram-based data traffic is not flow controlled
Congestion Control	performed on a virtual basis through the use of choke packets	packets are discarded as needed in order to relieve congestion
Connection Establishment and Termination Procedures	asymmetric procedures involving 3-parties: the primary and secondary logical units, and the SSCP -- with a distinction made between same-domain, and cross-domain sessions	symmetric procedures involving 2-parties: the two system element processes; no distinction is made between intra- and inter-network connections
Internetworking Philosophies	no architected internetworking approach per se; however concatenated virtual circuits and address mapping are used to facilitate SNA network interconnection	a common internetwork protocol based upon datagrams that uses a global addressing scheme

be addressed by a network interconnection approach are the following:

1. Preservation of Local Network Independence. Both IBM and Xerox have independently developed a sizable number of network products and services that are based upon a proprietary network end-use interface. In addition, both vendors have have implemented a number of internal network control protocols and products to facilitate resource management and network administration.

2. Provision of Data Transport Services. A primary objective of network interconnection is the extension of resource sharing capabilities across network boundaries. In order to do so, a comprehensive set of data transport services must be provided that addresses the following:

   - data transport between SNA and XNS networks
   - data transport between two SNA networks through an intermediate XNS network
   - data transport between two XNS networks through an intermediate SNA network

3. Application-Level Cooperation. The facilitating of interprocess communication and cooperation across network boundaries is an equally important objective of network interconnection. Since the provision of an end-to-end data transport service is not sufficient for achieving this goal, protocol translation will have to be performed at one or more higher protocol levels in addition.

SOLVING THE SNA/XNS NETWORK INTERCONNECTION PROBLEM

The literature discusses four network interconnection approaches that use gateways [1,6,8,16,20]. Each of these approaches differ with respect to their internetworking capability, which is a function of the level of network interconnection:

- Bridges are used to interconnect homogeneous networks.
- Subnetwork Media Conversion Gateways are used to interconnect networks that are homogeneous at OSI Layer 3 and above, but differ with respect to physical transmission media.
- Subnetwork Access Gateways are used to interconnect networks that are homogeneous at OSI Layer 4 and above, but differ with respect to subnetwork access interface.
- Protocol Translation Gateways are used to interconnect networks that are heterogeneous above OSI Layer 4.

Of the four alternative approaches, the protocol translation gateway is the only one suited for solving the SNA/XNS network interconnection problem since the first three gateway approaches require the interconnected networks to employ a homogeneous set of protocols at OSI Layer 4 and above. In fact, seven kinds of incompatibilities can exist when network interconnection is attempted, one kind for each OSI Reference Model layer. Thus, a protocol translation gateway must be employed to resolve these differences, subject to constraints placed upon it by protocol conversion.

The technique of protocol conversion described by Francois and Potocki [9], entails expressing the semantics carried by one protocol system in the message format and protocol syntax of another. Four protocol conversion principles, which are relevant to gateways, are stated by the authors:

1. Protocol conversion can be effectively used to support communication between two entities belonging to different protocol architectures if, and only if, the protocols pertaining to these entities carry the same semantics and are therefore functionally compatible.

2. When two protocols belonging to different architectures support different sets of functions, protocol conversion can be applied only to the functional subset that is common to both protocols.

3. When two protocols are not convertible, the only solution for communicating with the other network entity is to actually implement the other entity's protocol.

4. When two protocols differ only in some respects, a mixed solution can be employed. Protocol conversion can be applied to the common subset and this in turn can be complemented by implementing the missing function(s) on the corresponding side.

Dalal [7] asserts that the design of a protocol translation gateway can be greatly simplified by separating the seven layers of the OSI Reference Model into two functional groupings. The lower four layers (i.e., the physical, data link, network, and transport) can be viewed as providing a reliable end-to-end data transport service; whereas the upper three layers can be viewed as defining ways to achieve cooperation between processes, independent of where they may reside. In particular, the OSI session and presentation layers perform functions which are directly used and driven by application requirements. Therefore instead of having seven different kinds of protocol incompatibility, there are now just two:

- a data transport incompatibility
- an application-level incompatibility

Thus, the SNA/XNS network interconnection problem can be functionally decomposed into a data transport problem and an application-level cooperation problem. The following two subsections describe how these problems can be solved by an SNA/XNS protocol translation gateway.

Solving the SNA/XNS Data Transport Problem

A primary objective of network interconnection is the extension of resource sharing capabilities across network boundaries. In order to do so, a comprehensive set of data transport services must be provided. Specifically, an SNA/XNS gateway should facilitate the following categories of data transport service: (1) data transport between adjacent homogeneous networks; (2) data transport between adjacent heterogeneous networks; and (3) data transport between non-adjacent homogeneous networks.

The need for the first two categories of data transport service is immediately obvious. However, the need for the third category stems from the fact that heterogeneity exists at the data transport level only. Therefore, six data transport services must be architected into the design of the gateway:

1. Data Transport Between Adjacent SNA Networks. The SNA/XNS gateway should support the transportation of SNA message traffic between logical units residing in adjacent SNA networks. The requirement for an SNA-SNA data transport service stems from the situation where the SNA/XNS gateway will be used to interconnect two SNA networks in lieu of using IBM's SNA Network Interconnection facility [2].

2. Data Transport Between Non-Adjacent SNA Networks. The SNA/XNS gateway should also facilitate the transportation of SNA message traffic across an XNS subnetwork. The requirement for an SNA-XNS-SNA data transport service stems from the situation where two logical unit processes residing in non-adjacent SNA networks, are physically separated by one or more interconnected XNS networks. In this scenario, two cooperating gateways will be required to facilitate this data transport service, one at each SNA/XNS network boundary.

3. Data Transport Between Adjacent XNS Networks. From the perspective of an attached XNS network, an SNA/XNS gateway should appear as an internetwork router host. That is, the gateway should perform the functions associated with an XNS store-and-forward packet switching by virtue of the gateway's juxtaposition to the XNS network itself.

4. Data Transport Between Non-Adjacent XNS Networks. The SNA/XNS gateway should also support the transportation of XNS message traffic across an SNA subnetwork. The requirement for an XNS-SNA-XNS data transport service stems from the situation where two XNS-based system element processes residing in non-adjacent networks, are separated by one or more interconnected SNA networks. Like the SNA-XNS-SNA data transport scenario, two gateways will be required to facilitate this data transport service, one at each SNA/XNS network boundary.

5. Data Transport Between Adjacent SNA and XNS Networks. The SNA/XNS gateway should facilitate the transportation of message traffic between communicating SNA and XNS-based processes. The requirement for an SNA-XNS data transport service stems from the situation where an SNA logical unit requires transport-level access to an XNS-based system element process.

6. Data Transport Between Adjacent XNS and SNA Networks. The SNA/XNS gateway should reciprocally support the transportation of message traffic between communicating XNS and SNA-based processes. The requirement for an XNS-SNA data transport service stems from the situation where an XNS logical unit requires transport-level access to an SNA-based system element process.

Solving the SNA/XNS Application Cooperation Problem

A second objective of internetworking is to facilitate application-level cooperation between functionally related processes residing in heterogeneous networks. In order to fulfill this requirement, the SNA/XNS gateway should provide an open-ended set of application protocol conversion services that are built upon the previously discussed SNA-XNS and XNS-SNA data transport services. The following application-oriented utility functions are examples: bulk data transfer, distributed data base management, document transfer, document printing, file transfer, and virtual terminal.

Since a general transformation of all application-level protocols is not possible for the reasons cited by Francois and Potcoki [9], the gateway's open-ended set of application protocol conversion services must be complemented by an application protocol mapping service. This service, as its name implies, will selectively map SNA protocol message units into XNS messages and vice versa.

The need for application protocol mapping stems from the situation where two application processes residing in heterogeneous networks are capable of processing each other's application messages, but are lacking the data transport means to achieve application-level cooperation. The gateway's application protocol mapping service will fulfill this need by: (1) providing a suitable data transport mechanism and (2) allowing application processes to choose how protocol message units will be presented to each other (i.e., formatted into messages). Thus, protocol mapping will enable the gateway's limited set of application protocol conversion services to be augmented through the use of host-based application protocol conversion programs.

The SNA/XNS Gateway Solution Compared to Others

Three alternative approaches for solving the SNA/XNS network interconnection problem are briefly considered below:

1. Device-Oriented Interfaces. The simulation of device-oriented interfaces is a frequently used method for achieving connectivity between functionally similar processes across network boundaries. For example, Dalal [7] describes how an XNS-based terminal user can use an XNS-provided service to access an SNA-based application host. In this example, the XNS service emulates an SNA-supported terminal device by performing protocol conversion on the XNS device data stream.

   The drawback to this approach is that it does not provide the XNS resource with the same set of capabilities that an SNA resource might be afforded. That is, device-oriented SNA network resources are relegated "secondary logical unit" status whereas host-based resources are given either "primary logical unit" or "secondary logical unit" status. The implication of this approach is that it provides SNA and XNS resources with an asymmetric set of session capabilities. Thus, the use of device-oriented interfaces is not viewed as being particularly desirable for solving the SNA/XNS network interconnection problem.

2. Application Relay Programs. A second alternative approach entails using application relay programs to achieve connectivity between functionally similar processes across network boundaries. An application relay program is a message translation and transportation mechanism whose purpose is to perform application-level protocol conversion upon the messages conveyed between two session partners. This approach is well suited for applications such as virtual terminal, file transfer, and electronic mail.

   The principal drawback to this approach is that application relay programs do not address the issue of transport-level connectivity. Relay programs are intended to perform application-specific protocol conversion; they are not well suited for performing general data transport functions, such as store-and-forward packet switching.

3. Meta-Protocol Approaches. The network interconnection approach described by Einert and Glas [8] proposes the use of a protocol translation gateway to interconnect (IBM-based) SNA and (Siemens-based) TRANSDATA networks. The SNATCH gateway as it is called, incorporates an OSI-like meta-protocol architecture through which it performs protocol translation (i.e., SNA<=>meta-protocols<=>TRANSDATA). In so doing, SNATCH provides a limited set of application protocol conversion services (i.e., virtual terminal, remote job entry, and batch job data transfer).

   Although the SNATCH gateway is very well suited for solving the SNA/TRANSDATA network interconnection problem, two major deficiencies exist that preclude the SNATCH approach from being used to solve the SNA/XNS network interconnection problem. First, the SNATCH approach addresses a number of internetworking issues that are typically associated with WAN-WAN (wide-area network) interconnection, namely data transport between adjacent heterogeneous networks. The SNATCH approach however does not address data transport between adjacent and non-adjacent homogeneous networks, issues that are especially germane to LAN-LAN and LAN-WAN interconnection. Second, the SNATCH approach does not provide a protocol mapping service. Through the mapping of session, presentation, and application-level protocol message units, a gateway would be capable of supporting a wide variety of host-based application relay programs (each with its own set of protocol mapping requirements). Thus, the provision of protocol mapping would enable a gateway's limited set of application protocol conversion services to be augmented through the use of host-based application-relay programs.

Needless to say, the functions performed by an application relay program and a protocol translation gateway overlap to a large extent. Therefore, the SNA/XNS gateway should embellish both the strong points of the "pure" protocol translation gateway approach along with the strong points of the application relay program approach. The documented shortcomings of the SNATCH gateway support the motivation for developing a multi-level, multi-function protocol translation gateway to solve the SNA/XNS network interconnection problem.

DESCRIPTION OF AN SNA/XNS GATEWAY

The proposed SNA/XNS gateway [24] shown in Figure 4, will comprise three functional parts:

- The SNA Protocol Element will serve as the gateway's interface to an attached SNA network. The protocol element will process SNA protocols and provide the gateway with SNA functionality.
- The XNS Protocol Element will similarly serve as the gateway's interface to an attached XNS network. Like its SNA counterpart, the XNS Protocol Element will process XNS protocols and provide the gateway with XNS functionality.
- The Protocol Translation Element will perform two functions. It will convert the semantics carried in an SNA protocol data unit into the message format and syntax of an XNS protocol data unit, and vice versa. Secondly, it will convert SNA network addresses into XNS network addresses through address mapping and vice versa.

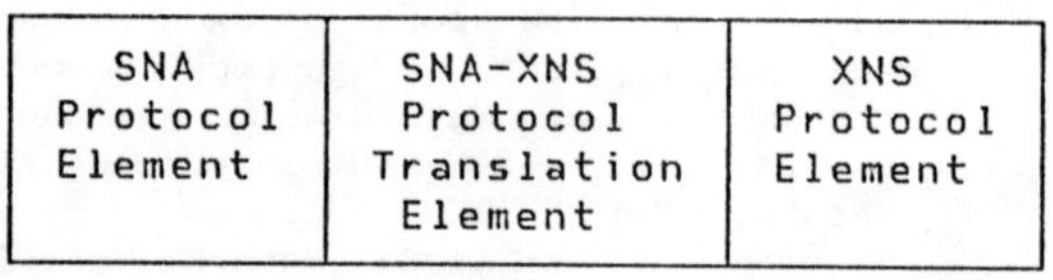

Figure 4. Structure of the SNA/XNS Gateway

The SNA Protocol Element

Architecturally, the SNA Protocol Element will appear as a subarea node containing a System Services Control Point (SSCP). The internal structure of the SNA Protocol Element shown in Figure 5, is described in the text that follows.

The SSCP component will facilitate cross-network SNA session establishment and termination between logical units residing in non-adjacent SNA networks as well as between an SNA logical unit and an XNS system element process. In the former case, the gateway SSCP will function as a message relay between the originating logical unit's SSCP and the destination logical unit's SSCP. In the latter case, the gateway SSCP will function as the owning cross-domain resource manager for all XNS system element sockets participating in SNA LU-LU sessions. In order to carry out these two functions, the SSCP must establish and maintain SSCP-SSCP sessions with neighboring SSCP's, residing in both adjacent SNA networks and non-adjacent SNA/XNS gateways.

The gateway's physical unit component will perform the activation, deactivation, and monitoring functions associated with the management of virtual routes, explicit routes, transmission groups, and link stations as described in [12]. The physical unit will also manage a pool of network addresses, which will be used to represent cross-network logical unit session partners.

Logical units will be used to facilitate a number of gateway-to-gateway functions, such as the transmission of XNS internet datagrams and routing information across an SNA network. In addition, a pool of logical units will be used to facilitate cross-network LU-LU data transport and SNA-to-XNS application protocol conversion.

Components of the gateway's path control layer will perform four inter-related path control functions: virtual route error control, virtual

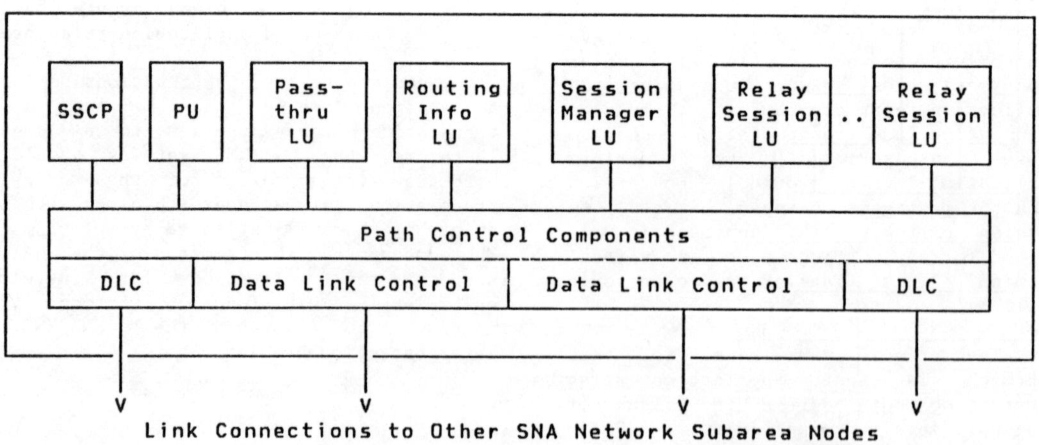

Figure 5. Structure of the SNA/XNS Gateway as Viewed by SNA

route flow control, packetization and reassembly, and store-and-forward packet routing.

A data link control element will serve as a line driver for each of the subarea links that attach the gateway to another SNA network gateway or subarea routing node. Each line driver will perform SDLC framing and point-to-point (i.e., SDLC frame-level) error detection and correction (through retransmission) as described in [11].

The XNS Protocol Element

Architecturally, the XNS Protocol Element will appear as an internetwork router host. The internal structure of the XNS Protocol Element shown in Figure 6, is described in the text which follows.

Three well-known sockets will be used to implement the following Level-2 protocols: routing information, echo, and error. In particular, the routing information socket will be used by the gateway's internetwork router to send and receive routing information packets as required by the XNS-specified dynamic routing algorithms. An echo socket will be used to receive and send echo request and response packets. Finally, a router error socket will be used to facilitate the sending of routing and transmission-related error packets by the gateway's internetwork router.

Service listener sockets will be used to facilitate the connection-time mapping of XNS gateway service requests to available service process socket resources. Service process sockets on the other hand, will be used to facilitate SNA LU-LU data transport across an XNS internet as well as XNS-to-SNA application protocol conversion.

The gateway's internetwork router will perform the functions of message routing and congestion control as described in [22].

A network driver will serve as the interface between the gateway's internet router and an attached packet transmission medium (network), such as an Ethernet. Each network driver will perform transmission medium-dependent framing, error detection or correction, as well as transmission medium-specific packet fragmentation and retransmission.

The SNA/XNS Protocol Translation Element

Figure 7 illustrates the internal structure of the SNA/XNS gateway as viewed from the protocol translation element.

The Session Management component will be used to provide a logical communication path between the gateway's SSCP and gateway-based logical unit and system element processes. The primary purpose of this component is to facilitate the passing of request and response indications during connection establishment and termination processing.

The Route Management component will be used to provide a logical communication path between the (SNA) physical unit and the (XNS) routing error socket. This component will convert SNA virtual route inoperative reports into corresponding XNS routing error packets and vice versa.

The Routing Information component (and its associated logical unit) will facilitate the maintenance of XNS dynamic routing tables across intermediate SNA networks.

The remaining components of the SNA/XNS protocol translation element will enable the

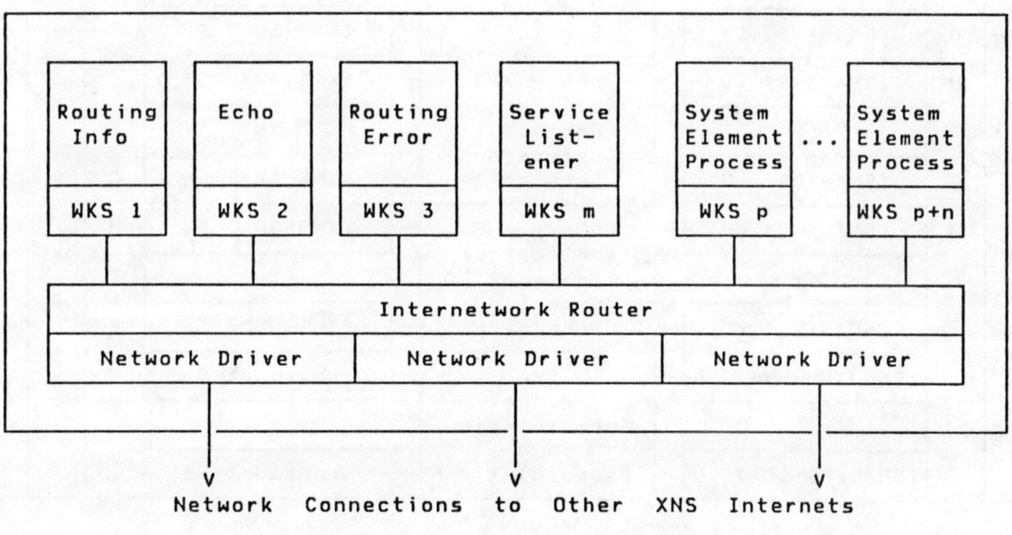

Figure 6. Structure of the SNA/XNS Gateway as Viewed by XNS

gateway to provide a comprehensive set of data transport and application-oriented services.

The XNS Packet Encapsulation / Decapsulation Element will be used to support XNS-SNA-XNS data transport.

The Transport Connection Protocol Translation Element will interface with the SNA and XNS Transport Connection Servers and the XNS Name and Address System Element in order to facilitate the establishment, management, and takedown of SNA-SNA, SNA-XNS, and XNS-SNA transport connections.

A number of SNA-XNS protocol translation elements will be used to facilitate application protocol conversion and application protocol mapping.

Finally, an SNA Packet Encapsulation / Decapsulation Element will be used to support SNA-XNS-SNA data transport.

Gateway Functions

This section describes nine functions that will be performed by the SNA/XNS gateway in support of the data transport and application support services described earlier.

Internetwork Resource Naming. The data transport requirements that have been imposed upon the design of the gateway raise a number of issues that are germane to the subject of naming and addressing:

1. Discerning Between SNA-SNA and SNA-XNS Transport Requests. The information that is conveyed within an SNA

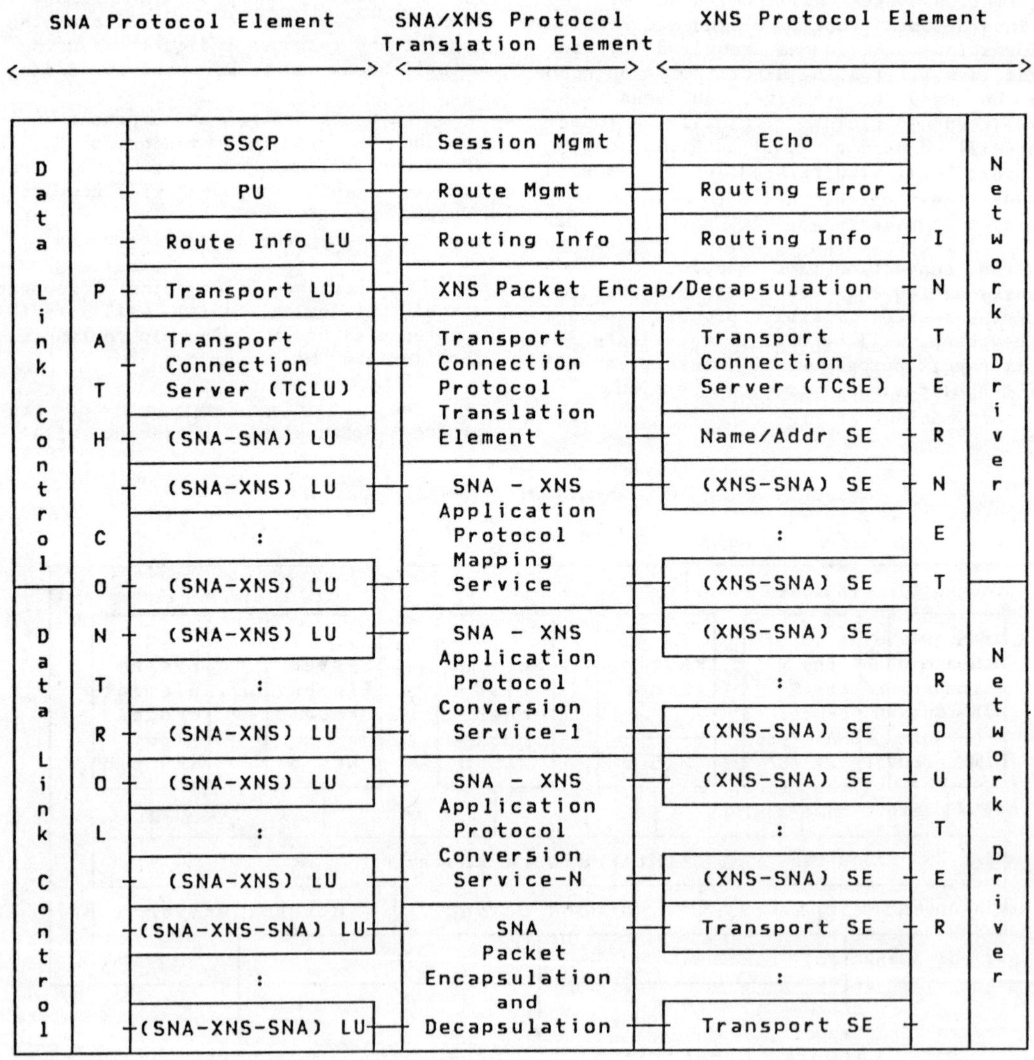

Figure 7. Detailed Structure of the SNA/XNS Gateway

Cross-Domain-Initiate (CDINIT) session establishment message, is not sufficient to differentiate between requests for SNA-SNA and SNA-XNS data transport. This problem results from the fact that the IBM SNA Network Interconnection (SNI) facility does not use network-qualified logical unit names (i.e., Network-Name.LU-Name). Although the IBM SNA Network Interconnection (SNI) facility has implemented a number of means for overcoming the cross-network resource location problem [18], these approaches are not suitable for locating resources in a mixed SNA/XNS networking environment for the following reasons: (1) the SNA name space cannot adequately express XNS resource names; and (2) an implicit SNA-XNS session setup request would necessitate sending connection requests on a trial-and-error basis to every SSCP in the interconnected SNA networking environment.

In order to remedy this problem, the SNA/XNS gateway will implement a logical unit server, whose purpose is to provide an SNA-compatible mechanism for explicitly requesting SNA-XNS transport sessions. The Transport Connection Logical Unit (TCLU) will enable a requesting logical unit to specify a fully or partially qualified XNS resource name as well as the session parameters which should be used in establishing the requested session. Since the requirement for application protocol support is so closely related to the issue of SNA-XNS data transport, the TCLU will also be used to capture application protocol support options at transport connection request time.

2. <u>Locating LU Resources in Other SNA Networks</u>. Rentschler et al. [18] describe the tables and procedures which are employed in IBM's current SNA Network Interconnection (SNI) implementation that enable an SSCP to locate logical unit resources residing in other networks. It is assumed that the gateway's SSCP will either implement the SNI procedure or use an alternative procedure for locating cross-network SNA resources through "CDINIT rerouting". In either case, the SSCP which owns the named resource would actually perform name/address binding.

3. <u>Locating XNS Resources From SNA</u>. XNS resources will be located through the XNS Clearinghouse [15] as always. In this situation however, a gateway-implemented (i.e., an XNS-based) name and address process will serve as the interface between the gateway's TCLU and the internetwork's composite Clearinghouse.

4. <u>Locating SNA Resources From XNS</u>. The SNA/XNS gateway will implement a transport connection system element (TCSE) process, whose purpose (like the TCLU) is to provide an XNS-compatible means of explicitly requesting XNS-SNA data transport sessions. This approach will be used since it would be impractical to dynamically maintain a copy of the composite SSCP's resource definition table within the Clearinghouse environment. However, the Clearinghouse could be used to maintain the following 3-tuple relationship, which would be useful for requesting SNA transport sessions through the XNS transport connection service: LU name, LU's network name, and LU's owning SSCP name.

<u>Internetwork Addressing</u>. Four internetwork addressing options were considered in the design of the SNA/XNS gateway:

The first option, assigning unique addresses from a flat address space, is presently employed by XNS. Accordingly, the gateway will support the use of internetwork-unique addresses for XNS-XNS and XNS-SNA-XNS data transport.

The second option, creating a new hierarchical addressing structure, was not considered because it violates one of the requirements imposed upon the design of the gateway, namely the preservation of local network independence.

The third option, moving one address space into the unused portion of another, was also not considered. SNA addresses are not statically assigned or uniquely specified across all networks.

The fourth option, mapping one address space into the unused portion of another on a demand basis, is presently employed by the SNA Network Interconnection (SNI) facility. Accordingly, the gateway will use this addressing technique for SNA-SNA and SNA-XNS-SNA data transport.

Address mapping will also prove to be helpful in bridging the differences that exist between SNA and XNS resource addressing for SNA-XNS and XNS-SNA data transport. Therefore, the gateway will represent SNA resources to XNS as being an XNS-based system element process residing in the gateway host. Conversely, XNS resources will be represented to SNA as being a (same-network) cross-domain logical unit resource owned by the gateway's SSCP.

<u>Packet Header and Request Data Translation</u>. In order to maintain the mapped resource relationships just described, the gateway's protocol translation element must perform name and address translations as follows.

For SNA-SNA and SNA-XNS-SNA data transport, the gateway must translate packet header addresses from one address space into another as a packet flows across network boundaries. In addition, the gateway must also translate resource names and addresses that are contained in session setup and takedown messages (e.g., CDINIT) in order to preserve each network's name/address space.

For SNA-XNS and XNS-SNA data transport, the gateway must translate SNA packet header addresses into corresponding XNS addresses and vice versa.

For XNS-XNS and XNS-SNA-XNS data transport, no packet header translations are required.

Internetwork Routing. Since both SNA and XNS addressing will be employed within the gateway, it follows that the gateway's internetwork routing mechanism should comprise an SNA path control router [12] and an XNS internetwork router [22]. Each component will perform routing in a manner that is consistent with its underlying network.

Implementation of the XNS Routing Information Protocol. A topic of special concern is the manner in which XNS routing tables will be maintained in a heterogeneous SNA/XNS networking environment. In XNS, the present routing table maintenance algorithms mandate that all internetwork router hosts gratuitously broadcast complete routing information once every thirty seconds. Since broadcasting is not supported by SNA and the point-to-point simulation of broadcasting would be very expensive to implement from a resource utilization viewpoint, another approach must be used to maintain XNS routing information in an SNA/XNS networking environment.

The SNA/XNS gateway will solve this problem by confining the broadcasting of XNS routing information to the XNS subnetwork, and transmitting "delta-XNS" (i.e., changed) routing information across the SNA subnetwork on a change-demand basis. The gateway's internetwork router will be used to perform the former function while paired routing information logical units will be used to perform the latter.

Message Encapsulation and Decapsulation. The SNA-XNS-SNA and XNS-SNA-XNS data transport services require that the gateway provide a means for transparently moving data between homogeneous networks across an intermediate (heterogeneous) data transport network. The technique of message wrapping [10] or message encapsulation [19] will be employed by the gateway.

For SNA-XNS-SNA data transport, SNA packets (arriving at the SNA-XNS network boundary) will be encapsulated within an XNS internet datagram packet prior to transmission across XNS. This will entail prefixing the SNA packet with a Level-2 (SPP) and a Level-1 (IDP) message header. Conversely, the embedded SNA packet will be decapsulated from its XNS packet at the XNS-SNA network boundary and presented to the SNA-side of the gateway for header translation and destination routing.

For XNS-SNA-XNS data transport, XNS packets (arriving at the XNS-SNA network boundary) will be encapsulated within an SNA path information unit (packet) prior to transmission across SNA. This will entail prefixing the XNS internet datagram packet with a request header (RH) and a transmission header (TH). Conversely, the embedded XNS IDP will be decapsulated from the SNA packet at the SNA-XNS network boundary, and presented to the XNS-side of the gateway for destination routing.

Packet Fragmentation and Reassembly. Another data transport-related function that will be performed by the gateway is packet fragmentation and reassembly.

For SNA-SNA and XNS-XNS data transport, fragmentation and reassembly are considered to be non-issues [18,22]

For SNA-XNS-SNA data transport, it is assumed that packet fragmentation will be required. Since SNA does not support fragmentation, an alternative approach will be used instead: SNA packets (containing whole or segmented messages) will be stored, resequenced, and then resegmented into 512 byte sized packets by the gateway's path control layer.

For XNS-SNA-XNS data transport, it is assumed that packet fragmentation will not generally be required due to the nominal maximum size of an XNS packet (i.e., 576 bytes). However, if fragmentation should need to be performed at an SNA/XNS network boundary, the approach described for SNA-XNS-SNA data transport can be employed.

For SNA-XNS and XNS-SNA sessions, it is assumed that the gateway will enforce the specification of a maximum SNA message size that can be readily accommodated within an XNS packet.

Error Control. Since concatenated virtual circuits will be employed by many of the data transport scenarios, a very important issue that must be addressed is the preservation of end-to-end message integrity. Accordingly, the gateway will maintain the significance of end-to-end message acknowledgements through the technique of cascading described by Gien and Zimmerman [10].

Flow and Congestion Control. Since the data transmission speeds employed by SNA and XNS typically differ by two to three orders of magnitude, it is imperative that the gateway closely monitor the availability of message buffer resources as a function of the data traffic flowing through the gateway. On the SNA-side of the gateway, this is accomplished by implementing the SNA virtual route flow and congestion control protocols. Since XNS has no corresponding network-level flow control mechanism, the gateway must control the composite XNS message flow through the control of individual XNS connection-oriented session flows.

The gateway must also bridge the differences that exist between SNA and XNS session-level flow control protocols. The gateway must monitor both flow-control protocols and try to keep both receiving-end windows permanently open, while at the same time providing enough buffer capacity within the gateway to prevent the loss of data when one receiving (session) end keeps its flow control window closed.

SUMMARY

The CCITT, DARPA/DoD, and ISO/ECMA internetworking models are three of the commonly used approaches for interconnecting computer resource sharing networks. Many of these approaches make the assumption that all interconnected networks must "subscribe" to the same internetworking standard. However, many computer networks that exist today are based upon vendor-proprietary, closed-system designs that do not support the aforementioned internetworking standards.

This paper has described an approach for internetworking such closed-systems that is based upon the solution of two problems: a data transport problem and an application-level cooperation problem. This internetworking approach has been applied to the pragmatic problem of interconnecting private data networks that are based upon IBM Corporation's System Network Architecture (SNA) and Xerox Corporation's Network System (XNS).

Paramount to the solution of the "SNA/XNS internetworking problem" has been the identification and definition of a number of gateway-provided data transport and application protocol support services, which were incorporated into a multi-function, multi-level, protocol translation gateway. Six specific data transport services were required to solve the data transport problem whereas the application-level cooperation problem was solved through the provision of an open-ended set of application protocol conversion services that were augmented by a single application protocol mapping function.

This paper has described a protypical gateway for interconnecting SNA and XNS networks in particular and closed-system networks in general. One obvious set of future work will focus on the design and implementation of an SNA/XNS gateway. A second area of future work will address the notion of "opening" vendor-proprietary closed-system network architectures through an OSI-based gateway.

REFERENCES

[1] Benhamou, E. and Estrin, J. "Multilevel Internetworking Gateways: Architecture and Applications", Computer. Vol. 16, No. 9, September 1983. pp. 27-34.

[2] Benjamin, J.H., Hess, M.L., Weingarten, R.A., and Wheeler, W.R. "Interconnecting SNA Networks", IBM Systems Journal. Vol. 22, No. 4, 1983. pp. 344-366.

[3] Boggs, D.R., Shoch, J.F., Taft, E.A., and Metcalfe, R.M. "Pup: An Internetwork Architecture", IEEE Transactions on Communications. Vol. 28, No. 4, April 1980. pp. 612-624.

[4] Callon, R. "Internetwork Protocol", Proceedings of the IEEE. Vol. 71, No. 12, December 1983. pp. 1388-1393.

[5] Cerf, V.G. and Cain, E. "The DoD Internet Architecture Model", Computer Networks. Vol. 7, No. 5, September 1983. pp. 307-318.

[6] Cerf, V.G. and Kirstein, P.T. "Issues in Packet-Network Interconnection", Proceedings of the IEEE. Vol. 66, No. 11, November 1978. pp. 1386-1408.

[7] Dalal, Y.K. "Use of Multiple Networks in the Xerox Network System", Computer. Vol. 15, No. 10, October 1981. pp. 82-92.

[8] Einert, D. and Glas, G. "The SNATCH Gateway: Translation of Higher Level Protocols", Journal of Telecommunication Networks. Vol 2. No. 1., Spring 1983. pp. 83-102.

[9] Francois, P. and Potocki, A. "Some Methods for Providing OSI Transport in SNA", IBM Journal of Research and Development. Vol. 27, No. 5, September 1983. pp. 452-463.

[10] Gien, M. and Zimmerman, H. "Design Principles for Network Interconnection", Proceedings of the Sixth Data Communications Symposium. Pacific Grove, California. November 1979. pp. 104-119.

[11] IBM Corporation. IBM Synchronous Data Link Control General Information. Document No. GA27-309302. International Business Machines Corporation, Systems Communication Division. Research Triangle Park, North Carolina. March 1979.

[12] IBM Corporation. Systems Network Architecture - Format and Protocol Reference Manual: Architectual Logic. Document No. SC30-3112-2. International Business Machines Corporation, Data Processing Division. White Plains, New York. November 1980.

[13] IBM Corporation. Systems Network Architecture Concepts and Products. Document No. GC30-3072-0. International Business Machines Corporation, Data Processing Division. White Plains, New York. January 1981.

[14] IBM Corporation. Systems Network Architecture Technical Overview. Document No. GC30-3073-0. International Business Machines Corporation, Data Processing Division. White Plains, New York. March 1982.

[15] Oppen, D.C. and Dalal, Y.K. The Clearinghouse: A Decentralized Agent for Locating Named Objects in a Distributed Environment. Document No. OPD-T8103. Xerox Corporation, Office Products Division. Palo Alto, California. October 1981.

[16] Postel, J.B. "Internetwork Protocol Approaches", IEEE Transactions on Communications. Vol. 28, No. 4, April 1980. pp. 604-611.

[17] Postel, J.B., Sunshine, C.A., and Cohen, D. "The ARPA Internet Protocol", Computer Networks. Vol. 5, No. 4, July 1981. pp. 261-271.

[18] Rentschler, K. Stoner, A., and Kaufmann, E. SNA Network Interconnection Planning and Design. Document No. GG24-1630-00. IBM Corporation, International Systems Center. Research Triangle Park, North Carolina. April 1984.

[19] Shoch, J.F., Cohen, D., and Taft, E.A. "Mutual Encapsulation of Internetwork Protocols", Computer Networks. Vol. 5, No. 4, July 1981. pp. 287-300.

[20] Sunshine, C.A. "Interconnection of Computer Networks", Computer Networks. Vol. 1, No. 3, January 1977. pp. 175-195.

[21] Weir, D.F., Holmblad, J.B., and Rothberg, A.C. "An X.75 Based Network Architecture", Proceedings of the 5th International Conference on Computer Communications. Atlanta, Georgia. October 1980. pp. 741-750.

[22] Xerox Corporation. Internet Transport Protocols. Document No. XSIS 028112. Stamford, Connecticut. December 1981.

[23] Xerox Corporation. Courier: The Remote Procedure Call Protocol. Document No. XSIS 038112. Stamford, Connecticut. December 1981.

[24] Zoline, K.O. "An Approach for Interconnecting SNA and XNS Networks", Master's Thesis, Illinois Institute of Technology. Chicago, Illinois. December 1984.

CNMGRAF - GRAPHIC PRESENTATION SERVICES FOR NETWORK MANAGEMENT

Richard S. Gilbert and Wolfgang B. Kleinöder*

IBM Thomas J. Watson Research Center, Yorktown Heights, NY 10598
* IBM Zurich Research Laboratory, 8803 Rüschlikon, Switzerland

Abstract: Two key problems in the design of systems and network management applications are: how to efficiently manage large quantities of related data, and how to allow the application user to view this data at a user-defined level of detail. This paper describes an architecture for network management facilities that may help solve these problems. It also describes CNMgraf (Communications Network Management Graphics Facility), which was developed to test the presentation service model defined by this architecture.

1. Introduction

Two key problems in the design of systems and network management applications are: how to efficiently manage large quantities of related data, and how to allow the application user to view this data at a user-defined level of detail. This paper describes an architecture for network management facilities that may help solve these problems. It also describes CNMgraf (Communications Network Management Graphics Facility), which was developed to test the presentation service model defined by this architecture.

CNMgraf combines the flexibility of a host-based relational database (SQL/DS) [1] with the power of a PC/AT-based graphic workstation. The prototype database is filled with information about different SNA (Systems Network Architecture) [2] networks. By using a GKS (Graphical Kernel System) [3] application running on the workstation as the interface to the data, a CNMgraf user may view the configuration of a network and ask questions about the network in general or the network nodes in particular (e.g. ask for routes between two nodes, or for the speed or utilization of links).

In current SNA network management products, all data and presentation services are host-based and provided as separate tasks within each network management product [4]. In the CNMgraf model, all presentation services are handled by the workstation, while all database queries are formatted and performed on the host. The CNMgraf data and presentation services may also be shared by several different network management applications so that these applications may be implemented as simple database utilities. The SQL database allows the user extraordinary flexibility in the type and format of the data displayed. In fact, a particular user may view a network configured to his own taste and experience. The intelligent workstation presentation application also allows the user to perform many functions locally (such as picture editing, zooming, or panning), thus eliminating the overhead of display management from the host. By using a multi-tasking operating system on the PC/AT, it is also possible to optimize the graphic and host interaction processes as well as allow for asynchronous alerts from the host.

2. The Use of Graphics for Network Management

The basic design principle of CNMgraf is that network data should be viewed in an appropriate and comfortable manner. Until recently, most network management tools have forced users to view all kinds of data in the same way - usually via alphanumeric lists. Our approach has been to show alphanumeric data (e.g. lists of component serial numbers) on alphanumeric panels and graphically oriented data on graphic panels. There seem to be several distinct classes of graphic data.

One obvious type of graphically oriented data are the statistical distributions that abound in network activities. These distributions are very easy to analyze when displayed using standard business graphic techniques. We were not concerned with these types of displays as they have been described elsewhere [5].

Also obvious, but much more difficult to implement, is the display of two-dimensional graphs representing network configurations. Such graphs may be produced from the node list and the adjacency matrix by using various heuristic algorithms. Once a network has been represented as a graph, the graph can be *colored* to indicate a wide variety of state related data such as: component status, utilization, link speed, routes, and active sessions.

Another way to use a graphic representation of a network is to allow the user to *pick*, from the image, a particular network component or set of components. Then the image may be used as a data directory to retrieve network related data which could then be displayed as an graphic overlay or on another panel. For example, the user could ask for the recent change history of a network node, by picking it, and the history would be displayed, in list form, on the alphanumeric panel.

A third use for an image of a network is as a driver for other network management processes. The user could interactively edit a network image and then submit the modified network to an application that processes the data and performs some computation intensive task such as selecting routes or selecting link speeds. It is possible for a user to perform the same function by editing an alphanumeric representation of the network node list and adjacency matrix. However, the user would lose the natural sense of relative position that comes from using the graphic representation.

Finally, given the capability to quickly create a graphic representation of a network, it is possible to view historical data in a pseudo-cinematic way so that many hours of history could be compressed into a few seconds of real time. The user could operate the system as he would a video tape player to selectively view important periods of a recorded history log.

3. CNMgraf System Architecture

The CNMgraf prototype is an implementation of the presentation service task of a more general network management model. Consider the network and all its activities as a separate *black box*. Then network management may be considered as the processes of collecting, monitoring, displaying, and manipulating network data. In this model, there are three key elements to the network management system.

The first element is a network monitor process that, in effect, filters dynamic network data and decides where to route the data. Possible route destinations are: to a particular network management workstation or to a general network management database facility.

The second element, the database facility, should be viewed as a central repository for all network related data. It includes both static data (slowly changing information such as the network configuration) and dynamic data (such as current component state). The database should be relational to allow efficient recovery of specific information.

The third element is the graphic workstation that is used as a generic presentation service facility for all network management functions. The workstation receives most of its data from the central database, but it also receives some dynamic alerts directly from the network monitor. It should be noted that there is no reason why this process could not be implemented on the host using dumb terminals for display, but that implementation would increase host and communications overhead while losing the capability for local workstation functions.

In this model, all other classical network management tasks can be implemented as simple applications that use the central relational database facility and the common presentation services provided via the workstation and database combination. This structure allows great simplification of many communications network management products that currently include individual, and therefore redundant, data and presentation service processes.

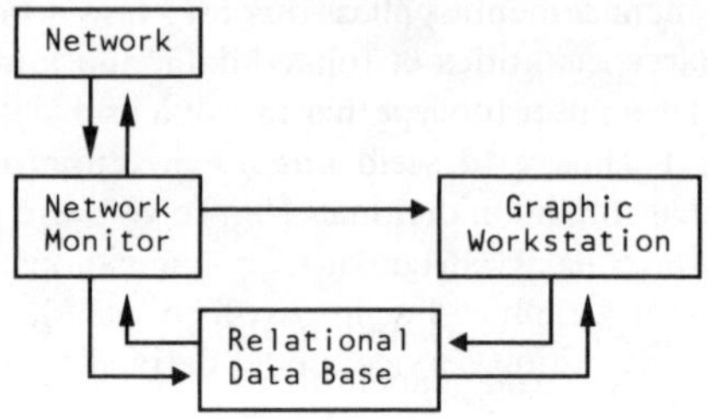

Fig. 1. Data flow in the CNMgraf architecture

In the CNMgraf prototype, the network monitor process has not yet been implemented. So there is no access to dynamic network data. The workstation has been completed as well as the workstation to database communications. Some static data has been stored in the database and a database application has been developed to extract specific data on workstation requests.

There were three distinct problem areas in the implementation of the CNMgraf prototype system; the graphic workstation, communication between the workstation and the host, and the relational database on the host. The rest of this paper will describe these three problem areas in more detail.

4. Graphic Workstation

The workstation application was designed as a, keystroke interactive, graphic editor which was limited to drawing two-dimensional representations of computer networks. We wanted to clearly separate *knowledge about networks* from *knowledge about drawing networks*. All network data was stored on the host and accessed via host-based database applications while all graphic data (such as graphic icons) was stored on the workstation. Hence, the host application would know of a physical or logical entity by name (e.g. 3081 or SDLC link) while the workstation would treat the same entities as shapes with colors (e.g. large red box or solid green line). The only graphically oriented data stored on the host are the positions of the network nodes. This was necessary, in CNMgraf, as the initial layout of a network is computed by an algorithm implemented on the host.

This separation of function allowed the development of a high-level protocol for communication between the workstation and the host. The workstation can, in effect, request the data required to draw a picture and the host will return three control blocks describing the network nodes, links, and text annotations. The workstation processes these control blocks by mapping object types and classes to graphic shapes and colors respectively, and then draws the appropriate icons. The fact that the workstation application is limited to drawing networks is important, as only a limited number of types and classes must be supported. There are about 30 different shapes (e.g. polygon types, line types, and text fonts) and 16 different colors supported by the prototype workstation application. The colors were chosen to yield images that appear flat and comfortable to most individuals [6], but a user has the option to custom design his color set by editing a profile.

The prototype workstation application supports two *panels*; a monochrome panel for user input/output and alphanumeric data display, and a color panel for network graphic displays. A CNMgraf *panel* should be viewed as a logically independent cache of information. We were more concerned with data organization than with the physical placement of these panels. Hence, although the prototype was developed on a workstation with two dedicated physical screens it could, as easily, have been implemented using multiple windows on a single high-resolution display.

The concept of the CNMgraf user interface is to allow the operator to select a presentation *level* from a large set of possibilities ranging from a geographic representation of a backbone network to a physical representation of a peripheral node. This hierarchically organized set of levels allows a user to view a network at his desired level of detail. Once the level has been selected, the user will select a *mode* that determines on what basis the network image is to be colored (e.g. state, speed, or utilization). The level/mode selection is sent to the host and defines the data required for the drawing. The host application then structures a semi-customized database query and returns the collected data to the workstation in the form of the three control blocks described above.

There are many keystroke interactive functions such as zoom or pan that do not require host interaction. These local functions are primarily used to accommodate user requests to manipulate an existing display. It should be noted however, that zooming and panning exist for user comfort only, they are not used as substitutes for the level/mode method of detail selection.

CNMgraf is implemented using the GKS (Graphical Kernel System) primitives that form an ISO/ANSI standard for vector oriented two-dimensional drawing. We found that GKS allowed an easy, device independent implementation of the graphic functions. However, the particular implementation of GKS that we used produced rather slow drawing functions. This led us to believe that a reasonable approach for developing graphic prototypes would be: to first develop a high-level GKS prototype and then refine the code by pushing functions to the lower level VDI (Virtual Device Interface) standard or even sacrifice device independence and implement functions using device specific hardware capabilities. The primary goal of our prototype was function, so GKS was sufficient. However, if CNMgraf were used in an operational environment it would also require interactive speed.

5. Network Graphic Communication Protocol

The design principle of using an independent graphic workstation for all presentation services enables us to offload the host. This is especially important during periods where the network may be in critical condition; during the time where the network operator will heavily interact with the network management service programs. On the other hand, by splitting the network management system onto multiple computers, we have to deal with the communication between those systems.

Our goal is to minimize the overhead introduced by the communication between the workstation and the host system. The chosen solution is based on the client/server model [7]. The host application is implemented as a database server machine. A high-level client/server protocol significantly reduces the amount of data which must be sent from the host to the workstation compared with a general solution which runs the graphic software on the host system and ships whole images to be displayed on the workstation. Instead of transmitting the description for complex symbols, only references to the symbols will be transmitted. The workstation can then efficiently generate the actual image by using locally stored icon definitions.

For heterogeneous systems, like personal workstations together with large mainframes, we get further improvements if another layer is put on top of the client/server model which shields the application programs in the workstation from the details of the communication. This is possible by *function shipping of service requests* [8]. The workstation application stores all parameters for a service request in a control block and the control block is shipped to the database server on the mainframe system. The application on the workstation then waits for the reply from the database server.

The server on the host provides services corresponding to levels 6 and 7 (presentation and application) of the ISO OSI model. First, the data will be translated from workstation format to host format. This service cannot be provided by any lower level communication software, because it requires knowledge about the syntax of the control block. Communication between a workstation based on the Intel 80286 and an IBM 370 mainframe requires a translation of text fields from ASCII to EBCDIC and an unscrambling of bytes in the binary data areas. One implementation of this sort of interface is known as the Virtual Service Interface (VSI) [9,10] as used in the IBM XT/370 product. We also used the VSI interface for the CNMgraf prototype.

The next step, which will be explained in the next section, is the data retrieval according to the parameters contained in the (translated) request control block. The results will then be translated back from mainframe format into workstation format and sent to the workstation as a set of control blocks.

An advantage of this design is that it can be used on top of various transportation services; 3270 datastreams, SNA application program to program communication, or ISO transport level services of either local-area networks or wide-area networks.

Using our prototype implementation, we could demonstrate that, even when slow channels were used for the communication between the workstation and the database server, the response times were quite reasonable because the requests and results can be represented in rather small data areas (1K - 15K bytes for complicated queries on large networks). The communication response delays, for our distributed system, are rather short (2 seconds).

6. Relational Database

The database application consists of two main parts. One set of routines supports data retrieval for the graphic workstation, while other routines deal with data input into the database. Our prototype database system was built using a relational database. This makes the application programming quite convenient as one advantage of the relational database system SQL/DS is that the access methods can be tuned after the system is developed. The table indices may be generated in order to optimize the most often used access patterns. Because data retrieval programs are executed more often that data input programs, the indices may be tuned to optimize the data retrieval operations. The rest of this section will explain how the data retrieval programs access the network information and describe the database input operation, in particular, how initial layouts for arbitrary networks are generated.

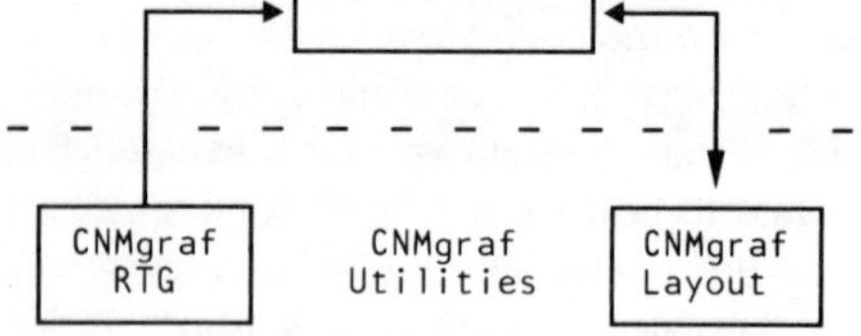

Fig. 2. CNMgraf prototype design

As explained in the last section, the communication routines take care of the data translation necessary between the workstation and the database server. The workstation prepares the request control block and executes the remote procedure call to the database application, then the database server retrieves the information from the database and sends back only the information that is necessary for the graphic application to construct the picture:

- node information: shape, color, and location
- link information: end nodes, line type, and color
- text information: text, font, and corresponding object (node or link)

The user gets an explanation what shape, line type, and color mean for each image in a dynamically generated key which can be shown on the graphic screen by request. By using this technique, it was possible to restrict the features that have to be handled by the workstation to a small but very general set. For example, all graphical interaction on a green line will be independent of whether the color green will represent the speed of the link, the utilization of the link, or even the fact that the link is contained in a physical route (SNA explicit route) which is represented by green.

For each workstation request, the database server executes the following operations:

Check logon information: The database can contain multiple networks and information for multiple operators. On the first access, the operator's password is checked. The network name and operator name are then translated into an internal representation that can be used more efficiently for further queries.

CNMG_NETWORKS

NETID	NETWORK
1	AMEX
16	NETHER
4	NYLIFE1
8	PENNY3
9	STF_FEIS
13	STL0107A
11	TRAVELER

CNMG_OPERATORS

OPID	OPERATOR
201	ANDY
-5	CNMGRAF
-9	CNMGRAF9
102	GILBERT
301	HOCKER
101	WBK

Select window: The viewing window may be selected either manually, or the database server can choose an optimal window using one of the following options:

- show whole network
- show all nodes which are connected to a selected node
- show all nodes which are in one site (i.e. connected by channels)

Retrieve nodes: The node information is retrieved from two different tables: The shape and color of a node is calculated from network specific data, such as node type (indicated by PULVL in the CNMG__SUBAREA table). The location of a node on the screen is determined from a table which is specific to the graphic presentation approach for network management. Different coordinates are stored for each node and each presentation level, also, if he chooses, an operator can have his own set of coordinates for any network/level combination. These operator specific coordinates may be defined via the SQL/DS interface, but are more often the result of interactive editing of the default image of a network.

The set of nodes often remains constant for several requests from the workstation. Therefore, a node cache was implemented so that the database retrieval functions have only to be executed when one of the parameters that defines the node set (network name, operator name, or presentation level) has changed since the last request. A reduction in the set of nodes necessary for the picture generation is sometimes possible after the link processing because, by then, it is known which nodes are connected to at least one link. Our experience is that, usually, only those nodes need to be shown.

CNMG__SUBAREA

NETID	SANUM	SANAME	PULVL
22	8	NEWYORK	NCP3
22	70	NEWYORK	VTAMV2

CNMG__SITE

NETID	SANUM	SITENUM
22	8	8
22	70	8

CNMG__COORDINATES

NETID	OPID	LEVEL	SANUM	X	Y
22	101	4	70	10059	9958
22	101	4	8	10039	9972
22	101	3	8	10040	9972

Retrieve links: Links are represented on the screen as lines connecting nodes. Only links where at least one end point is in the selected window are considered for presentation on the graphic screen. There are various methods to calculate the line type and color information. For the default network representation, the line type represents the transmission medium (e.g. channel = solid) and the color represents the speed (e.g. 48-56 kbps = blue). A similar method may be used to represent other numerical values mapped to links (e.g. utilization, availability, or length) where different ranges of possible values are represented by different colors. Small sets of non-numeric values, such as link state (e.g. active or inactive) can also be represented by colors.

CNMG__LINK

NETID	SA1	SA2	LNAME	TGN	MED	SPEED	PROTOCOL
22	5	8	5A8	1	T	19200	H
22	8	9	8A9	1	S	19200	H
22	8	70	8A70	1	C	1000000000	H

A different approach is taken if link coloring is used to mark routes. To retrieve the information for virtual routes, we have to access 3 tables: The table CNMG__VRN__ERN contains the mapping from virtual routes to physical routes, the table CNMG__ER contains which links are used in physical routes. The table CNMG__LINK is used to discover which medium (mapped to line type) a link has, and to find all the links that are not used in any of the selected routes (these are shown as background information in a dark grey color). Because it is possible that a link is used by multiple routes, it is sometimes necessary to show multiple lines between a given pair of nodes. After all values for a given link are retrieved, end point coordinates are calculated for a set of parallel lines representing the different routes. An SQL query for virtual routes is shown below as an example.

CNMG__ER

NETID	OSA	DSA	ER	SA1	SA2	TGN
22	70	4	3	7	4	1
22	70	4	3	70	8	1
22	70	4	3	8	9	1
22	70	4	3	9	7	1

CNMG__VRN__ERN

NETID	OSA	DSA	VRN	ER	RER	RL
22	70	4	0	3	6	4

SQL Query for Virtual Routes

```
EXEC SQL DECLARE CURSOR LINK__VR CURSOR FOR
SELECT DISTINCT A.SA1,A.SA2,A.TGN,A.MED,-1,A.LNAME
FROM CNMG__LINK A
WHERE A.NETWORKID=:SQL__NETI
UNION
SELECT DISTINCT A.SA1,A.SA2,A.TGN,A.MED,C.VRN+0,A.LNAME
FROM CNMG__LINK A, CNMG__ER B, CNMG__VRN__ERN C
WHERE A.NETWORKID=:SQL__NETI AND B.NETWORKID=:SQL__NETI
AND  C.NETWORKID=:SQL__NETI
AND  C.DSA=:SQL__DSA AND C.OSA=:SQL__OSA
AND  C.ER=B.ER
AND  B.DSA=:SQL__DSA AND B.OSA=:SQL__OSA
AND  ((B.SA1=A.SA1 AND B.SA2=A.SA2)
OR   (B.SA1=A.SA2 AND B.SA2=A.SA1)) AND A.TGN=B.TGN
ORDER BY 1,2,3,5 DESC
```

Text retrieval: During the node and link processing, object names are collected which can be shown, by request, on the graphic screen. Simultaneously, the database application program keeps track of which types of nodes and links actually appear on the screen and generates a customized key that can also be displayed upon request.

Update of graphic information

Network management presentation graphics are very well suited for keeping the network description up-to-date. If a new node is to be added to a network, the user would select a location for the node on the graphic screen using a locator device (joystick or mouse). He would then enter, on the alphanumeric screen, the required non-graphic information about the node such as the subarea number, subarea name, or node type. Finally, he would pick on the graphic screen all the nodes which have links to the new node. The new node and associated links would then appear on the graphic screen and, if the user desires, the addition can be saved in the host database.

In the prototype implementation, we have emphasized modification of the graphic information. Although the automatic layout algorithm (explained later in this section) produces, in most cases, very reasonable network layouts the user may sometimes want to change a layout. If so, he will pick a node to be moved and select a new position using the graphic cursor. The node will then be moved and any associated links will automatically move with the node. This process may be repeated as needed. After he has finished the modifications, the user may save the updated network in the database, or cancel the changes. Only the table for the coordinates CNMG__COORDINATES must be updated, because all other tables only reference the subarea numbers.

First experience

Our prototype implementation has proven that a host-based relational database system is very well suited for the retrieval operations necessary in a network management presentation system. Although the queries tend to be quite complicated, the response times are rather exciting. Even for large networks with more than 100 nodes in the backbone network, the retrieval time (on a time-shared host system) rarely exceed 3 - 5 seconds. These response times were made possible by generating the appropriate indices for all the queried tables. The small increase in the table update time, caused by using multiple indices for some of the tables is tolerable, because the retrieval process is performed significantly more often than the tables are updated.

Database input

Input into the CNMgraf database is supported by utility programs. It is very important for a new network management tool to be able to use existing network definitions. Because these network descriptions are often complex (e.g. more than 50000 lines just for the definition of the backbone network), it is not reasonable to ask a manager of a large network to redefine the network from scratch. Therefore, CNMgraf provides a utility which reads the input/output decks from the RTG (Route Table Generator) program [11], and loads the node, link, and route definitions into the database tables.

A similar approach is possible for other static (slowly changing) data such as the boundary network configuration data. Even the results of network design tools (e.g. calculated projected values for the throughput of links) can be treated similarly; a utility program simply dumps the data into the database.

Much more complicated, is the dynamic data (changing frequency in the order of minutes or less). An example of this category are alert messages. It is important that the network operator receives alerts as soon as possible and it is not possible to wait until he requests the alerts. In this case, the host system will act as client, and the workstation as a server.

In our current prototype system, we concentrated on static data for backbone networks. However, we are considering extending CNMgraf to include boundary networks as well as add some dynamic data. Note that the latter task will require a multi-tasking operating system on the workstation so that the workstation can, simultaneously, act as a client for the retrieval and presentation of static data and as a server for the dynamic data.

Layout Algorithm

The architecture of CNMgraf differs radically from other graphic network management systems in the way the network layout is constructed. Most existing systems were developed for use by one customer. Therefore, the network layout was often hard coded so that, if the network was modified, the layout definition had to be changed by hand and some of the system had to be recompiled. CNMgraf can not only store the information of multiple networks in the database but it also provides an automatic network layout algorithm.

The goal for this algorithm was that the user should have to supply as little additional information as possible before he can view the network image. Of course, it would be preferable to provide a geographic layout. Unfortunately, very few current network descriptions contain any geographical information. One exception is BITNET, which includes the latitude and longitude (and planet!) in each node definition.

BITNET Node Definition

```
:nick.DERRZE1 :alias.DERRZE1 :nodenum.0380
:abbr.Univ. Erlangen - Nuernberg
:system.VM/SP Release 3 :machine.IBM 4361
:country.D :longitude.011E 00 :latitude.049N 36
:linkfail.OPERATOR@DERRZE1 (09131/85-7037)
:planet.EARTH :linkspeed.9600 :linktype.LEASED
```

CNMgraf contains a logical network layout algorithm which is based only on a classification of links:

- links, which are short
- links, which may be long

If the network definition contains the length of a link, then this information can be used. Otherwise, the algorithm can use the speed of a link to guess the classification, very high speed links (e.g. channels) are usually short while lower speed connections may be longer.

1. The algorithm first calculates which nodes form a site by clustering all those nodes together which are channel (high speed) connected. This step attempts to guess which nodes are geographically close together.

   The next steps only consider the links between different sites. This reduces the complexity of the algorithm significantly.

2. Sites, which have connections to one other site, are called subsites. They will be shown close to their main site and need not be considered in the following steps.
3. Place all (main) sites on one large circle.
4. Reorder the sites to minimize the number of crossings between links.

   The next steps are repeated for each site.

5. Place all nodes of a site which only have connections to other nodes of the same site on a small circle.
6. Reorder the nodes to minimize the number of crossings between links inside the small site circle.

The experience from our prototype development shows that, for SNA networks, it is sufficient to distinguish between channels and other links in order to obtain very reasonable looking network layouts (see Fig. 3). In some cases, the layout can be further enhanced by moving a small number of the nodes using the interactive editing features of CNMgraf.

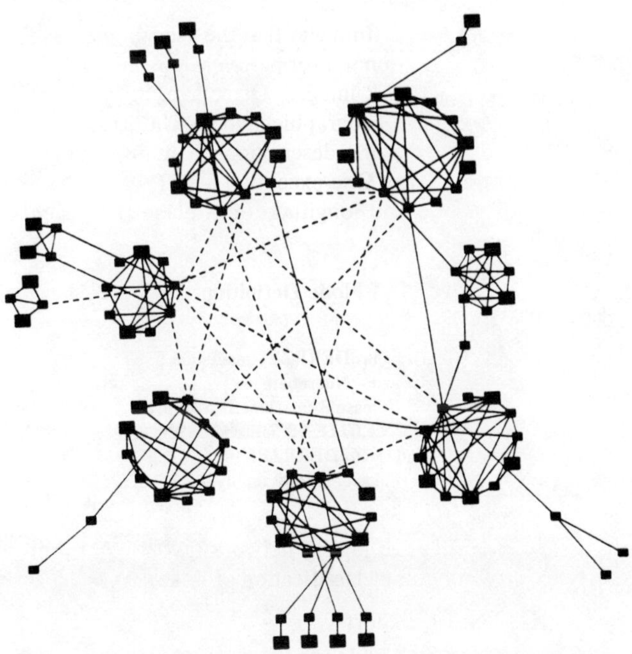

Fig. 3. A sample network layout produced by the algorithm described in this paper.

7. Summary

In this paper, we introduced an architecture for systems and network management facilities that consolidates all data management activities on a large host system and distributes all presentation service activities to small workstations. When compared with previous host-based network management applications, this architecture has the advantage of decreasing the host and communications overhead requirements because of its exploitation of the local processing power and keystroke interactive capabilities of intelligent workstations. The CNMgraf architecture should also allow for fast development of new network management utilities as such utilities are now reduced to simple applications that access a common database and use common presentation services via high-level interfaces.

Our work, designing and implementing CNMgraf, has convinced us that it is practical to build efficient network management presentation service facilities based on this architecture. The three problem areas encountered in the implementation of CNMgraf, the graphic workstation, the network graphic communication protocol, and the relational database, have been discussed and our solutions have been described. In conclusion, we believe that the inherent flexibility of the prototype CNMgraf presentation service facility allows the user consistent and fast access to the specific data he requires and the capability to display that data in the format that he desires.

Acknowledgments

The monochrome display manager used in the prototype, ADMS (Application Display Mangement System), was developed by Ed Iacobucci who was very helpful debugging this part of our code. The C interface to ADMS was developed by Dave Hocker. The authors would like to thank our colleagues, Josh Auerbach and Brian Carr, for their development and support of the communications software. We would also like to thank our management, Harold Anderson and Kiyoshi Maruyama, for their advice and support.

References

1. *SQL/Data System General Information*, GC24-5012; available through IBM branch offices.
2. *Systems Network Architecture - Concepts and Products*, GC30-3072; available through IBM branch offices.
3. F. R. A. Hopgood, D. A. Duce, J. R. Gallop, and D. C. Sutcliffe, *Introduction to the Graphical Kernel System (GKS)*, Academic Press, Inc., New York, NY, 1983. 200 pp. (ISBN 0-12-355570-1).
4. *Network Management Products General Information*, GC27-0657; available through IBM branch offices.

5. R. S. Gilbert and P. P. Wong, "NETS - A Network Evaluation and Tracking System," *IBM Technical Report TR 03.182*, IBM General Products Division, Santa Teresa Laboratory, CA 95105 (1982).
6. Gerald M. Murch, "Physiological Principles for the Effective Use of Color" *IEEE Computer Graphics and Applications*, Vol. 4, No. 11, Nov. 1984.
7. Liba Swobodova, "Client/Server Model of Distributed Processing," *IBM Research Report RZ 1350*, IBM Thomas J. Watson Research Center, Yorktown Heights, NY 10598 (1985).
8. F. N. Parr, J. S. Auerbach, and B. C. Goldstein, "Distributed Processing involving personal computers and mainframe hosts", *Journal of Selected Areas in Communications*, Vol. SAC-3, No. 3, May 1985.
9. B. C. Goldstein, A. R. Heller, F. H. Moss, and I. Wladawsky-Berger, "Directions in cooperative processing between workstations and hosts", *IBM Systems Journal*, Vol. 23, No. 3, 1984.
10. B. C. Goldstein, G. Trivett, and I. Wladawsky-Berger, "Distributed Processing in a Large System Environment", *IBM Research Report RC 9027*, IBM Thomas J. Watson Research Center, Yorktown Heights, NY 10598 (1976).
11. K. Maruyama, "Defining routing tables for SNA networks", *IBM Systems Journal*, Vol. 22 No. 4, 1983.

Panel 7

End User Applications

Chairperson
N. Shacham
SRI

Panelists
A. Maurer, *Rochester Telephone*
J. Schiller, *MIT*

Late Paper

ISDN ARCHITECTURE: A BASIS FOR NEW SERVICES

Dr. Daryl J. Eigen

Siemens Communication Systems, Inc.
5500 Broken Sound Blvd., N.W.
Boca Raton, FL 33431

Abstract

ISDN is an acronym for Integrated Services Digital Network. ISDN embodies world-wide standards that will engender the technology necessary to weave the fabric of a global communications network and provide the foundation for information age services. Siemens has a cost effective ISDN introduction strategy that will allow ISDN to interwork with existing switching systems, current transmission networks, and nonproprietary Customer Premises Equipment (CPE), as well as to interface with operational and administration systems. This strategy also entails the development of ISDN terminals, terminal adaptors and service capabilities so that the end user can quickly realize the benefits of ISDN. Siemens is forming partnerships with Bell Operating Telephone Companies, like Wisconsin Bell Telephone Co., to perform market research of the end users to better determine their ISDN needs and applications. ISDN architecture is briefly discussed to provide a basis for some examples on how ISDN and network components can be combined to meet customer needs.

ISDN Network Architecture

ISDN provides end-to-end digital connectivity with message level signaling[1,2]. The Basic Access Standard is 2x64 Kbps B channels for voice or data and 1x16 Kbps D channel for signaling data and packet data. With existing technology ISDN Basic Access can be implemented over the twisted wire pair(s) that now goes to one's home or business.

The Primary Access U.S. Standard is 23x64 Kbps B channels for voice or data and 1x64 Kbps D channel for signaling data. The Primary Access uses 4 wire digital carrier. Interoffice signaling is handled via Common Channel Interoffice Signaling Number 7 (CCSS#7) or Channel Associated Signaling (CAS), e.g. MF. Packet data is handled via B or D channels for Basic Access and B channels for Primary Access. Packet Data follows the X.25 standard for intra-packet switch access and X.75 protocol for inter-packet switch communications. Figure 1 illustrates the ISDN network architecture.

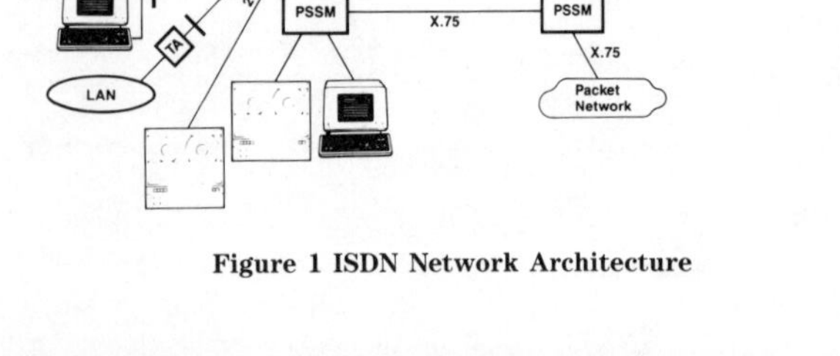

Figure 1 ISDN Network Architecture

Service Modules (SMs) provide a modular architecture for adding new ISDN service capabilities. Service Modules are loosely coupled and can be part of the EWSD, part of the ISDN network or a vendor node depending upon the interface used and where the SM is placed. There are several broad classes of Service Modules:

- Network Gateways
  - Packet Switch
  - Modem Pool
  - Network PAD
- Store and Forward Services
- Telemetry Monitoring
- Entrepreneurial Databases
- Special Features

The core of the ISDN network is the Central Office. The Siemens' EWSD is actually a family of switching applications which provide for ISDN as an add-on to a fully LSSGR (LATA Switching System General Requirements) compatible POTS (Plain Old Telephone Service) digital switch. Centrex is provided on either POTS lines or ISDN Basic Access lines. EWSD ISDN is achieved with a software generic change, the modular addition of ISDN service modules and pin-for-pin compatible ISDN line cards.

The EWSD ISDN provides access and interconnectivity to the Public Telephone Switched Network, the Public Packet Switched Network, Private and Channel Networks, the Signaling Network and an overlay 64 Kbps clear channel network. This ISDN capability provides the fundamental basis for a new generation of network services. It holds the promise of transport and terminal integration for voice, data and imaging.

Network Terminations, Terminal Adaptors and Terminals

Basic Access is achieved by a Network Termination (NT) device at the customer premises side of the loop and a companion line card at the Central Office (CO). The diagram in Figure 2 summarizes the ISDN Basic Access arrangement. The NT to be provided with EWSD is NT1. Thus the S(o) interface or T reference point is offered. This is sometimes called the T interface and provides a standard interface for ISDN compatible Customer Premises Equipment (CPE). The T interface and D channel protocols are defined in CCITT Recommendations I.440, I.441, I.450 and I.451. These ISDN standards are also being worked in the U.S. T1D1 Committee. Siemens' goal is to contribute to and be compatible with these standards. This is to enable the EWSD to interface with NTs, TAs, and terminals provided by other vendors and manufacturers.

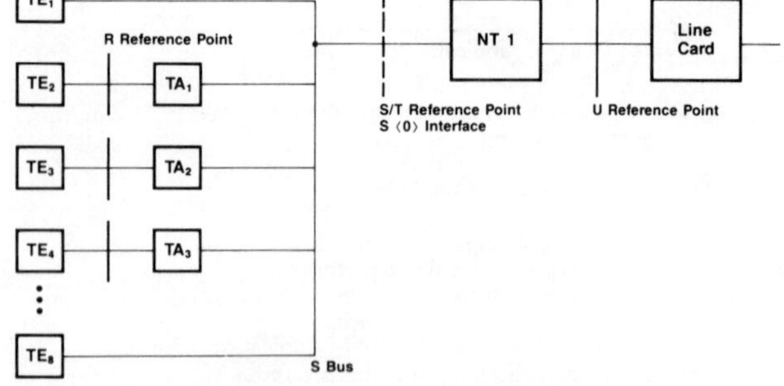

Figure 2 ISDN Basic Access Arrangement

Basic Access allows for simultaneous and independent voice or data calls on each B channel as well as simultaneous packet data transport. This is illustrated in Figure 3.

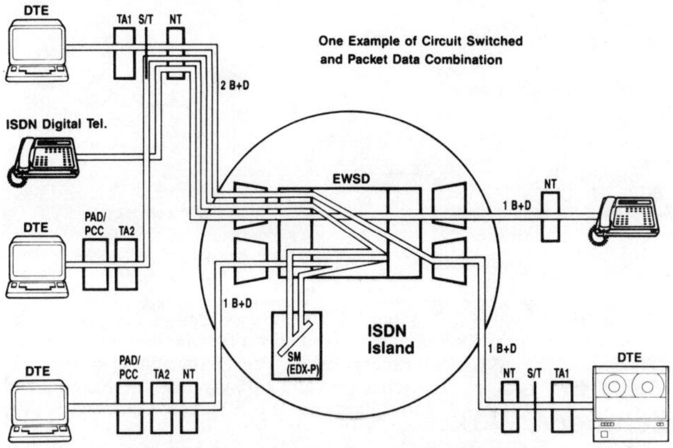

Figure 3 ISDN Simultaneous and Independent Voice Call, Data Call, and Packet Data

To enable interfacing with existing terminals, a number of Terminal Adapters (TAs) will be provided. The Basic Access TAs will electrically adapt terminals to the S/T interface, provide D channel protocol conversion, and perform bit rate adaption where necessary. Figure 4 summarizes the high priority TAs and terminals.

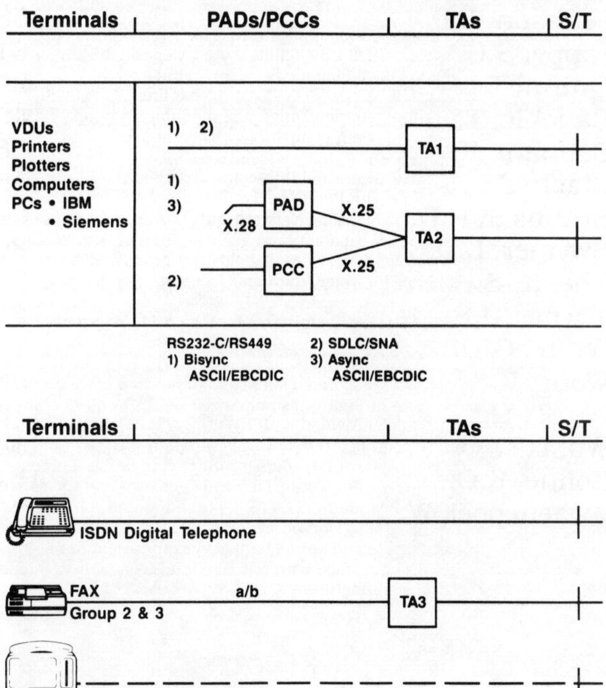

Figure 4 ISDN Terminals and Terminal Adaptors

Services

With ISDN there are fundamental transport services that, in and of themselves, should reduce costs and improve performance for the end user. These services include 64 Kbps (restricted or unrestricted) data transport, 64 Kbps digital telephony, and 9.6 Kbps and 64 Kbps X.25 packet data transport.

The ISDN Basic Access signaling standards are being expanded to provide for ISDN Centrex. Newly designed ISDN terminals will improve the human-machine interface and make the Centrex services much more powerful and easy to use. Centrex can be made to be a very competitive alternative to PBXs in terms of cost and features. Further, businesses will have the opportunity to be freed from the maintenance, administration and operational headaches that accompany owning PBX equipment.

But what about PBXs? With ISDN and CCSS#7, PBXs can take advantage of the intrinsic capabilities of ISDN and the network. The concept of ISDN Private Virtual Networks (PVNs) enables businesses to have control of a mix of data and voice, demand and semi-permanent (nailed), inter-office connections that can change as a function of time of day, day of week or date. Perhaps the greatest benefit to business communications will come from the integration of PBX and Centrex on a LATA-wide basis as shown in Figure 5.

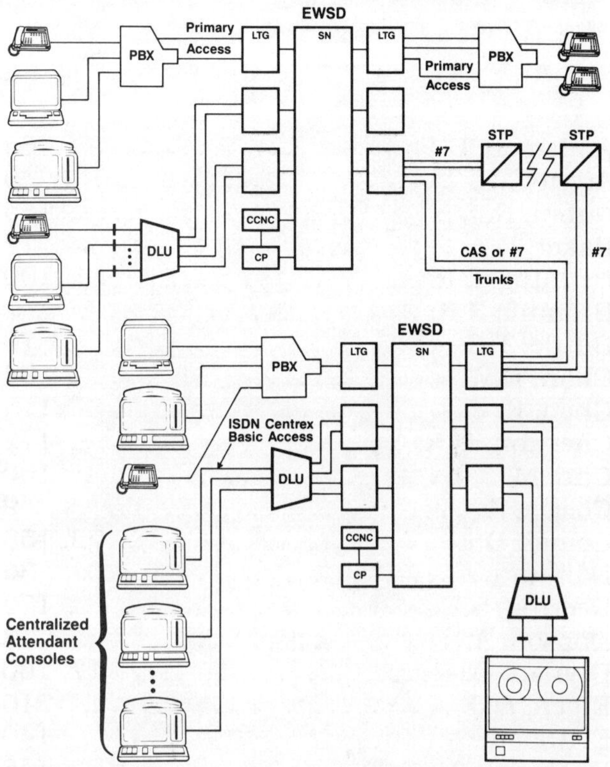

Figure 5 LATA-Wide Business Service

Another business telecommunication problem that can be better met by combining ISDN and network capabilities is the requirement for attendant teams to handle customer calls. ISDN, Centrex, CCSS #7, and 800 services can be combined with ISDN to solve this specific business need in an integrated manner. ISDN multifunctional terminals can be connected to the Central Office via ISDN Basic Access Centrex lines. Centrex call hunting can be used to select attendants who are available. ISDN Basic Access will allow these attendants to simultaneously query a computer from the same terminal (and line) from which they received the call. The query and information transfer can be made at 64 Kbps with a circuit switched B channel using one of the conventional data protocols such as BSC or SDLC/SNA. X.25 data can be sent via the D channel at 9.6 Kbps. Higher data rates will speed the transaction. The interconnectivity will allow the host computer to serve a number of attendant locations or allow an attendant to quickly access a number of host computers. In addition a 800 plus capability will provide toll free calling to the customer and more control over routing the call to an available attendant team.

These are just a few examples which show how ISDN and network capabilities can be orchestrated to meet real business needs.

References

(1) R. Blake, "Integrated networks: A Realistic Approach," *Telephone Engineer and Management*, P64 - 72, June 1, 1985.

(2) H. Suckfuell, "ISDN - The Universal Network for All Individual Communication Services" *Siemens Telecom Report*, Volume 8, April 1985.

Author Index